W9-ACU-976

# CONTROL AND DYNAMIC SYSTEMS

*Advances in Theory and Applications*

Volume 34

*CONTRIBUTORS TO THIS VOLUME*

*AMIT M. ATHALYE*
*M. S. CHEN*
*MARTIN CORLESS*
*HENRYK FLASHNER*
*WALTER J. GRANTHAM*
*RAMESH S. GUTTALU*
*STEFEN HUI*
*GEORGE LEITMANN*
*W. W. MELVIN*
*A. MIELE*
*SANDEEP PANDEY*
*JANISLAW M. SKOWRONSKI*
*M. TOMIZUKA*
*H. WANG*
*T. WANG*
*STANISLAW H. ŻAK*
*PEDRO J. ZUFIRIA*

# CONTROL AND DYNAMIC SYSTEMS

## ADVANCES IN THEORY AND APPLICATIONS

Edited by
*C. T. LEONDES*

School of Engineering and Applied Science
University of California, Los Angeles
Los Angeles, California
and
College of Engineering
University of Washington
Seattle, Washington

VOLUME 34: ADVANCES IN CONTROL MECHANICS
Part 1 of 2

ACADEMIC PRESS, INC.
**Harcourt Brace Jovanovich, Publishers**
San Diego New York Boston
London Sydney Tokyo Toronto

This book is printed on acid-free paper. (∞)



Academic Press, Inc.
San Diego, California 92101

United Kingdom Edition published by
Academic Press Limited
24-28 Oval Road, London NW1 7DX

Library of Congress Catalog Card Number: 64-8027

ISBN 0-12-012734-2 (alk. paper)

Printed in the United States of America
90 91 92 93 9 8 7 6 5 4 3 2 1

Dedicated to
Drs. George Leitmann and Angelo Miele
in appreciation for their many contributions over the years

# CONTENTS

# CONTRIBUTORS

Numbers in parentheses indicate the pages on which the authors' contributions begin.

Amit M. Athalye (205), *Department of Mechanical and Materials Engineering, Washington State University, Pullman, Washington 99164*

M. S. Chen (155), *Mechanical Engineering Department, National Taiwan University, Taipei, Taiwan 10764, Republic of China*

Martin Corless (125), *School of Aeronautics and Astronautics, Purdue University, West Lafayette, Indiana 47907*

Henryk Flashner (xiii), *Department of Mechanical Engineering, University of Southern California, Los Angeles, California 90089*

Walter J. Grantham (205), *Department of Mechanical and Materials Engineering, Washington State University, Pullman, Washington 99164*

Ramesh S. Guttalu (xiii, 279), *Department of Mechanical Engineering, University of Southern California, Los Angeles, California 90089*

Stefen Hui (175), *Department of Mathematical Sciences, and School of Electrical Engineering, San Diego State University, San Diego, California 92182*

George Leitmann (1), *Department of Mechanical Engineering, University of California, Berkeley, California 94720*

W. W. Melvin (81), *Delta Airlines, Atlanta, Georgia, and Airworthiness and Performance Committee, Air Line Pilots Association (ALPA), Washington, D.C.*

A. Miele (81), *Aero-Astronautics Group, Rice University, Houston, Texas 77001*

Sandeep Pandey (1), *Department of Mechanical Engineering, University of California, Berkeley, California 94720*

Janislaw M. Skowronski (xiii), *Department of Mechanical Engineering, University of Southern California, Los Angeles, California 90089*

M. Tomizuka (155), *Mechanical Engineering Department, University of California, Berkeley, California 94720*

H. Wang (81), *Aero-Astronautics Group, Rice University, Houston, Texas 77001*

T. Wang (81), *Aero-Astronautics Group, Rice University, Houston, Texas 77001*

Stanislaw H. Żak (175), *Department of Mathematical Sciences, and School of Electrical Engineering, Purdue University, West Lafayette, Indiana 47907*

Pedro J. Zufiria (279), *Department of Mechanical Engineering, University of Southern California, Los Angeles, California 90089*

# PREFACE

Modern technology makes possible amazing things in control and dynamic systems and will continue to do so to an increasing extent with the passage of time. It is a fact, however, that control systems go back over the millennia, and while rudimentary in earlier centuries, they were, generally, enormously effective. A classic example of this is the Dutch windmill. In any event, this is true today, and, as just noted, amazing things are now possible. As a result, this volume and the following one, Volumes 34 and 35, respectively, in the series *Control and Dynamic Systems,* are based on a National Science Foundation–Sponsored Workshop on Control Mechanics, i.e., control system development using analytical methods of mechanics and active control of mechanical systems. Publication of the presentations at this workshop in this Academic Press series has made it possible to expand them into a format which will facilitate the study and utilization of their significant results by working professionals and research workers on the international scene.

Ordinarily, the preface to the individual volumes in this series presents a summary of the individual contributions in the respective volumes. In the case of these two volumes, this is provided by the Introduction, which follows immediately.

# INTRODUCTION

The articles in this volume were presented at the National Science Foundation–Sponsored Second Workshop on Control Mechanics, held at the University of Southern California, Los Angeles, January 23–25, 1989. This workshop is the second in a series devoted to promoting control mechanics, i.e., control system development using analytical methods of mechanics and active control of mechanical systems.

Research in control mechanics is motivated by the demands imposed on modern control systems. The tasks of modern industry in areas like high technology manufacturing, construction and control of large space structures, and aircraft control can be accomplished only by precisely controlled and highly autonomous mechanical systems. These systems are inherently nonlinear due to simultaneous high speed large motions of multiple interconnected bodies and their complex interactions with the environment. In addition, weight limits imposed by space-based systems and power constraints lead to highly flexible structures. High speed operation combined with structural flexibility necessitates inclusion of both nonlinear effects and vibrational modes in control law development. Consequently, one needs to apply methods of analytical mechanics to develop an adequate mathematical representation of the system. Then the underlying global characteristics of the equations of motion must be employed to develop sophisticated multivariable, possibly nonlinear, control laws.

The theme of the above-mentioned promotion of control mechanics is covered by a wide scope of papers in mechanical systems control theory presented at the workshop. The topics are arranged in two volumes—seven papers in the first and thirteen papers in the second. The first two chapters of Part 1 deal with microburst, a severe meteorological condition pertinent to aircraft control. These two chapters consider the problem of stable control of an aircraft subjected to windshear caused by microburst. New results concerning control of an aircraft under windshear conditions are given by G. Leitmann and S. Pandey in the first chapter. The problem of control during windshear was first posed and has been extensively studied by A. Miele, who kindly accepted the invitation of C. T. Leondes to present a review of his results in the second chapter. In the third chapter, M. Corless addresses the issue of designing controllers for uncertain mechanical systems which are robust against unmodelled flexibility. Results concerning robust control design without matching conditions are presented by M. S. Chen and M. Tomizuka in the fourth chapter. Their results are applicable to SISO systems with disturbances or modelling errors that are either bounded or that have a cone-bounded growth rate. In the fifth chapter, S. Hui and S. H. Żak deal with the robust control problem using the variable structure control method. They present a methodology to design controllers and state observers and analyze their stability properties.

W. G. Grantham and A. M. Athalye study the control of chaotic systems in the sixth chapter. This article is concerned with the numerical chaotic behavior which can occur under feedback control, even with a stabilizing control law. Part 1 concludes with an article by R. S. Guttalu and P. J. Zufria that considers the problem of finding zeros of a nonlinear vector function, using methods of dynamical systems analysis. The role of singularities and their effect on the global behavior of dynamical systems is studied in detail.

Liapunov design method, an approach often used in control mechanics, is the first topic considered in Part 2. In the first chapter of Part 2, A. Olas studies the question of finding recursive Liapunov functions for autonomous systems. This is a sequel to the novel method of "converging series" presented by him at the First Control Mechanics Workshop in 1988. In the second chapter, J. E. Gayek discusses a new approach to verifying the existence of stabilizing feedback control laws for systems with time-delay by using Liapunov functionals. In the third chapter, K. Shamsa and H. Flashner use the notions of passivity and Liapunov stability to define a class of discrete-time control laws for mechanical systems. Their approach is based on the Hamiltonian structure of the equations of motion, a characteristic common to a wide class of mechanical systems. Reduction in dimensionality of models is discussed by R. E. Skelton, J. H. Kim, and D. Da in the fourth chapter. This problem is of interest for controlling high-order mechanical systems such as large space structures and robotic manipulators with structural flexibility. Regarding the systems with structural flexibility, a new method of control via active damping augmentation is introduced by T. L. Vincent, Y. C. Lin, and S. P. Joshi in the fifth chapter. This study is an extension to two-dimensional structures that the authors presented for beams in the First Control Mechanics Workshop. O. D. I. Nwokah, D. Afolabi, and F. M. Damra discuss the modal stability of imperfect cyclic systems in the sixth chapter. The paper is of particular value in resolving some of the disagreement in the literature concerning the qualitative behavior of cyclic systems. In the seventh chapter, W. E. Schmitendorf and C. Wilmers investigate the problem of developing reduced-order stabilizing controllers. They present a numerical algorithm for designing a minimum-order compensator to stabilize a given plant.

Most of the papers presented in this workshop are applicable to the analysis and design of robotic manipulators. This field has recently attracted attention in developing strategies for coordination control of multi-arm systems, adaptive control of robots, and control of manipulators with varying loads. A numerical step-by-step collision avoidance technique is proposed and demonstrated by R. J. Stonier in the eighth chapter. A theoretical basis for collision avoidance using Liapunov stability theory is investigated by G. Bojadziev in the ninth chapter. In the tenth chapter, a new method of using differential game approach to coordination is proposed by M. Ardema and J. M. Skowronski. A single arm problem using nonlinear Model Reference Adaptive Control studied by R. J. Stonier and C. N. Wheeler appears in the next chapter. A path-tracking method for control of mechanical systems is presented by H. A. Pak and R. Shieh in the twelfth chapter. In this chapter, a class of optimal feedforward tracking controllers have been proposed using preview and feedback control actions. Finally, our Australian participants presented a number of applications with this field, specifically on how to use the robotic manipulators in cane-sugar production analysis, sheep shearing, and the mining industry. The last-mentioned application is discussed in the final chapter by G. F. Shannon.

The participants of the workshop are in debt to Professor G. Leitmann for initiating this series of meetings; to Professor L. M. Silverman, Dean of the School of Engineering at the University of Southern California, for supporting and opening the workshop; the administrative staff of the Mechanical Engineering Department, Ms. G. Acosta and Ms. J. Givens, for their invaluable help; and to the editor of *Control and Dynamic Systems,* Professor C. T. Leondes, for inviting these proceedings for publication. The organizing committee gratefully acknowledges a grant from the National Science Foundation.

Janislaw M. Skowronski
Ramesh S. Guttalu
Henryk Flashner

# AIRCRAFT CONTROL UNDER CONDITIONS OF WINDSHEAR

GEORGE LEITMANN
AND SANDEEP PANDEY

Department of Mechanical Engineering
University of California
Berkeley, CA 94720

## 1. INTRODUCTION

The problem of control of aircraft flying through windshear has gained considerable importance since a 1977 FAA study revealed the presence or potential of low-level windshear as a factor in many accidents involving large aircraft, [1]. This condition is hazardous for an aircraft flying at low altitudes, e.g., during take-off or landing. Over the past 20 years, some 30 aircraft accidents have been attributed to windshear, [2]. A few of them are mentioned below, [1,3,4]:

Jan. 4, 1971: DC-3C at LaGuardia Airport, New York

Dec. 17, 1973: DC-10 at Logan International Airport, Boston

Jun. 24, 1975: B-727 at J.F. Kennedy International Airport, New York

Jul. 9, 1982: B-727 at New Orleans International Airport

Aug. 2, 1985: L-1011 at Dallas-Fort Worth International Airport

Windshear is caused by several different motions of atmospheric masses. The atmospheric boundary layer in its natural state always contains some degree of windshear. Thus, during most approach and take-off operations some degree of windshear is encountered. The strength of the shear and the degree to which it becomes hazardous is dependent upon the existing combination of meteorological conditions, [5]. Generally a windshear condition involves several factors such as horizontal shear, vertical shear, wind direction change and height of shear above ground level. In [6] windshear is defined as significant changes in wind speed and/or direction up to 500 m above the ground which may adversely affect the approach, landing or take-off of an aircraft.

To study this problem we must have accurate and reliable wind profiles. However, the existing windshear models used in computer and manned flight simulator studies are not very realistic. Existing mathematical models of windshear are spatially two-dimensional and based on limited data. None of them includes time dependence. Moreover, few of these models contain small-scale microburst type windshear, [3]. The aim of present work is, however, not to improve the existing windshear models, but to develop a control strategy which is effective for flying in different kinds of situations. For modeling apsects of windshear the reader is referred to [1,3,5-7].

Recently it was discovered that a small-scale but severe low-level thunderstorm wind, now referred to as a *microburst*, occurs with surprising frequency and can adversely affect an aircraft, [3]. During the investigation of the catastrophic JFK accident of June 1975, Dr. T.T. Fujita of the University of Chicago discovered this previously unidentified weather phenomenon associated with thunderstorms. He identifies two effects due to strong downdrafts: downburst and microburst. He defines downburst as a strong temporary localized downdraft with a speed of up to 30 m/s that hits the ground and spreads horizontally in a radial burst of damaging wind (he suggests that the JFK accident could have been caused by a series of downbursts bombarding the approach to the runway). He defines microburst as downflows which spread horizontally within a layer only 15-30 m above the ground and a horizontal extent of typically less than 5 km in length, and 1 to 5 min. in duration [1,3,8].

Most of the research done so far on this topic has been concerned with meteorology, instrumentation, aerodynamics, flight mechanics and stability and control, [9]. In take-off, once the aircraft becomes airborne, the pilot has no choice but to fly through a windshear. His only control is the angle of attack, assuming that the power setting, his other control, is already held at a value which results in maximum thrust, [9,15].

In one of his seminal papers, [16], Miele deduces optimal control strategies for take-off in the presence of a given windshear; that is, the controlled trajectories are optimum in the specified flow field. Several guidance strategies which achieve near optimum performance in the given windshear were studied in the pioneering investigations [9,17-19]. These strategies employ local information on windshear. Since local information is difficult to obtain on present day aircraft, several piloting strategies were studied in [4,19]. The so-called *simplified gamma guidance* strategy, which yields a quick transition to horizontal flight in a windshear, was shown to have very good survival capabilities in the prescribed windshear. This strategy is particularly suitable for flight in a severe windshear of the type considered; it is based on the philosophy of trying to avoid altitude loss while simultaneously containing velocity loss. A possible disadvantage of this kind of guidance scheme is that it is based on a particular model of windshear and may not be effective when the aircraft flies through other windshear distributions.

In [20] a control scheme is deduced by considering the so-called pseudo-energy as a measure of aircraft altitude and speed relative to the wind. A set of linearized equations about equilibrium level flight conditions together with a linear feedback scheme are employed to obtain prescribed closed-loop eigenvalues. There is no assurance that this feedback control will yield stable trajectories under varying windshear conditions of relatively high intensity.

The investigation reported in [21] is the first attempt to apply a theory of deterministic control of uncertain systems ([13,14]) to the problem of aircraft guidance during conditions of windshear. The resulting control scheme utilizes the full state of the aircraft in order to assure the convergence of the aircraft state to a desired end state.

In the present work, the problem of aircraft take-off under windshear conditions is considered as a problem of controlling an uncertain dynamical system: the mathematical model of the aircraft dynamics involves uncertainty due to the unknown windshear. Here, an upper bound of the uncertainty must be known to assure effective control.

The control of dynamical systems, whose mathematical models contain uncertainties, has been studied extensively, e.g., see [10-14]. These uncertainties may be due to parameters, constant or varying, which are unknown or imperfectly known, or uncertainties due to unknown or imperfectly known inputs into the system. Taking a deterministic point of view, controllers have been designed which, utilizing only information about the possible magnitudes of the uncertainties, guarantee desirable system performance such as some form of stability (see Appendix A). The design of these control schemes is based on a constructive use of Lyapunov stability theory.

The problem of controlling an aircraft through windshear falls into a general class of control problems of nonlinear systems. In the following we use the theory developed in [10-14] to design a deterministic controller to stabilize the relative path inclination which is one of the states of the system. Then we test this controller on different windshear models of [7,17,20]. The simulations indicate that it is a promising strategy which can work in a very general windshear condition.

## 2. AIRCRAFT EQUATIONS OF MOTION

Following Miele's lead, we employ equations of motion for the center of mass of the aircraft, in which the kinematic variables are relative to the ground ("inertial reference frame") while the dynamic ones are taken relative to a moving but non-rotating reference frame translating with the wind velocity at the aircraft center of mass ("wind based reference frame").

We employ the following notation:

*Notation*

$ARL \triangleq$ aircraft reference line;

$D \triangleq$ drag force, $lb$;

$g \triangleq$ gravitational force per unit mass (= constant), $ft\ \sec^{-2}$;

$h \triangleq$ vertical coordinate of aircraft center of mass (altitude), $ft$;

$L \triangleq$ lift force, $lb$;

$m \triangleq$ aircraft mass, $lb\ ft^{-1}\sec^{2}$;

$O \triangleq$ mass center of aircraft;

$S \triangleq$ reference surface, $ft^2$;

$t \triangleq$ time, sec;

$T \triangleq$ thrust force, $lb$;

$V \triangleq$ aircraft speed relative to wind based reference frame, $ft\ \sec^{-1}$;

$V_e \triangleq$ aircraft speed relative to ground, $ft\ \sec^{-1}$;

$W_x \triangleq$ horizontal component of wind velocity, $ft\ \sec^{-1}$;

$W_h \triangleq$ vertical component of wind velocity, *ft* $sec^{-1}$;

$x \triangleq$ horizontal coordinate of aircraft center of mass, *ft*;

$\alpha \triangleq$ relative angle of attack, *rad*;

$\gamma \triangleq$ relative path inclination, *rad*;

$\gamma_e \triangleq$ path inclination, *rad*;

$\delta \triangleq$ thrust inclination, *rad*;

$\rho \triangleq$ air density, *lb* $ft^2$ $sec^2$ ;

Dot denotes time derivative.

*Assumptions*

1) The rotational inertia of the aircraft is neglected.

2) The aircraft mass is constant.

3) Flight is in the vertical plane.

4) Maximum thrust is used.

In view of Assumption 1, we consider only the equations of motion of the center of mass (see Figure 1).

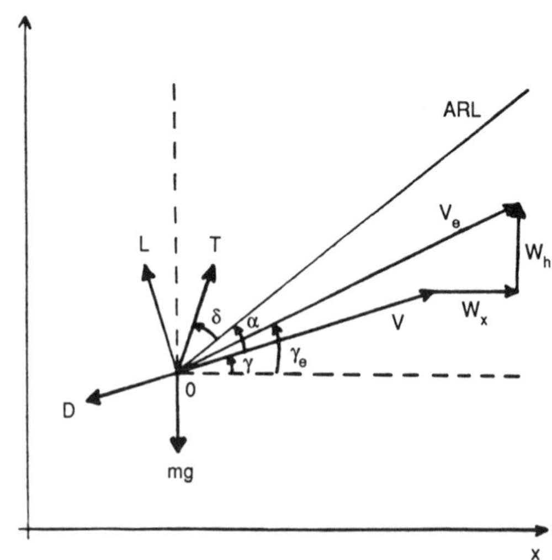


Fig. 1 Coordinate System and Free Body Diagram

The kinematical equations[1] are

$$\dot{x} = V\cos\gamma + W_x \tag{1}$$

$$\dot{h} = V\sin\gamma + W_h \tag{2}$$

and the dynamical equations are

$$m\dot{V} = T\cos(\alpha + \delta) - D - mg\sin\gamma - m(\dot{W}_x\cos\gamma + \dot{W}_h\sin\gamma) \tag{3}$$

$$mV\dot{\gamma} = T\sin(\alpha + \delta) + L - mg\cos\gamma + m(\dot{W}_x\sin\gamma - \dot{W}_h\cos\gamma) \tag{4}$$

These equations must be supplemented by specifying the thrust force $T = T(V)$, the drag $D = D(h, V, \alpha)$, the lift $L = L(h, V, \alpha)$, the horizontal windshear $W_x = W_x(x, h)$ or $W_x(t)$ and the vertical windshear $W_h = W_h(x, h)$ or $W_h(t)$. For a given value of the thrust inclination $\delta$, the differential equation system (Eqs. (1)-(4)) involves four state variables - the horizontal distance $x(t)$, the altitude $h(t)$, the relative speed $V(t)$, and the relative path inclination $\gamma(t)$ - and one control variable - the angle of attack $\alpha(t)$, since maximum thrust is employed according to Assumption 4.

## A. BOUNDED QUANTITIES

In order to account for the aircraft capabilities, we shall assume that there is a maximum attainable value of the relative angle of attack, $\alpha$; that is,

$$\alpha \in [0, \alpha_*] \tag{5}$$

where $\alpha_* > 0$ depends on the specific aircraft and generally is taken to be the *stick shaker* angle of attack.

In view of Assumption 1, the rotational inertia of the aircraft is neglected. To account for the neglected dynamics of rotation we bound the attainable magnitude of the rate of change of the relative angle of attack, $\dot{\alpha}$; that is,

$$|\dot{\alpha}| \le C \tag{6}$$

where $C > 0$ depends on the specific aircraft.

Furthermore, the range of practical values of the relative aircraft speed, $V$, is limited; that is,

$$\underline{V} \le V \le \overline{V} \tag{7}$$

[1] For the sake of brevity we shall delete the arguments of functions whenever this does not entail loss of clarity.

where $\underline{V} > 0$ and $\overline{V} > \underline{V}$ depend on the specific aircraft. These bounds correspond, for instance, to the relative stall speed and the maximum attainable relative speed, respectively.

The bounds (5) and (6) on $\alpha$ and $|\dot{\alpha}|$ will be neglected in deducing the proposed aircraft guidance scheme; however, they will be taken into account in the numerical simulations. On the other hand, the bounds (7) on the relative speed, $V$, will be employed in the construction of the proposed guidance system.

### B. APPROXIMATIONS FOR THE FORCE TERMS

*Thrust.*

$$T = A_0 + A_1 V + A_2 V^2 , \tag{8}$$

where the coefficients $A_0$, $A_1$, and $A_2$ depend on the altitude of the runway, the ambient temperature and the engine power setting.

*Drag.*

$$D = \frac{1}{2} C_D \rho S V^2, \tag{9}$$

where $C_D = B_0 + B_1\alpha + B_2\alpha^2$ . The coefficients $B_0$, $B_1$ and $B_2$ depend on the flap setting and the under-carriage position.

*Lift.*

$$L = \frac{1}{2} C_L \rho S V^2, \tag{10}$$

where

$$C_L = C_0 + C_1\alpha \quad \text{if} \quad \alpha \le \alpha_{**} < \alpha_* \tag{11}$$

$$C_L = C_0 + C_1\alpha + C_2(\alpha - \alpha_{**})^2 \quad \text{if} \quad \alpha_{**} \le \alpha \le \alpha_*. \tag{12}$$

The coefficients $C_0$, $C_1$ and $C_2$ depend on the flap setting and the undercarriage position.

## 3. PROPOSED CONTROL

### A. A NOMINAL CONTROL

We shall consider a *nominal control*, $\alpha = \alpha_n$, which corresponds to a constant relative path inclination, $\gamma$, in the absence of windshear. Of course, in the absence of windshear, $\gamma \equiv \gamma_e$.

Under no windshear conditions Eq. (4) becomes

$$\dot{\gamma} = \frac{T\sin(\alpha + \delta)}{mV} + \frac{L}{mV} + \frac{g\cos\gamma}{V} \tag{13}$$

so that for constant $\gamma$ we have

$$T\sin(\alpha + \delta) + L - mg\cos\gamma = 0. \tag{14}$$

Since $\alpha_*$ is relatively small, we make the following approximation:

$$\sin(\alpha + \delta) \approx \alpha\cos\delta + \sin\delta. \tag{15}$$

For $\alpha \leq \alpha_{**}$, using Eqs. (8), (10, (11) and (15) in Eq. (14), we obtain

$$\alpha = \alpha_n = \frac{D_0}{D_1} \tag{16}$$

where

$$D_0 \overset{\Delta}{=} mg\cos\gamma - (A_0 + A_1V + A_2V^2)\sin\delta - \frac{1}{2}C_0\rho SV^2 \tag{17}$$

$$D_1 \overset{\Delta}{=} (A_0 + A_1V + A_2V^2)\cos\delta + \frac{1}{2}C_1\rho SV^2. \tag{18}$$

For $\alpha_{**} \leq \alpha \leq \alpha_*$ , using Eqs. (8), (10), (12) and (15) in Eq. (14), we obtain

$$\alpha = \alpha_n = \frac{(-E_1 + \sqrt{E_1^2 - 4E_0E_2})}{2E_0} \tag{19}$$

where

$$E_0 \overset{\Delta}{=} \frac{1}{2}C_2\rho SV^2 \tag{20}$$

$$E_1 \overset{\Delta}{=} D_1 - C_2a_{**}\rho SV^2 \tag{21}$$

$$E_2 \overset{\Delta}{=} -D_0 + \frac{1}{2}C_2a_{**}{}^2\rho SV^2. \tag{22}$$

This nominal control will be employed in numerical simulations to illustrate the need for a guidance scheme which accounts for the possible presence of windshear.

## B. A STABILIZING CONTROL

The general problem of controlling an aircraft flying through windshear falls into the class of problems of stabilizing a so-called mismatched uncertain nonlinear system (see [10-14] for definitions, and [21] for the first application to the windshear case). Moreover, the system does not satisfy the conditions required for the controllers of [13,14]. Hence, in the present investigation we make some simplifications which allow direct utilization of the controllers proposed in [13] and [14] and in many earlier papers referenced there.

First of all, we consider the stabilization problem of only the relative path inclination, $\gamma$. Although it is desirable to stabilize all the states of the system (see [21]), the selection of only one state reduces the problem to a so-called matched one. Since the successful stabilization of relative path inclination about a nominal relative path inclination should ensure avoidance of a crash, it was chosen as the state to be stabilized.

Next we make the approximation (see Eqs. (10), (11), (12)),

$$C_L \approx C_0 + C_1\alpha \quad \text{for} \quad \text{all} \quad \alpha \in [0, \alpha_*] \tag{23}$$

in order to satisfy condition (B.5)(see Appendix B) for the design of a controller. From the plot of $C_L$ versus $\alpha$, [17], we observe that this is a valid approximation. Moreover, the contribution of $C_L$ to lift (see Eq. (2.9)) is small compared to other terms. As mentioned earlier, we consider bounds on the relative speed when deriving the controller, that is, we impose

$$\underline{V} \leq V \leq \overline{V}$$

where $\underline{V}$ and $\overline{V}$ depend on the specific aircraft.

A major assumption in the theory of deterministic control is that uncertain terms are bounded. In our case, it suffices to assume a known bound on the magnitude of the time rate of change of the wind; that is,

$$\sqrt{\dot{W}_x^2 + \dot{W}_h^2} \leq q \tag{24}$$

where $q > 0$ is assumed to be known from previous flights through various windshear conditions.

Now we consider the stabilization of the relative path inclination. Let $\gamma_R(t)$ be the reference value of $\gamma(t)$ with respect to which we want to stabilize $\gamma(t)$. Let us define,

$$\Delta\gamma(t) \triangleq \gamma(t) - \gamma_R(t) \tag{25}$$

Then from Eqs. (4), (8), (10), (23) and (25) we have

$$\dot{\Delta\gamma} = -K\Delta\gamma + \frac{(A_0 + A_1 V + A_2 V^2)\sin(\alpha + \delta)}{mV}$$

$$+ \frac{(C_0 + C_1\alpha)\rho S V^2}{2mV} - \frac{g\cos(\Delta\gamma + \gamma_R)}{V}$$

$$+ \frac{\dot{W}_x \sin(\Delta\gamma + \gamma_R)}{V} - \frac{\dot{W}_h \cos(\Delta\gamma + \gamma_R)}{V} - \dot{\gamma}_R + K\Delta\gamma \quad (26)$$

where a linear portion $K\,\Delta\gamma$, $K > 0$, has been added and subtracted in order to facilitate the controller design (see *Remark* of Appendix B). Now the above equation falls into the class of systems described by Eqs. (A.1) and (B.3) of Appendices A and B, repsectively. Comparing Eq. (26) with these equations, we have

$$y = \Delta\gamma \quad (27)$$

$$f^s = -K\Delta\gamma \quad (28)$$

$$B = 1 \quad (29)$$

$$z = \frac{(A_0 + A_1 V + A_2 V^2)\sin(\alpha + \delta)}{mV} + \frac{(C_0 + C_1\alpha)\rho S V^2}{2mV} - \frac{g\cos(\Delta\gamma + \gamma_R)}{V}$$

$$+ \frac{\dot{W}_x \sin(\Delta\gamma + \gamma_R)}{V} - \frac{\dot{W}_h \cos(\Delta\gamma + \gamma_R)}{V} - \dot{\gamma}_R + K\Delta\gamma \quad (30)$$

Referring to the *Remark* in Appendix B, since Eq.(26) has a stable linear time-invariant nominal part we use Eq. (B.10) (see Appendix B) with $Q = 1$ to obtain

$$\phi = B^T P y = \frac{1}{2K}\Delta\gamma. \quad (31)$$

Now we can choose the functions $\beta_1$ and $\beta_2$ as follows so that Eq. (B.5) (see Appendix B) is satisfied:

$$\beta_1 = \frac{|A_1|}{m} + \frac{g}{V} - \frac{\dot{W}_x \sin(\Delta\gamma + \gamma_R)}{V} + \frac{\dot{W}_h \cos(\Delta\gamma + \gamma_R)}{V} + \dot{\gamma}_R \quad (32)$$

$$\beta_2 = \frac{C_1 \rho S \underline{V}}{2m}. \quad (33)$$

The choice of the following functions satisfies conditions (B.6), (B.7) and (B.11) (see Appendix B):

$$\kappa^c = \kappa = \frac{|A_o| + |A_1|\overline{V} + |A_2|\overline{V}^2}{m\underline{V}} + \frac{g}{\underline{V}} + \frac{q}{\underline{V}} + \overline{\dot{\gamma}_R} + K|\Delta\gamma| \tag{34}$$

$$\xi^c = \xi = \frac{2m\kappa^c}{|C_1|\rho S\underline{V}} \tag{35}$$

where

$$\dot{\gamma}_R(t) \leq \overline{\dot{\gamma}_R} \quad \text{for} \quad \text{all } t \in [0,t_c\,] \tag{36}$$

and $[0,t_c\,]$ is the duration of the control period.

Then we use the stabilizing controller (B.17) of Appendix B, namely,

$$\alpha = p^\varepsilon = \begin{cases} -\xi^c\varepsilon^{-1}\eta & \text{if } |\eta| \leq \varepsilon \\ -\xi^c|\eta|^{-1}\eta & \text{if } |\eta| > \varepsilon \end{cases} \tag{37}$$

where $\eta = \kappa^c\phi$ and $\varepsilon > 0$ is a design constant.

## 1. WINDSHEAR MODELS

For purposes of testing the proposed controller (guidance scheme) we present simulation results for flight in four different windshear situations.

### A. MODEL 1 (Figure 2)

This model is a discretized approximation of the one used in [17]. The horizontal windshear is given by

$$W_x = \begin{cases} -k, & x \leq a, \\ -k + 2k(x-a)/(b-a), & a \leq x \leq b,, \\ k, & x \geq b \end{cases} \tag{38}$$

and vertical windshear is given by

$$W_h = \begin{cases} 0, & x \le a, \\ -8kh(x-a)/(50h^*(c-a)), & a \le x \le c, \\ [-8kh(e-c) - 34kh(x-c)]/[50h^*(e-c)], & c \le x \le e, \\ [-42kh(i-e) - 8kh(x-e)]/[50h^*(1-e)], & e \le x \le i, \\ -kh/h^* & i \le x \le j, \\ [-50kh(f-j) + 8kh(x-j)]/[50h^*(f-j)], & j \le x \le f, \\ [-42kh(d-f) + 34kh(x-f)]/[50h^*(d-f)], & f \le x \le d, \\ [-8kh(b-d) + 8kh(x-d)]/[50h^*(b-d)], & d \le x \le b, \\ 0, & x \le b, \end{cases} \tag{39}$$

where $a,b,c,d,e,f,i,j$ are various horizontal distances measured from the initial position, $h^*$ is a reference altitude and $k$ is a measure of windshear intensity.

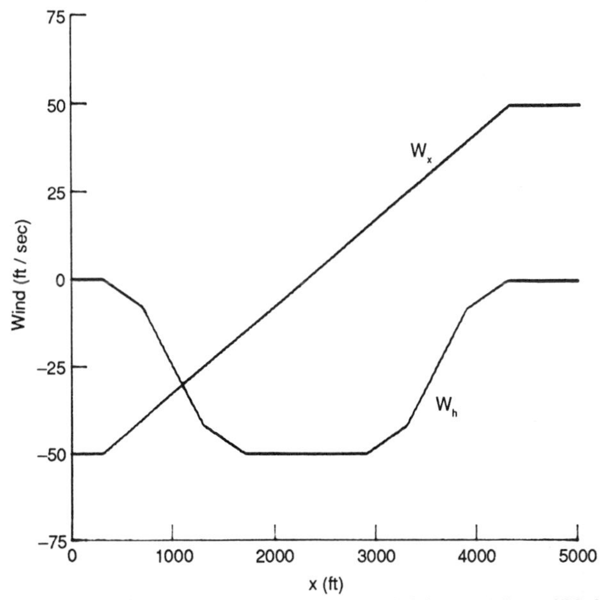


Fig. 2 Windshear Model 1 for $k$ = 50 ft/sec and $h$ = 1000 ft.

B. MODEL 2[1] (Figure 3)

This model is the one used in [20]. The horizontal windshear is given by

$$W_x = W_{x0} sin(2\pi t/T_0). \tag{40}$$

[1] Note that this is the only model in which $W_x$ and $W_h$ are given as functions of the time rather than position.

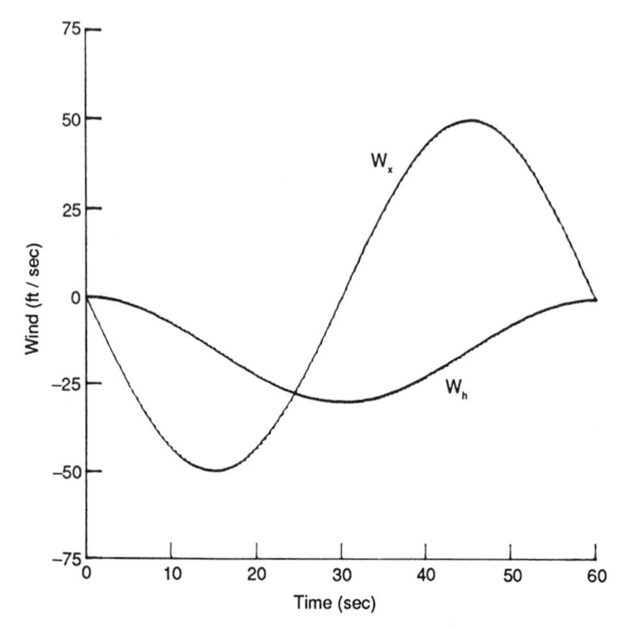


Fig. 3 Windshear Model 2 for $W_{x0}/W_{h0} = 50/30$.

The vertical windshear is given by

$$W_h = W_{h0}[1 - cos(2\pi t/T_0)]/2 \tag{41}$$

where $W_{x0}$ and $W_{h0}$ are given constants reflecting the windshear intensity, and $T_0$ is the total time to fly through the downburst.

## C. MODEL 3 (Figure 4)

This is a single vortex model, [7]. This model is described with respect to a polar coordinate system located at the center of the vortex. For $r \leq R$, where $R$ is the core radius, the transverse speed, $V_\theta$, varies linearly from zero at the center to a maximum $V_0$ at $r = R$. Outside the core ($r \geq R$), $V_\theta$ decreases asymptotically to zero as $r$ increases. The speed induced at a point is

$$V_\theta = \begin{cases} V_0 r/R, & 0 \leq r \leq R, \\ V_0 R/r, & r \geq R. \end{cases} \tag{42}$$

## D. MODEL 4 (Figure 5)

This is a double vortex model, [7]. A pair of counter-rotating vortices of equal strength, each of them modeled as in Eq. (42), is considered. The parameters which define the flowfield resulting from such a vortex pair are: the core radius $R$, the horizontal distance $L$ (half the distance between centers of the two cores), the maximum transverse speed $V_0$ and the directions of rotation.

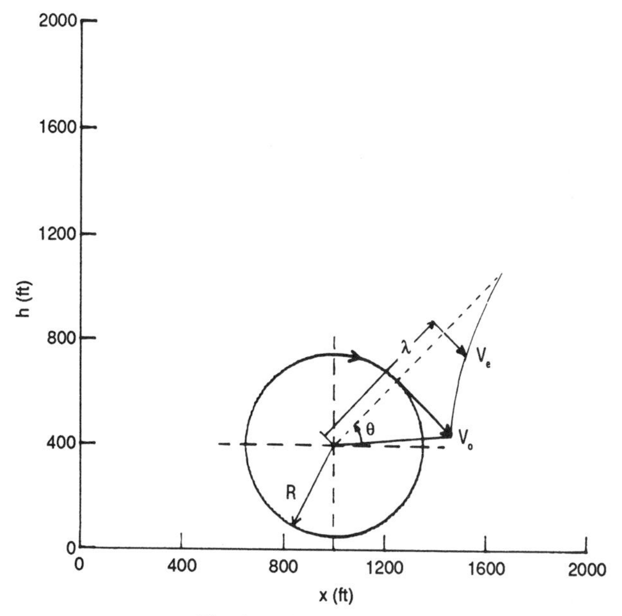


Fig. 4 Windshear Model 3.

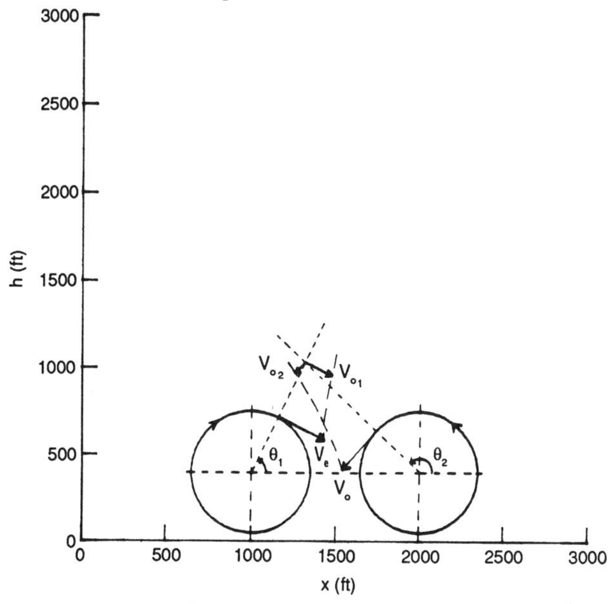


Fig. 5 Windshear Model 4.

## 1. NUMERICAL DATA

### A. AIRCRAFT DATA

The numerical data pertain to a Boeing-727 aircraft with three JT8D-17 turbofan engines. It is asssumed that the aircraft has become airborne from a runway located at sea-level. We use the same data as those used by Miele:

$$\alpha_* = 16^\circ, \quad \alpha_{**} = 12^\circ,$$

$$C = 3^\circ/sec\ ,$$

$$A_0 = 44564.0\ lb\ ,$$

$$A_1 = -\ 23.98\ lb\ ft^{-1}\mathrm{sec}\ ,$$

$$A_2 = 0.01442\ lb\ ft^{-2}\mathrm{sec}^2\ ,$$

$$\delta = 2^\circ\ ,$$

$$\rho = 0.002203\ lb\ ft^{-4}\mathrm{sec}^2\ ,$$

$$S = 1560\ ft^2\ ,$$

$$B_0 = 0.07351\ ,$$

$$B_1 = -\ 0.08617\ ,$$

$$B_2 = 1.996\ ,$$

$$C_0 = 0.1667\ ,$$

$$C_1 = 6.231\ ,$$

$$C_2 = -\ 21.65\ ,$$

$$mg = 180000\ lb\ ,$$

$$\underline{V} = 184\ ft\ \mathrm{sec}^{-1}, \quad \overline{V} = 422\ ft\ \mathrm{sec}^{-1}\ .$$

### B. WINDSHEAR MODELS

#### 1. MODEL 1

The following values yield a model which is very close to the model used by Miele (see Fig.2 and [17] for comparison):

$a = 300,\ c = 700,\ e = 1300,\ i = 1700,\ j = 2900, f = 3300,\ d = 3900,\ b = 4300,\ h^* = 1000;$

(all distances are in ft).

We consider four windshear intensities, namely, $k = 40,50,55,60$ ft/sec.

2. MODEL 2

We use $T_0 = 60$ sec and consider two cases: $W_{x0}/W_{h0} = 80/48$ and $W_{x0}/W_{h0} = 50/30$ (the values $W_x$ and $W_h$ given in ft/sec). These are the same values as those considered in [20].

3. MODEL 3

We consider the vortex to be located at a horizontal distance of 1000 ft from the location of initial position and at an altitude of 400 ft. We choose $R = 350$ ft and consider two cases: $V_0 = 100$ ft/sec and $V_0 = 140$ ft/sec. The direction of wind rotation is taken to be clockwise.

4. MODEL 4

This model consists of two counterrotating vortices, each of the type described in Model 3, with half separation distance $L = 500$ ft. The center of the vortex pair is located at a horizontal distance of 1500 ft from the initial aircraft position and an altitude of 400 ft. We consider the same two cases for $V_0$ as in Model 3.

C. INITIAL CONDITIONS

We consider the same initial conditions as Miele.

$$x(0) = 0\,ft, \quad h(0) = 50\,ft, \quad V(0) = 276.8\,ft/sec, \quad \gamma(0) = 6.989^\circ.$$

For the simulations in which a bound on derivative of $\alpha$ is considered we use $\alpha(0) = 10.36^\circ$.

D. BOUND ON WINDSHEAR RATE OF CHANGE

The bound on the rate of change of windshear, which is required for the controller design (see Eq. (24)), is computed for the worst case from the windshear Model 1. We use $q = 75.46$ $ft$ sec$^{-2}$ for all the simulations. The worst case value of $q$ is lower for Model 2 but higher for Models 3 and 4.

E. THE DESIGN PARAMETERS

The value of the gain $K$ is chosen as 500 sec$^{-1}$ and $\varepsilon$ is taken as 1.0.

## 2. SIMULATION RESULTS

In order to compare the robustness of various guidance schemes with respect to windshear distribution and intensity, we present simulation results for take-off in windshear modelled by Models 1-4 subject to three guidance schemes: the proposed *stabilizing control* of Section III.B., the *simplified gamma guidance* of [19] and the *nonlinear feedback controller* of [20]. However, before presenting these results, we illustrate the inadequacy of a feedback controller which is predicated on the absence of windshear, namely, the *nominal control* $\alpha_n$ (Section III.A.); see Figures 6-11. Whereas, in the absence of windshear, the *nominal control* assures a constant rate of altitude gain, the presence of even moderate windshear (the least severe case of Model 1) leads to a crash when the nominal control scheme is employed.

Next we examine the performance of each of the three guidance schemes for windshear models 1,2,3 and 4, respectively. We begin with Model 1, a discretized version of the model considered in [19]. Figures 12-18 and 19-25, repectively, show simulation results for the proposed *stabilizing control*; the first set of figures pertain to the case of $\alpha \le \alpha_*$ but $\dot{\alpha}$ not subject to a bound, while the second set of figures is for the case of $\alpha \le \alpha_*$ and $|\dot{\alpha}| \le C$. The reason for considering the first case, that of $\dot{\alpha}$ not subject to a bound, is to demonstrate the effect of bounding $\dot{\alpha}$. As can be seen by comparing Figures 18 and 25, the primary effect of enforcing a bound on $\dot{\alpha}$ is the introduction of severe oscillations in the relative angle of attack, $\alpha$; however, bounding $\dot{\alpha}$ has little effect on performance, that is, survivability, as can be seen by comparing altitude plots, namely, Figures 15 and 22. We conclude that employment of the proposed *stabilizing control* results in survival except for the most severe windshear intensity considered, k = 60.

Next we utilize the *simplified gamma guidance* for Model 1. The simulation results are presented in Figures 26-31. As seen in Figure 28, use of this control leads to a crash for two windshear intensities, k = 55 and k = 60. Here it must be pointed out that *simplified gamma guidance* results in survival for intensity k = 55 for flight in a windshear modelled by the windshear model employed in [19].

Finally, considering Model 1, we show simulation results for the *nonlinear feedback control* of [20] in Figures 32-37. We present only two cases of windshear intensity, k = 40 and k = 50, since this guidance scheme results in survival only for the least severe case, k = 40, as seen in Figure 34. However, it should be noted that, among the three control schemes considered, *nonlinear feedback control* is the smoothest (see Figure 37).

We now turn to Model 2, recalling that in this model the windshear velocity components are given as functions of the time rather than position. Figures 38-40 are plots of windshear velocity components and windshear acceleration magnitude, respectively. As before, we first present simulation results for the proposed *stabilizing controller* , Figures 41-44 for only $\alpha$ subject to a bound and Figures 45-48 for both $\alpha$ and $\dot{\alpha}$ limited. Here, suvival occurs only for the less severe situation, $W_{x0}/W_{h0}$ = 50/30 (see Figures 41 and 45).

Next we give simulation results for Model 2 with the *simplified gamma guidance* in Figures 49-52. As shown in Figure 49, again the guidance scheme fails in the more severe case, $W_{x0}/W_{h0}$ = 80/48.

Finally, for Model 2, Figures 53-56 present simulation results employing the *nonlinear feedback control* . As seen in the Figure 53, this guidance scheme, too, fails in the more severe case.

The next set of figures pertain to Model 3, the single vortex. Figures 57-64 (for $\alpha \le \alpha_*$) and Figures 65-72 (for $\alpha \le \alpha_*$ and $|\dot{\alpha}| \le C$) show simulation results utilizing the proposed *stabilizing control* . As seen in Figures 61 and 69, this guidance scheme results in survival for both vortex intensities, $V_0 = 100$ and $V_0 = 140$. As always, however, imposition of a bound on $\dot{\alpha}$ leads to oscillations in $\alpha$ (see Figures 64 and 72).

Figures 73-79 present simulation results for Model 3 and *simplified gamma guidance* . As seen in Figure 76, survival occurs for both vortex intensities.

Finally, for Model 3, Figures 80-86 show simulation results for the case of *nonlinear feedback control* . As seen in Figure 83, survival is assured for both vortex intensities.

Finally, we present simulation results for Model 4, a vortex pair. For the proposed *stabilizing control* , these are shown in Figures 87-94 (only $\alpha$ limited) and Figures 95-102 (both $\alpha$ and $\dot{\alpha}$ limited). As seen in Figures 91 and 99, this guidance scheme results in survival for both vortex intensities.

Simulation results for Model 4 and *simplified gamma guidance* are presented in Figures 103-109. As seen in Figure 106, this guidance scheme results in a crash for the more intense vortex pair.

For Model 4 with the *nonlinear feedback control* , Figures 110-116 show the relevant simulation results. As seen in Figure 113, survival occurs for both vortex intensities.

Although not shown here, it should be noted that the survival capabilities of the three control schemes considered here can be enhanced by adjusting the controller parameters (the design parameters $K$ and $\varepsilon$ in the proposed stabilizing controller, and the gains in the other two guidance schemes) for the various windshear distributions and intensities. However, such "tuning" employs unavailable information, since neither the windshear distribution nor its intensity are known *a priori* . Once the controller parameters are selected, they must be retained throughout the whole flight regime.

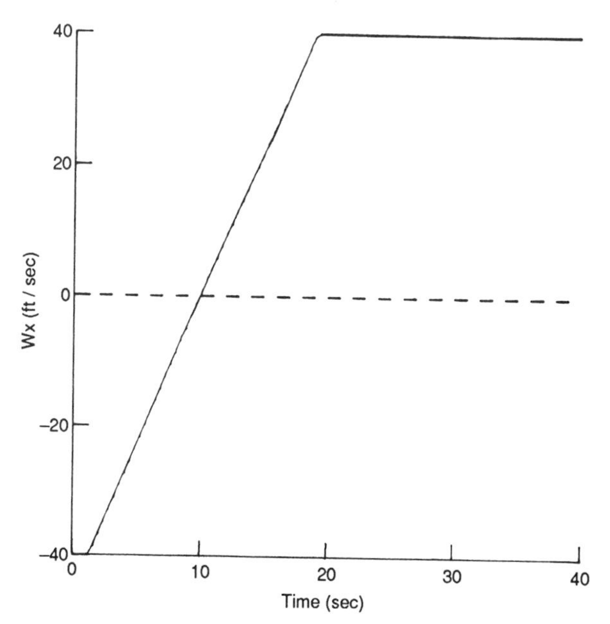


Fig. 6 Nominal Control: Model 1. Horizontal Windshear $W_x$ for $k = 0$ (dash), $k = 40$ (solid)

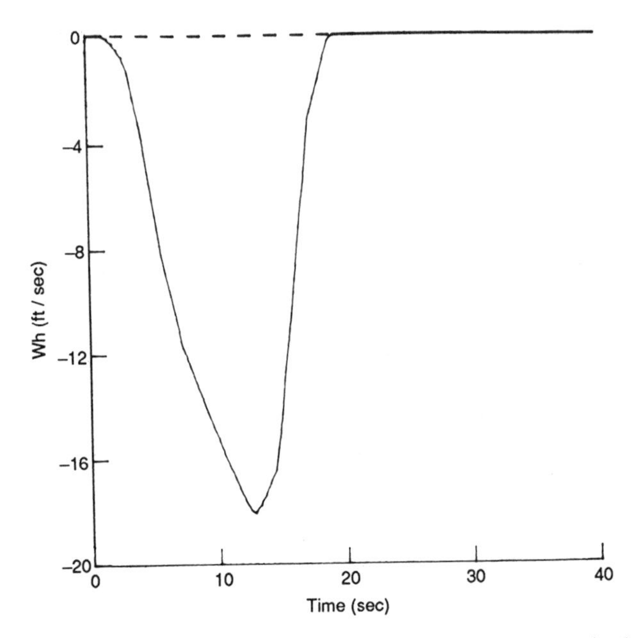


Fig. 7 Nominal Control: Model 1. Vertical Windshear $W_h$ for $k$ = 0 (dash), $k$ = 40 (solid).

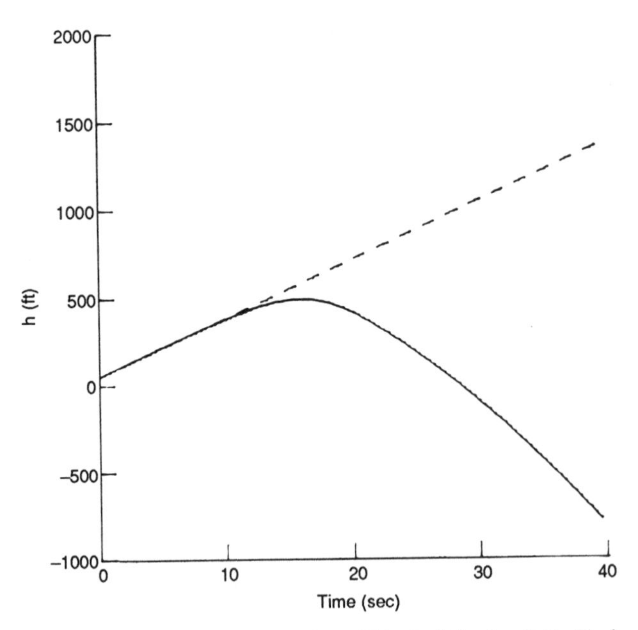


Fig. 8 Nominal Control: Model 1. Altitude $h$ for $k$ = 0 (dash), $k$ = 40 (solid).

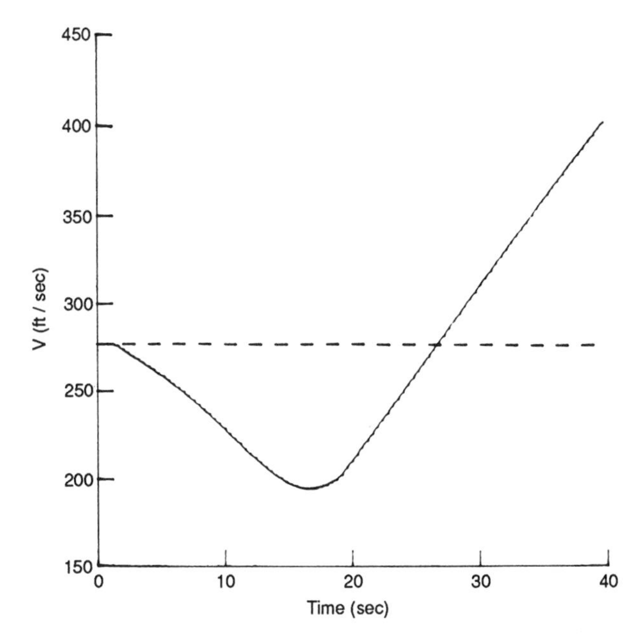


Fig. 9 Nominal Control: Model 1. Relative Speed $V$ for $k = 0$ (dash), $k = 40$ (solid).

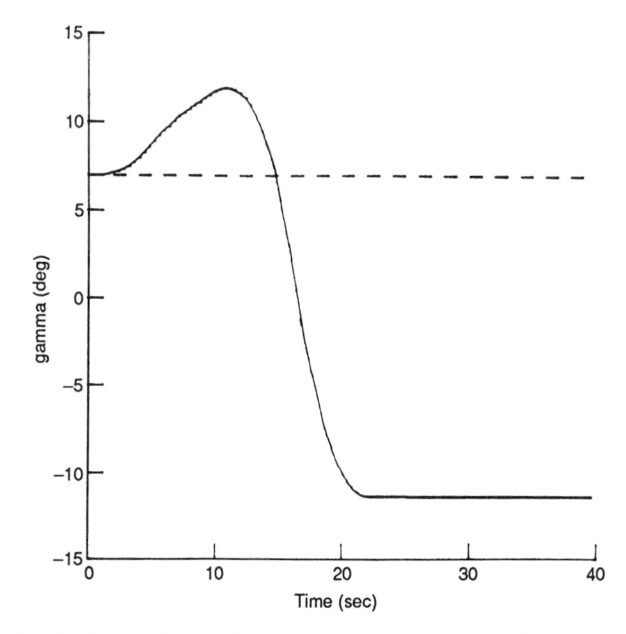


Fig. 10 Nominal Control: Model 1. Relative Path Inclination $\gamma$ for $k = 0$ (dash), $k = 40$ (solid).

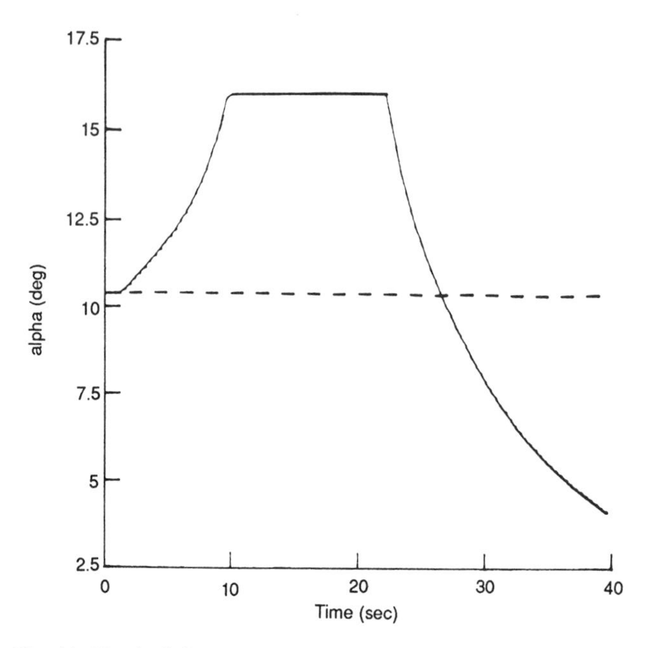


Fig. 11 Nominal Control: Model 1. Relative Angle of Attack $\alpha$ for $k = 0$ (dash), $k = 40$ (solid).

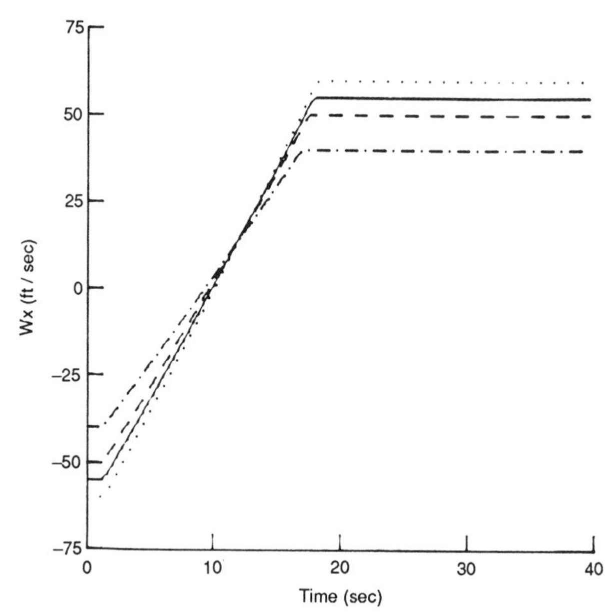


Fig. 12 Proposed Stabilizing Control, $\alpha \leq \alpha_*$: Model 1. Horizontal Windshear $W_x$ for $k = 40$ (dash-dot), $k = 50$ (dash), $k = 55$ (solid), $k = 60$ (dot).

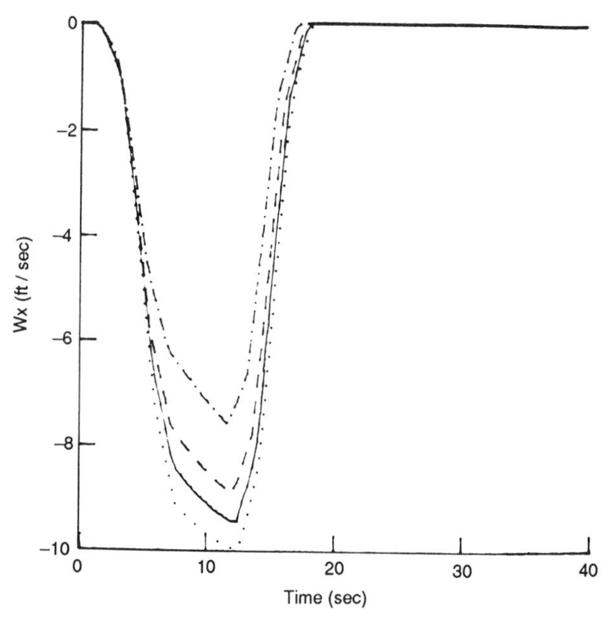


Fig. 13 Proposed Stabilizing Control, $\alpha \leq \alpha_*$: Model 1. Vertical Windshear $W_h$ for $k$ = 40 (dash-dot), $k$ = 50 (dash), $k$ = 55 (solid), $k$ = 60 (dot).

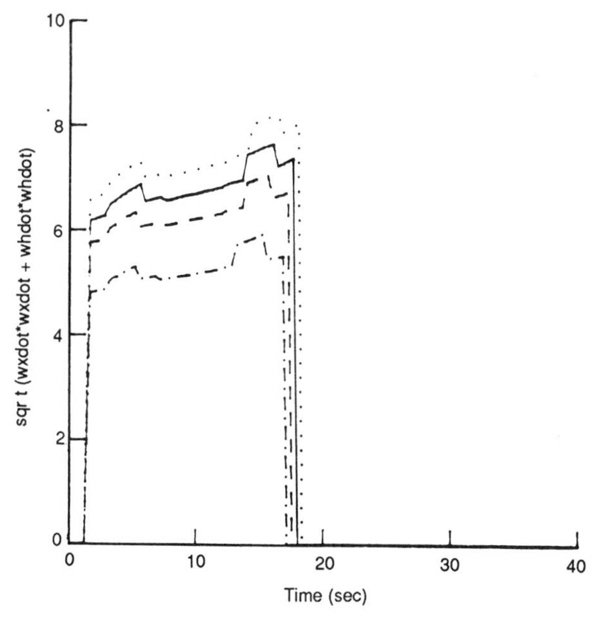


Fig. 14 Proposed Stabilizing Control, $\alpha \leq \alpha_*$: Model 1. Magnitude of Windshear Acceleration $\sqrt{\dot{W}_x^2 + \dot{W}_h^2}$ for $k$ = 40 (dash-dot), $k$ = 50 (dash), $k$ = 55 (solid), $k$ = 60 (dot).

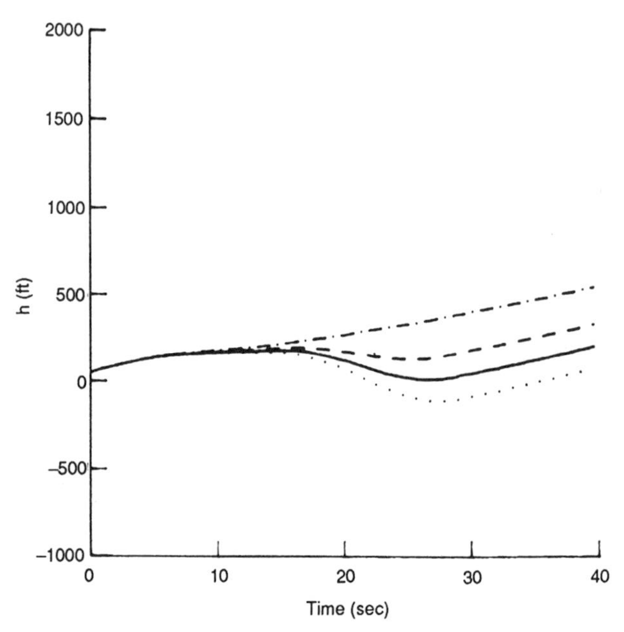


Fig. 15 Proposed Stabilizing Control, $\alpha \leq \alpha_*$: Model 1. Altitude $h$ for $k$ = 40 (dash-dot), $k$ = 50 (dash), $k$ = 55 (solid), $k$ = 60 (dot).

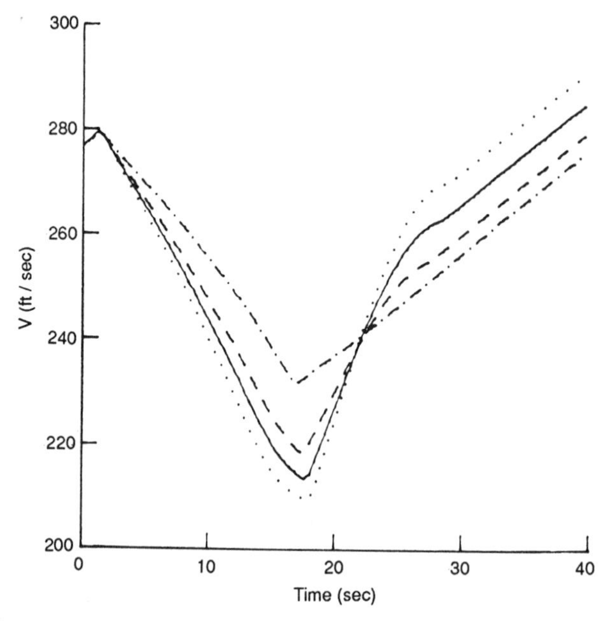


Fig. 16 Proposed Stabilizing Control, $\alpha \leq \alpha_*$: Model 1. Relative Speed $V$ for $k$ = 40 (dash-dot), $k$ = 50 (dash), $k$ = 55 (solid), $k$ = 60 (dot).

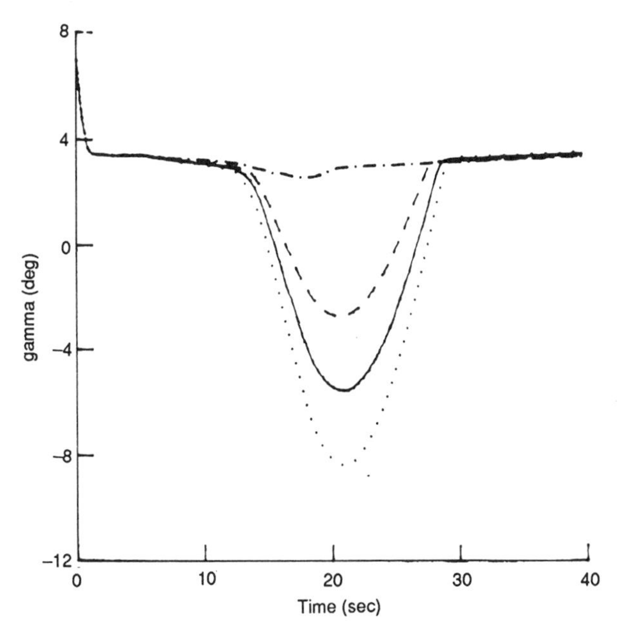


Fig. 17 Proposed Stabilizing Control, $\alpha \leq \alpha_*$: Model 1. Relative Path Inclination $\gamma$ for $k = 40$ (dash-dot), $k = 50$ (dash), $k = 55$ (solid), $k = 60$ (dot).

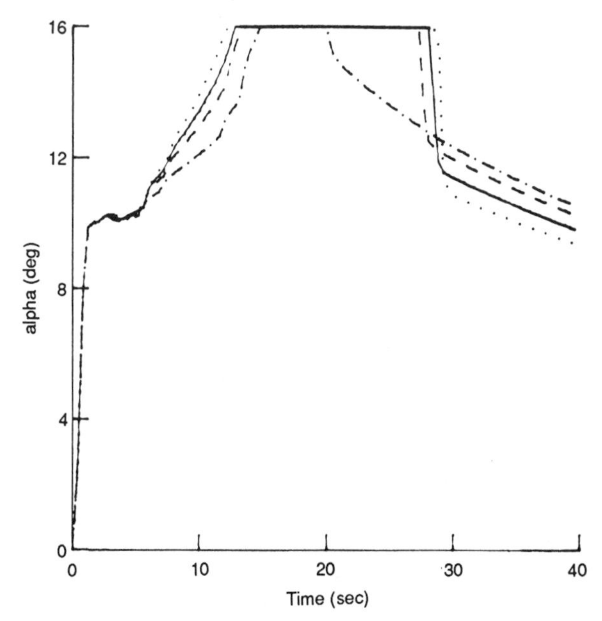


Fig. 18 Proposed Stabilizing Control, $\alpha \leq \alpha_*$: Model 1. Relative Angle of Attack $\alpha$ for $k = 40$ (dash-dot), $k = 50$ (dash), $k = 55$ (solid), $k = 60$ (dot).

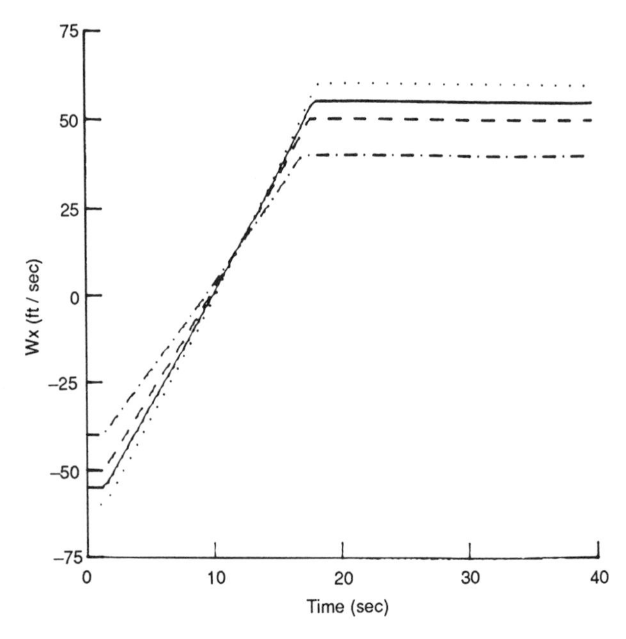


Fig. 19 Proposed Stabilizing Control, $\alpha \leq \alpha_*$ and $|\dot{\alpha}| \leq C$: Model 1. Horizontal Windshear $W_x$ for $k = 40$ (dash-dot), $k = 50$ (dash), $k = 55$ (solid), $k = 60$ (dot).

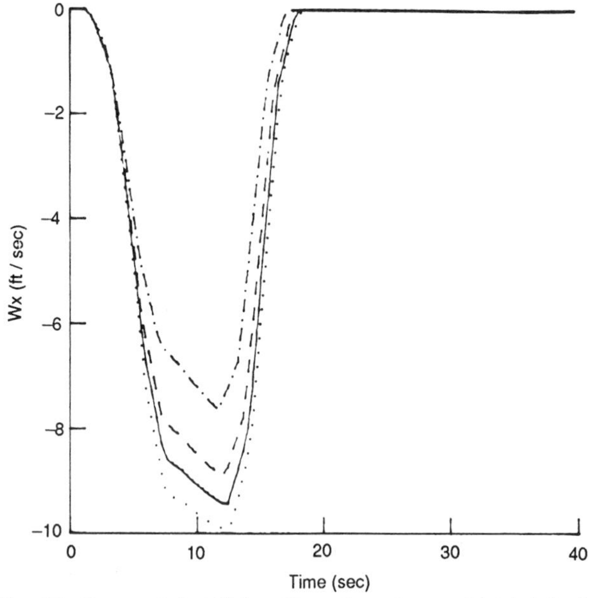


Fig. 20 Proposed Stabilizing Control, $\alpha \leq \alpha_*$ and $|\dot{\alpha}| \leq C$: Model 1. Vertical Windshear $W_h$ for $k = 40$ (dash-dot), $k = 50$ (dash), $k = 55$ (solid), $k = 60$ (dot).

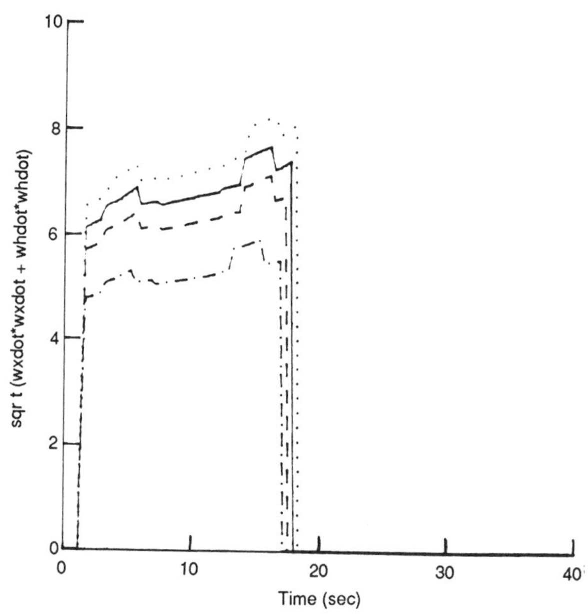


Fig. 21 Proposed Stabilizing Control, $\alpha \leq \alpha_*$ and $|\dot{\alpha}| \leq C$: Model 1. Magnitude of Windshear Acceleration $\sqrt{\dot{W}_x^2 + \dot{W}_h^2}$ for $k = 40$ (dash-dot), $k = 50$ (dash), $k = 55$ (solid), $k = 60$ (dot).

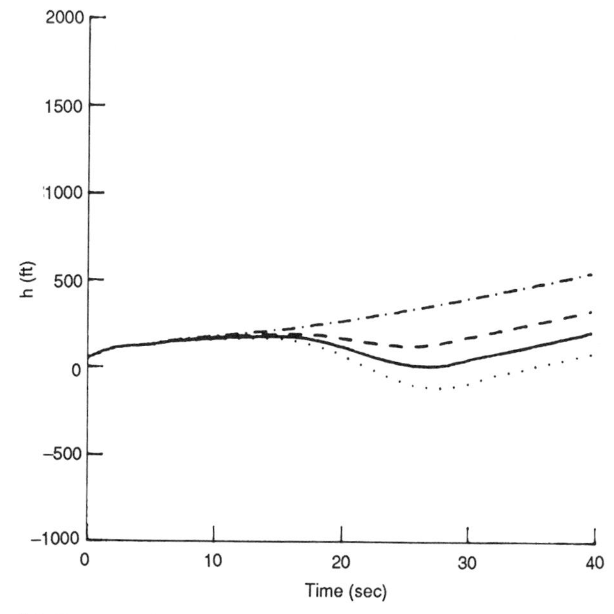


Fig. 22 Proposed Stabilizing Control, $\alpha \leq \alpha_*$ and $|\dot{\alpha}| \leq C$: Model 1. Altitude $h$ for $k = 40$ (dash-dot), $k = 50$ (dash), $k = 55$ (solid), $k = 60$ (dot).

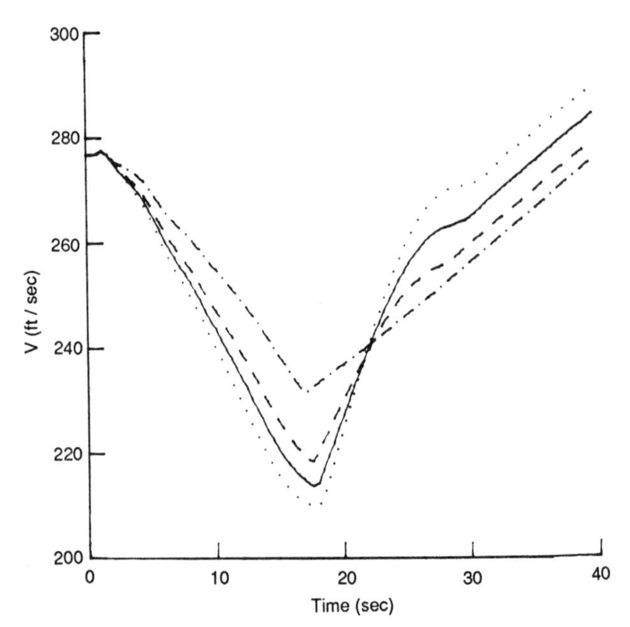


Fig. 23 Proposed Stabilizing Control, $\alpha \le \alpha_*$ and $|\dot{\alpha}| \le C$: Model 1. Relative Speed $V$ for $k = 40$ (dash-dot), $k = 50$ (dash), $k = 55$ (solid), $k = 60$ (dot).

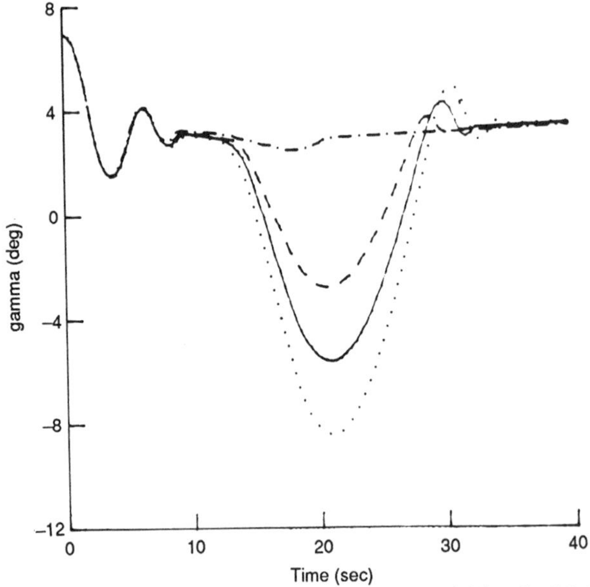


Fig. 24 Proposed Stabilizing Control, $\alpha \le \alpha_*$ and $|\dot{\alpha}| \le C$: Model 1. Relative Path Inclination $\gamma$ for $k = 40$ (dash-dot), $k = 50$ (dash), $k = 55$ (solid), $k = 60$ (dot).

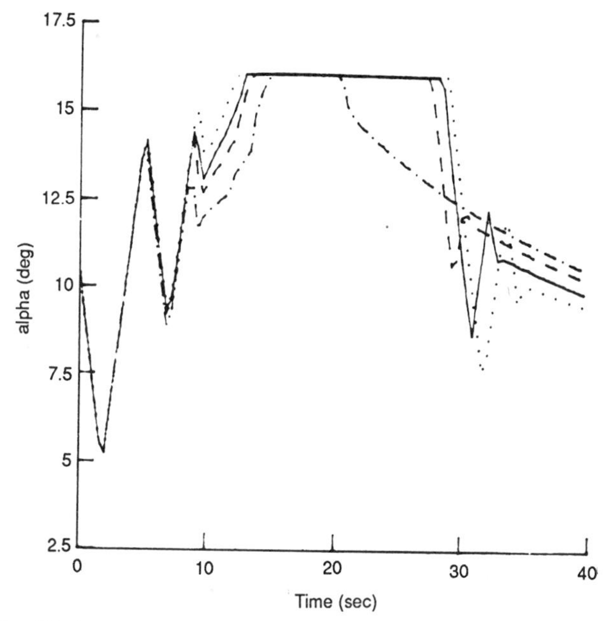


Fig. 25 Proposed Stabilizing Control, $\alpha \leq \alpha_*$ and $|\dot{\alpha}| \leq C$: Model 1. Relative Angle of Attack $\alpha$ for $k = 40$ (dash-dot), $k = 50$ (dash), $k = 55$ (solid), $k = 60$ (dot).

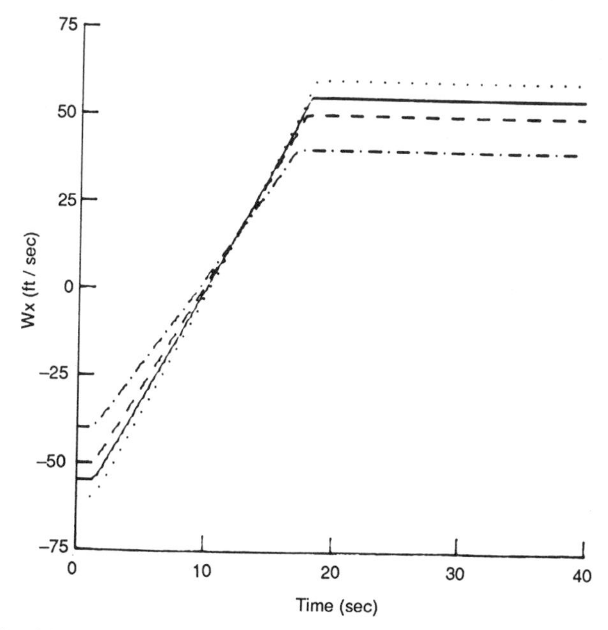


Fig. 26 Simplified Gamma Guidance, [19], $\alpha \leq \alpha_*$ and $|\dot{\alpha}| \leq C$: Model 1. Horizontal Windshear $W_x$ for $k = 40$ (dash-dot), $k = 50$ (dash), $k = 55$ (solid), $k = 60$ (dot).

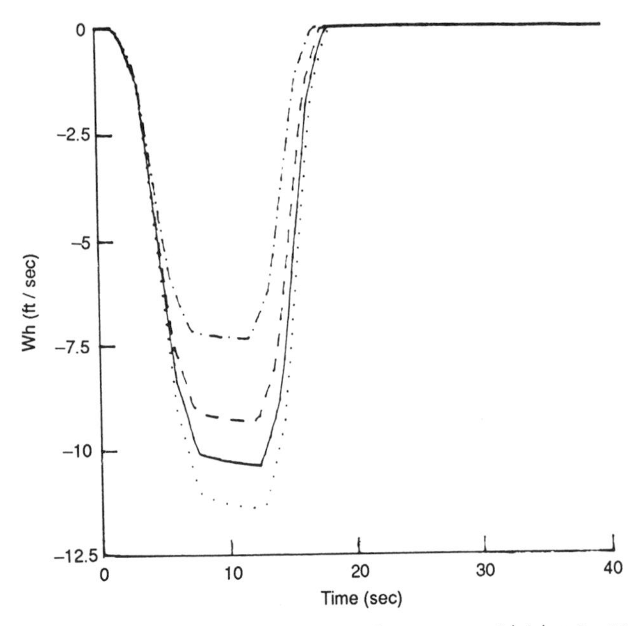


Fig. 27 Simplified Gamma Guidance, [19], $\alpha \leq \alpha_*$ and $|\dot{\alpha}| \leq C$: Model 1. Vertical Windshear $W_h$ for $k = 40$ (dash-dot), $k = 50$ (dash), $k = 55$ (solid), $k = 60$ (dot).

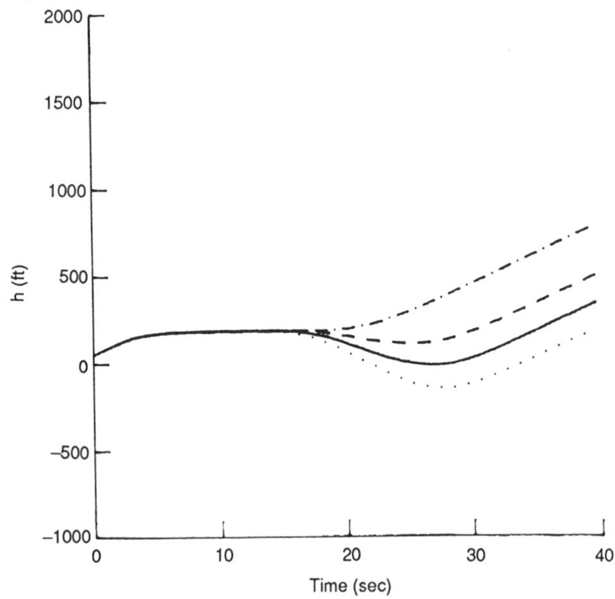


Fig. 28 Simplified Gamma Guidance, [19], $\alpha \leq \alpha_*$ and $|\dot{\alpha}| \leq C$: Model 1. Altitude $h$ for $k = 40$ (dash-dot), $k = 50$ (dash), $k = 55$ (solid), $k = 60$ (dot).

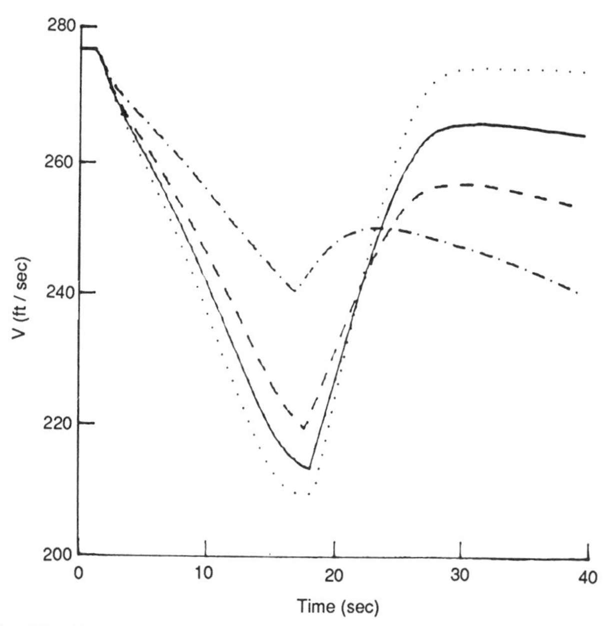


Fig. 29 Simplified Gamma Guidance, [19], $\alpha \leq \alpha_*$ and $|\dot{\alpha}| \leq C$: Model 1. Relative Speed $V$ for $k = 40$ (dash-dot), $k = 50$ (dash), $k = 55$ (solid), $k = 60$ (dot).

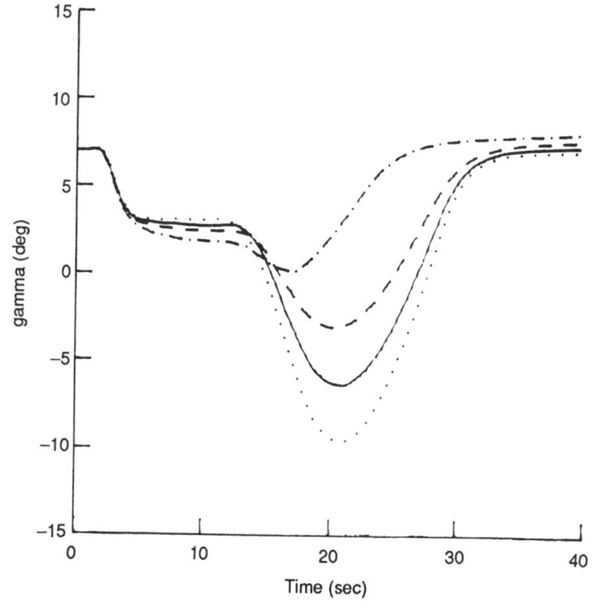


Fig. 30 Simplified Gamma Guidance, [19], $\alpha \leq \alpha_*$ and $|\dot{\alpha}| \leq C$: Model 1. Relative Path Inclination $\gamma$ for $k = 40$ (dash-dot), $k = 50$ (dash), $k = 55$ (solid), $k = 60$ (dot).

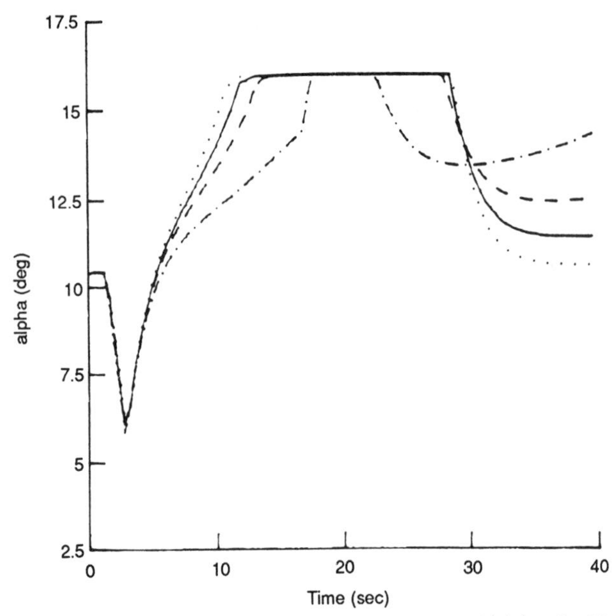


Fig. 31 Simplified Gamma Guidance, [19], $\alpha \leq \alpha_*$ and $|\dot{\alpha}| \leq C$: Model 1. Relative Angle of Attack $\alpha$ for $k = 40$ (dash-dot), $k = 50$ (dash), $k = 55$ (solid), $k = 60$ (dot).

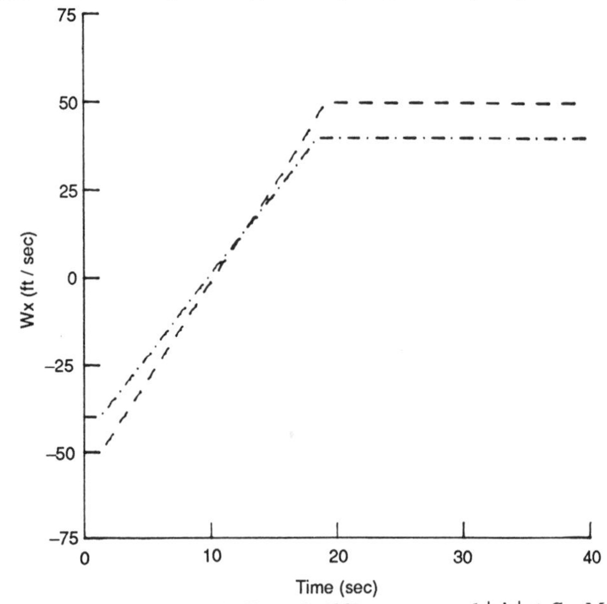


Fig. 32 Nonlinear Feedback Control, [20], $\alpha \leq \alpha_*$ and $|\dot{\alpha}| \leq C$: Model 1. Horizontal Windshear $W_x$ for $k = 40$ (dash-dot), $k = 50$ (dash).

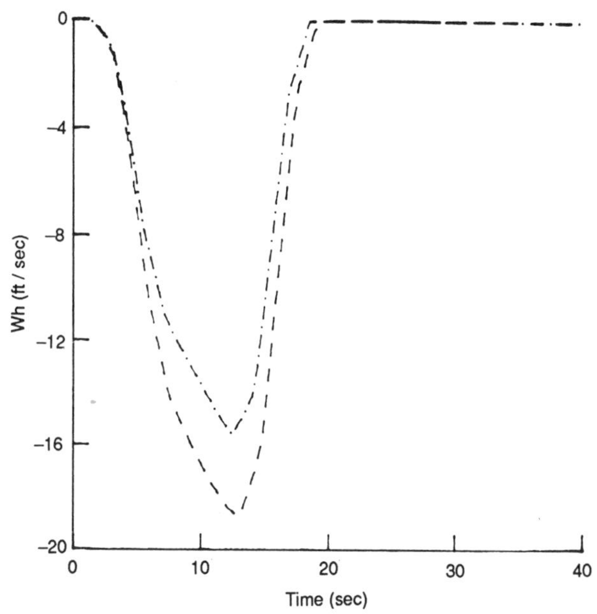


Fig. 33 Nonlinear Feedback Control, [20], $\alpha \leq \alpha_*$ and $|\dot{\alpha}| \leq C$: Model 1. Vertical Windshear $W_h$ for $k = 40$ (dash-dot), $k = 50$ (dash).

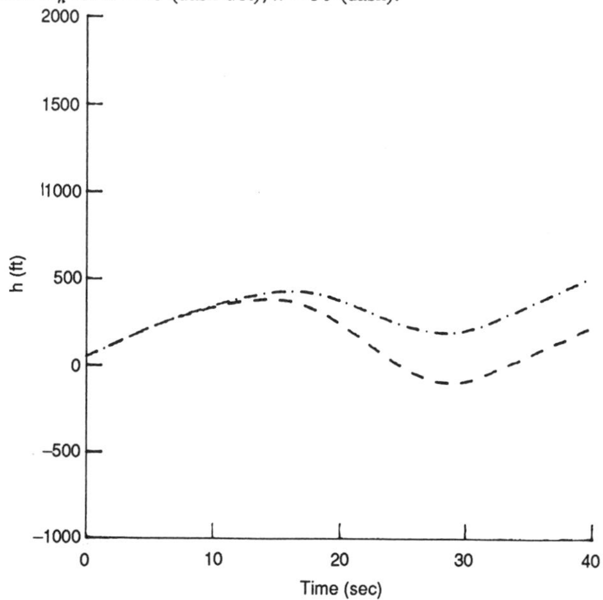


Fig. 34 Nonlinear Feedback Control, [20], $\alpha \leq \alpha_*$ and $|\dot{\alpha}| \leq C$: Model 1. Altitude $h$ for $k = 40$ (dash-dot), $k = 50$ (dash).

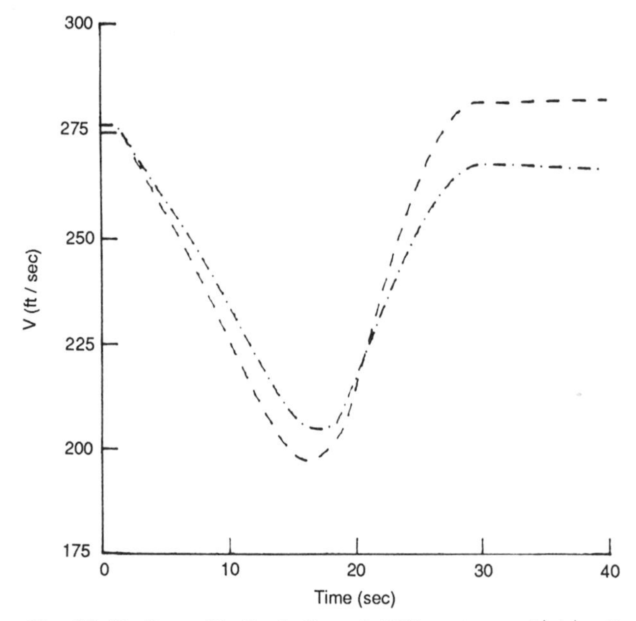


Fig. 35 Nonlinear Feedback Control, [20], $\alpha \leq \alpha_*$ and $|\dot{\alpha}| \leq C$: Model 1. Relative Speed $V$ for $k = 40$ (dash-dot), $k = 50$ (dash).

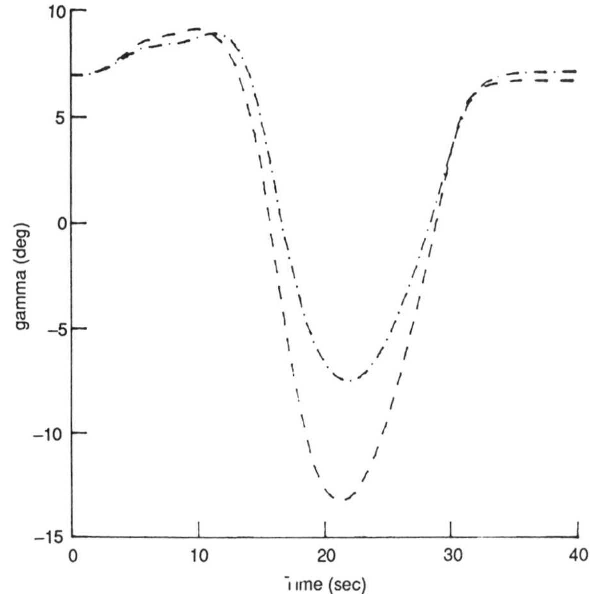


Fig. 36 Nonlinear Feedback Control, [20], $\alpha \leq \alpha_*$ and $|\dot{\alpha}| \leq C$: Model 1. Relative Path Inclination $\gamma$ for $k = 40$ (dash-dot), $k = 50$ (dash).

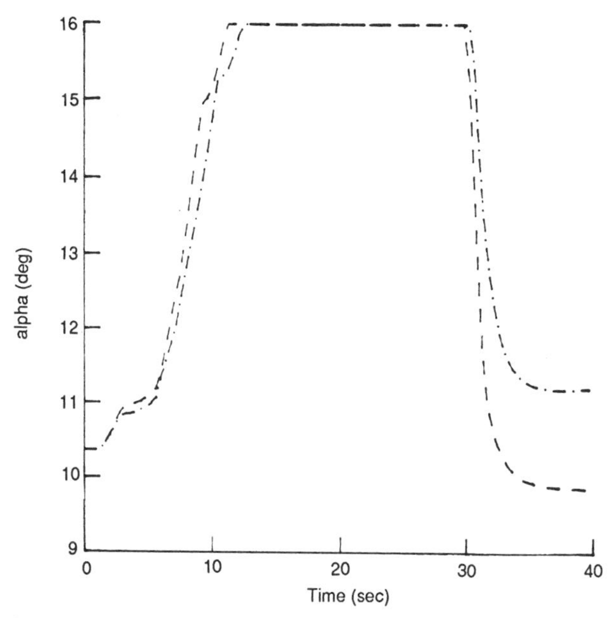


Fig. 37 Nonlinear Feedback Control, [20], $\alpha \leq \alpha_*$ and $|\dot{\alpha}| \leq C$: Model 1. Relative Angle of Attack $\alpha$ for $k = 40$ (dash-dot), $k = 50$ (dash).

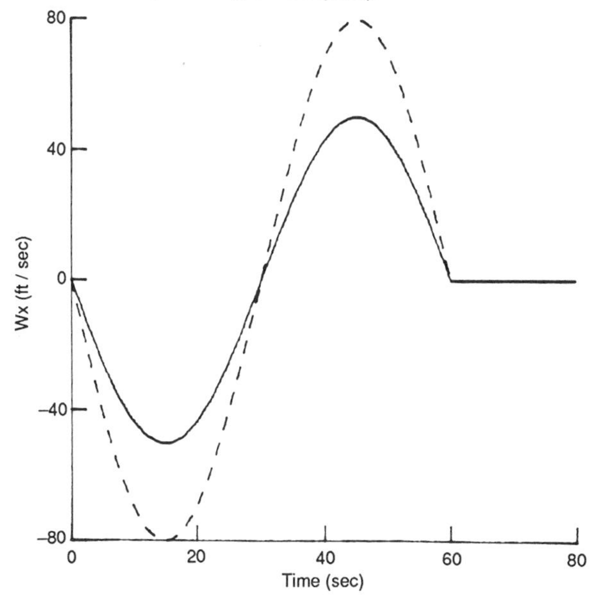


Fig. 38 Model 2. Horizontal Windshear $W_x$ for $W_{x0} = 50$ (solid), 80 (dash).

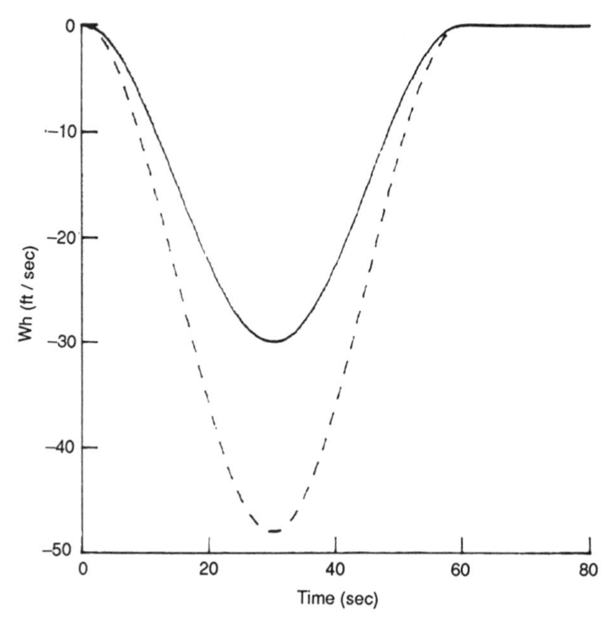


Fig. 39 Model 2. Vertical Windshear $W_h$ for $W_{h0}$ = 30 (solid), 48 (dash).

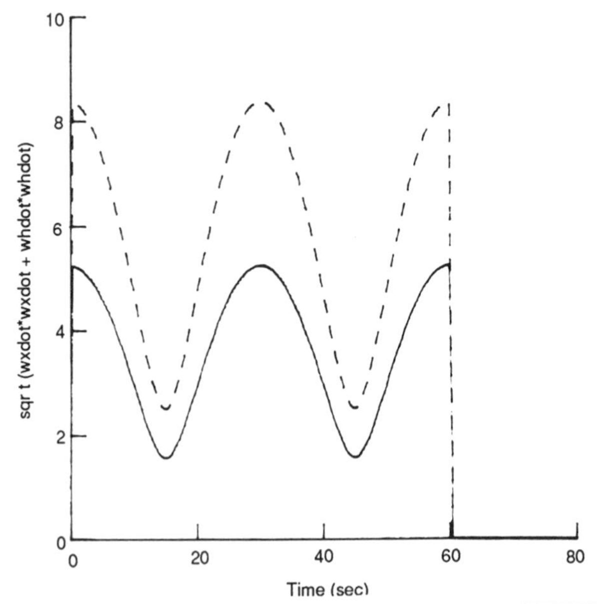


Fig. 40 Model 2. Magnitude of Windshear Acceleration $\sqrt{\dot{W}_x^2 + \dot{W}_h^2}$ for $W_{x0}/W_{h0}$ = 50/30 (solid), $W_{x0}/W_{h0}$ = 80/48 (dash).

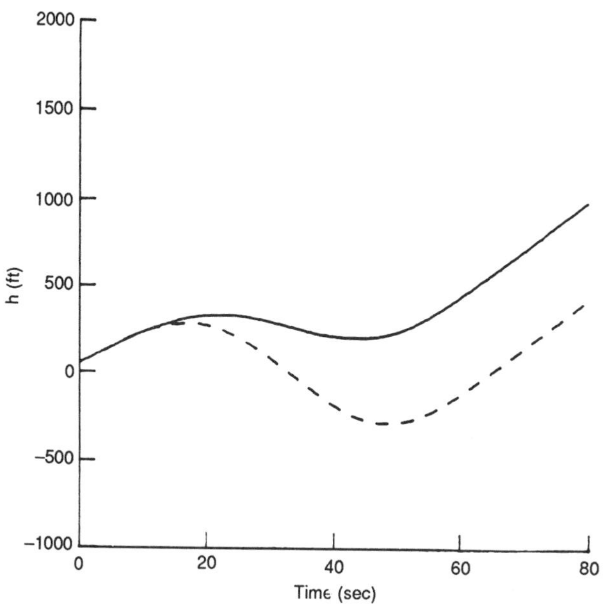


Fig. 41 Proposed Stabilizing Control, $\alpha \leq \alpha_*$: Model 2. Altitude $h$ for $W_{x0}/W_{h0}$ = 50/30 (solid), $W_{x0}/W_{h0}$ = 80/48 (dash).

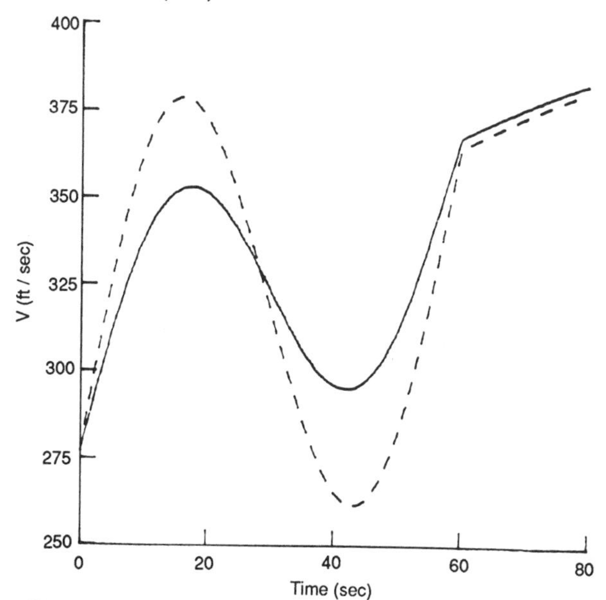


Fig. 42 Proposed Stabilizing Control, $\alpha \leq \alpha_*$: Model 2. Relative Speed $V$ for $W_{x0}/W_{h0}$ = 50/30 (solid), $W_{x0}/W_{h0}$ = 80/48 (dash).

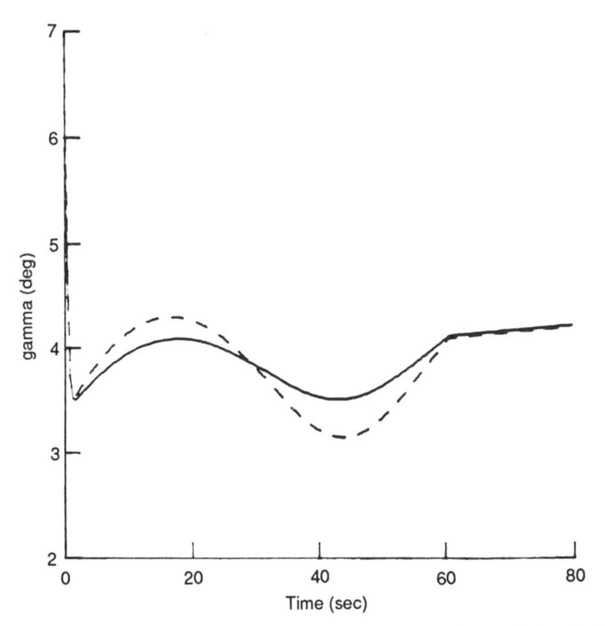


Fig. 43 Proposed Stabilizing Control, $\alpha \le \alpha_*$: Model 2. Relative Path Inclination $\gamma$ for $W_{x0}/W_{h0}$ = 50/30 (solid), $W_{x0}/W_{h0}$ = 80/48 (dash).

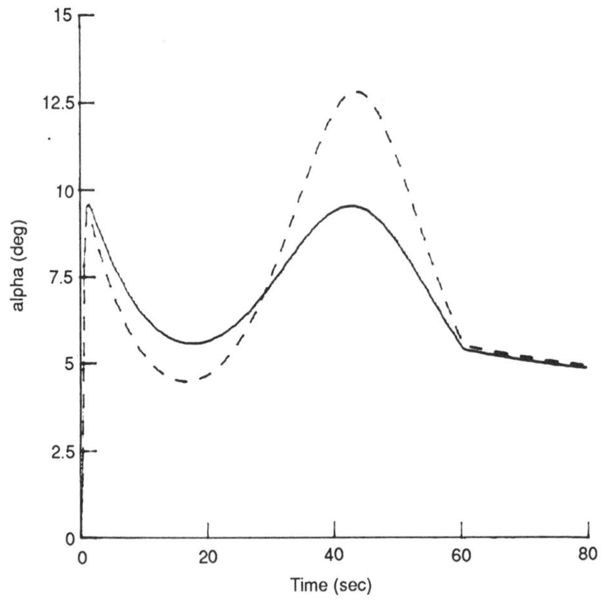


Fig. 44 Proposed Stabilizing Control, $\alpha \le \alpha_*$: Model 2. Relative Angle of Attack $\alpha$ for $W_{x0}/W_{h0}$ = 50/30 (solid), $W_{x0}/W_{h0}$ = 80/48 (dash).

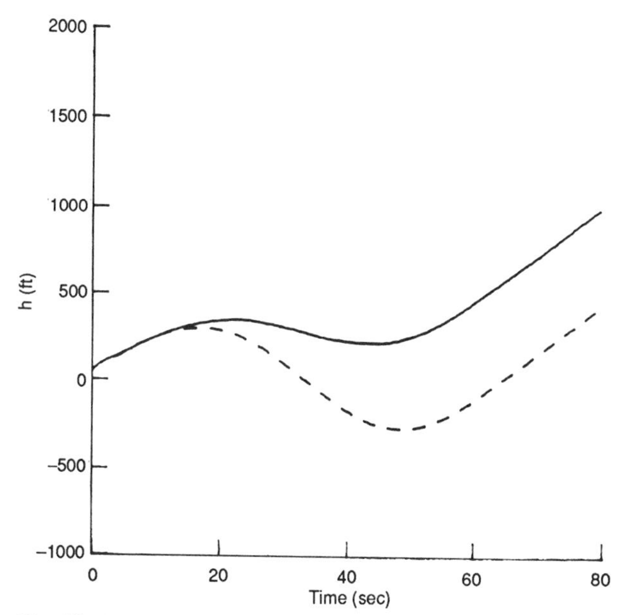


Fig. 45 Proposed Stabilizing Control, $\alpha \leq \alpha_*$ and $|\dot{\alpha}| \leq C$: Model 2. Altitude $h$ for $W_{x0}/W_{h0} = 50/30$ (solid), $W_{x0}/W_{h0} = 80/48$ (dash).

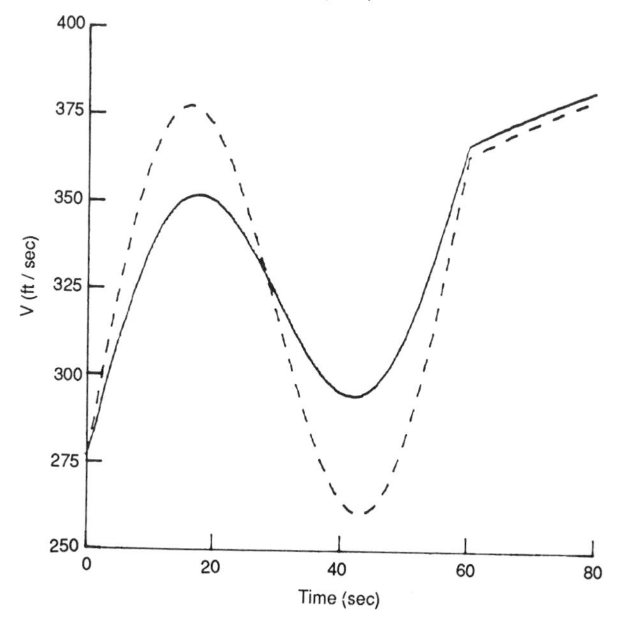


Fig. 46 Proposed Stabilizing Control, $\alpha \leq \alpha_*$ and $|\dot{\alpha}| \leq C$: Model 2. Relative Speed $V$ for $W_{x0}/W_{h0} = 50/30$ (solid), $W_{x0}/W_{h0} = 80/48$ (dash).

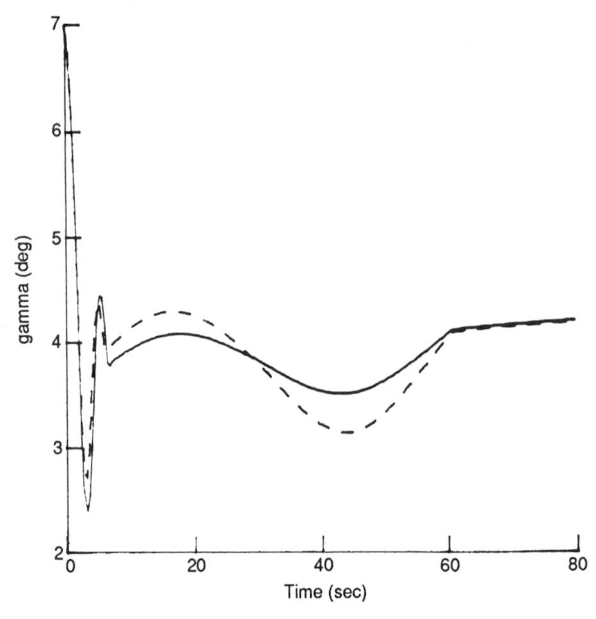


Fig. 47 Proposed Stabilizing Control, $\alpha \leq \alpha_*$ and $|\dot{\alpha}| \leq C$: Model 2. Relative Path Inclination $\gamma$ for $W_{x0}/W_{h0} = 50/30$ (solid), $W_{x0}/W_{h0} = 80/48$ (dash).

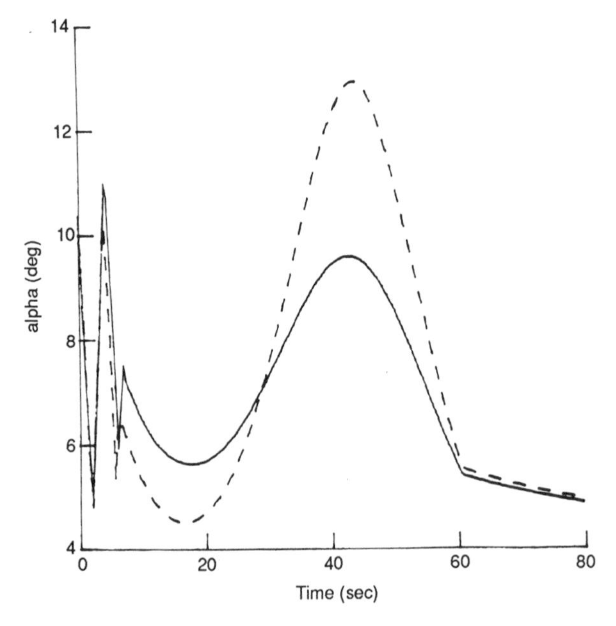


Fig. 48 Proposed Stabilizing Control, $\alpha \leq \alpha_*$ and $|\dot{\alpha}| \leq C$: Model 2. Relative Angle of Attack $\alpha$ for $W_{x0}/W_{h0} = 50/30$ (solid), $W_{x0}/W_{h0} = 80/48$ (dash).

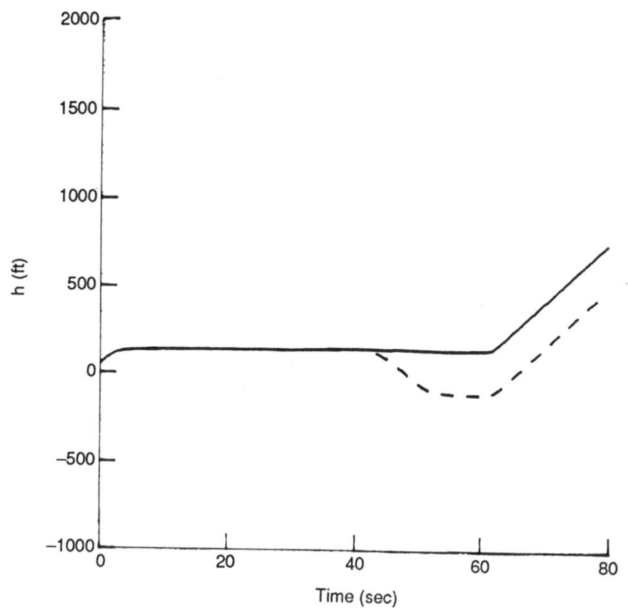


Fig. 49 Simplified Gamma Guidance, [19], $\alpha \le \alpha_*$ and $|\dot{\alpha}| \le C$: Model 2. Altitude $h$ for $W_{x0}/W_{h0} = 50/30$ (solid), $W_{x0}/W_{h0} = 80/48$ (dash).

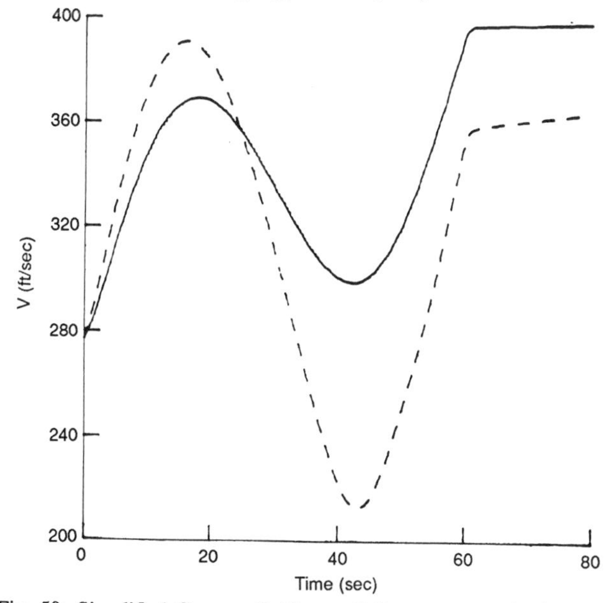


Fig. 50 Simplified Gamma Guidance, [19], $\alpha \le \alpha_*$ and $|\dot{\alpha}| \le C$: Model 2. Relative Speed $V$ for $W_{x0}/W_{h0} = 50/30$ (solid), $W_{x0}/W_{h0} = 80/48$ (dash).

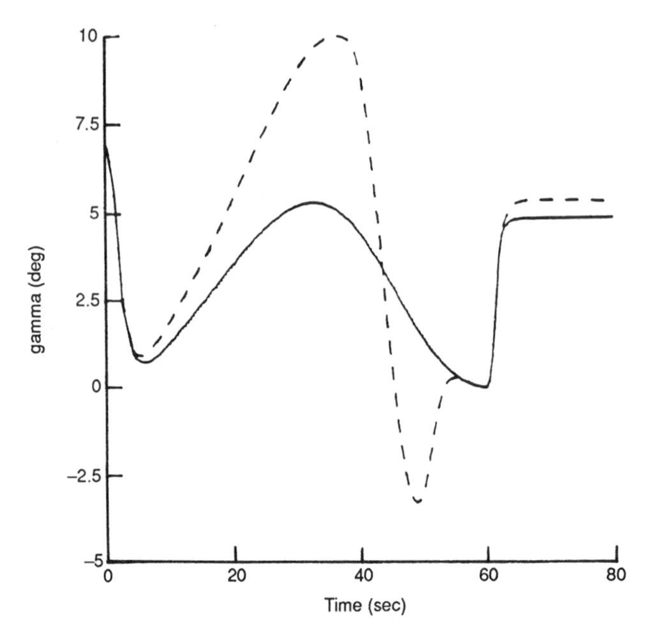


Fig. 51 Simplified Gamma Guidance, [19], $\alpha \le \alpha_*$ and $|\dot{\alpha}| \le C$: Model 2. Relative Path Inclination $\gamma$ for $W_{x0}/W_{h0} = 50/30$ (solid), $W_{x0}/W_{h0} = 80/48$ (dash).

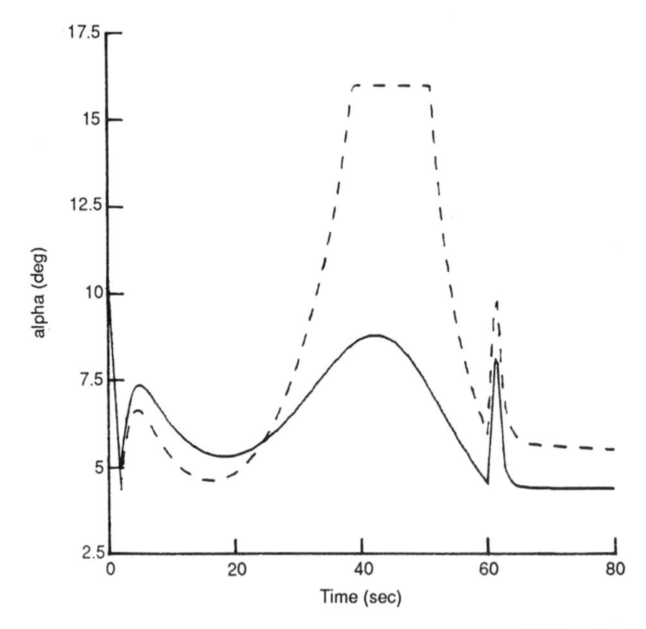


Fig. 52 Simplified Gamma Guidance, [19], $\alpha \le \alpha_*$ and $|\dot{\alpha}| \le C$: Model 2. Relative Angle of Attack $\alpha$ for $W_{x0}/W_{h0} = 50/30$ (solid), $W_{x0}/W_{h0} = 80/48$ (dash).

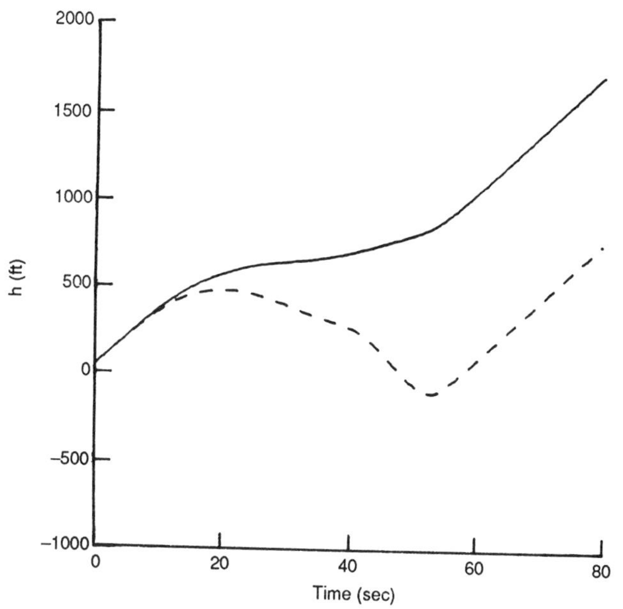


Fig. 53 Nonlinear Feedback Control, [20], $\alpha \leq \alpha_*$ and $|\dot{\alpha}| \leq C$: Model 2. Altitude $h$ for $W_{x0}/W_{h0} = 50/30$ (solid), $W_{x0}\ W_{h0} = 80/48$ (dash).

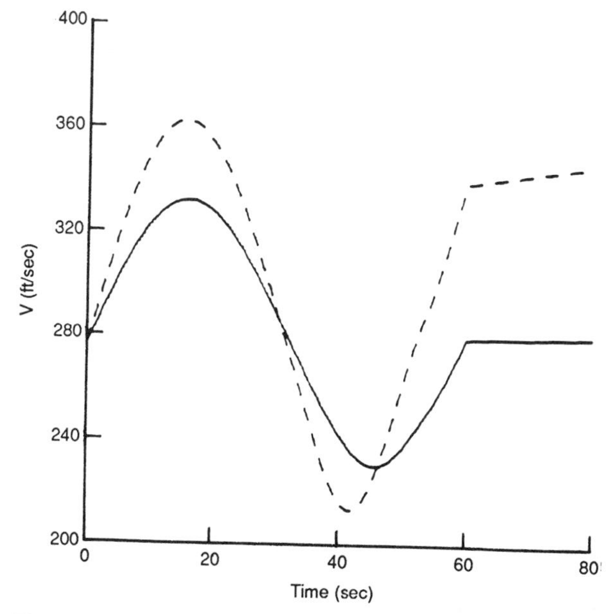


Fig. 54 Nonlinear Feedback Control, [20], $\alpha \leq \alpha_*$ and $|\dot{\alpha}| \leq C$: Model 2. Relative Speed $V$ for $W_{x0}/W_{h0} = 50/30$ (solid), $W_{x0}/W_{h0} = 80/48$ (dash).

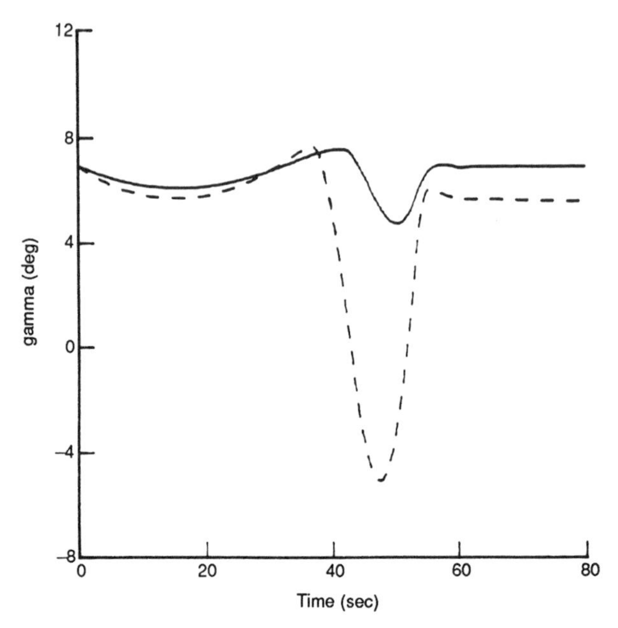


Fig. 55 Nonlinear Feedback Control, [20], $\alpha \le \alpha_*$ and $|\dot{\alpha}| \le C$: Model 2. Relative Path Inclination $\gamma$ for $W_{x0}/W_{h0} = 50/30$ (solid), $W_{x0}/W_{h0} = 80/48$ (dash).

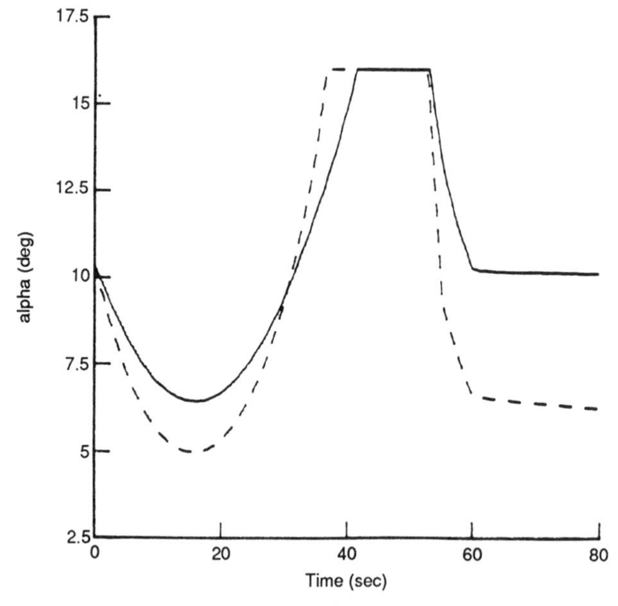


Fig. 56 Nonlinear Feedback Control, [20], $\alpha \le \alpha_*$ and $|\dot{\alpha}| \le C$: Model 2. Relative Angle of Attack $\alpha$ for $W_{x0}/W_{h0} = 50/30$ (solid), $W_{x0}/W_{h0} = 80/48$ (dash).

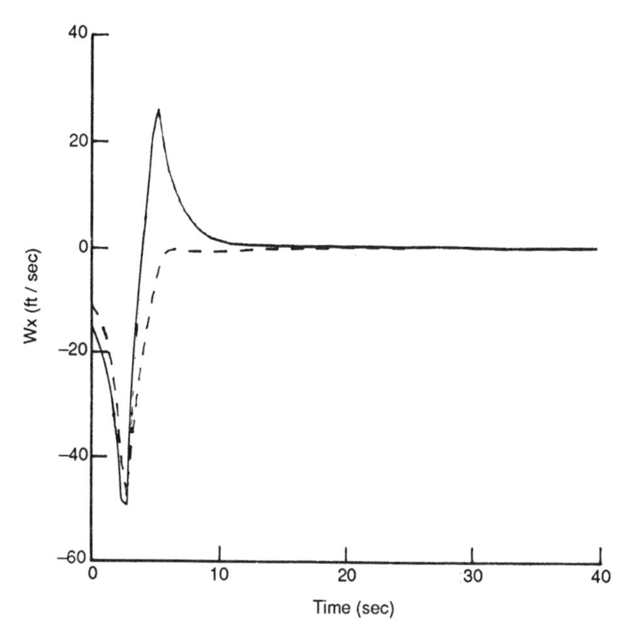


Fig. 57 Proposed Stabilizing Control, $\alpha \leq \alpha_*$: Model 3. Horizontal Windshear $W_x$ for $V_0 = 100$ (dash), $V_0 = 140$ (solid).

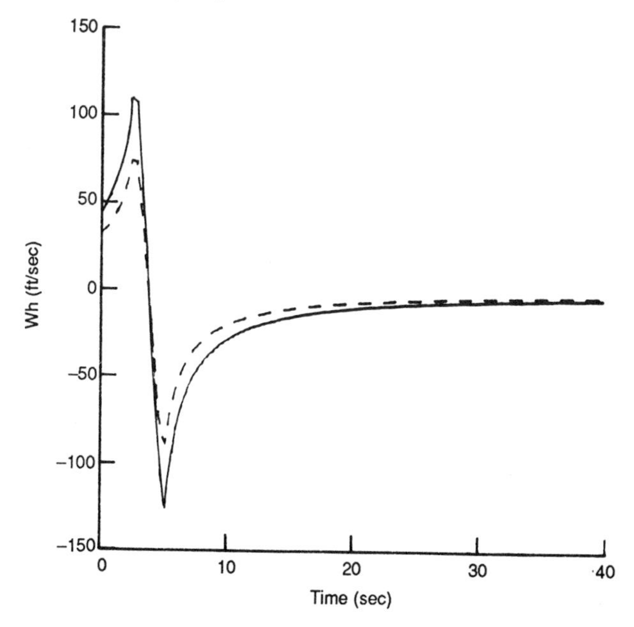


Fig. 58 Proposed Stabilizing Control, $\alpha \leq \alpha_*$: Model 3. Vertical Windshear $W_h$ for $V_0 = 100$ (dash), $V_0 = 140$ (solid).

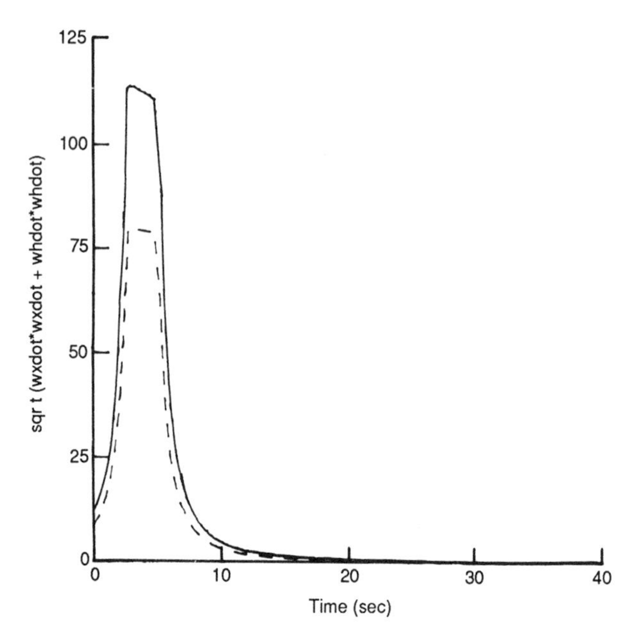


Fig. 59 Proposed Stabilizing Control, $\alpha \leq \alpha_*$: Model 3. Magnitude of Windshear Acceleration $\sqrt{\dot{W}_x^2 + \dot{W}_h^2}$ for $V_0 = 100$ (dash), $V_0 = 140$ (solid).

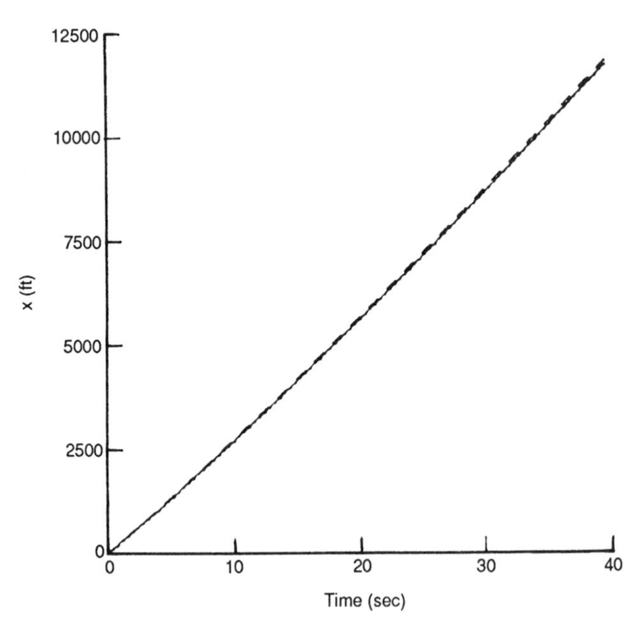


Fig. 60 Proposed Stabilizing Control, $\alpha \leq \alpha_*$: Model 3. Horizontal Distance $x$ for $V_0 = 100$ (dash), $V_0 = 140$ (solid).

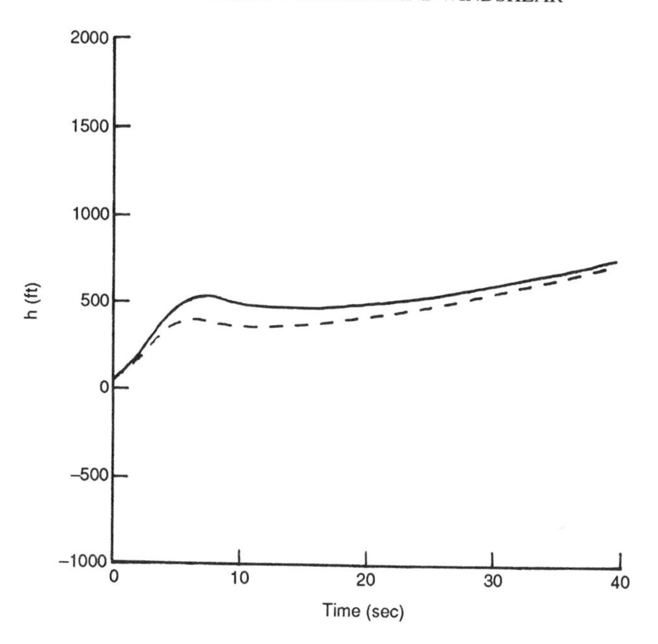


Fig. 61 Proposed Stabilizing Control, $\alpha \leq \alpha_*$: Model 3. Altitude $h$ for $V_0$ = 100 (dash), $V_0$ = 140 (solid).

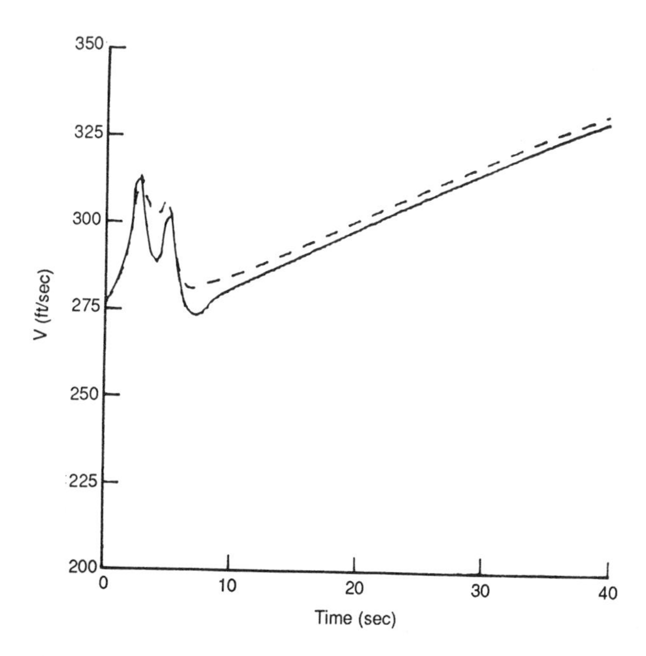


Fig. 62 Proposed Stabilizing Control, $\alpha \leq \alpha_*$: Model 3. Relative Speed $V$ for $V_0$ = 100 (dash), $V_0$ = 140 (solid).

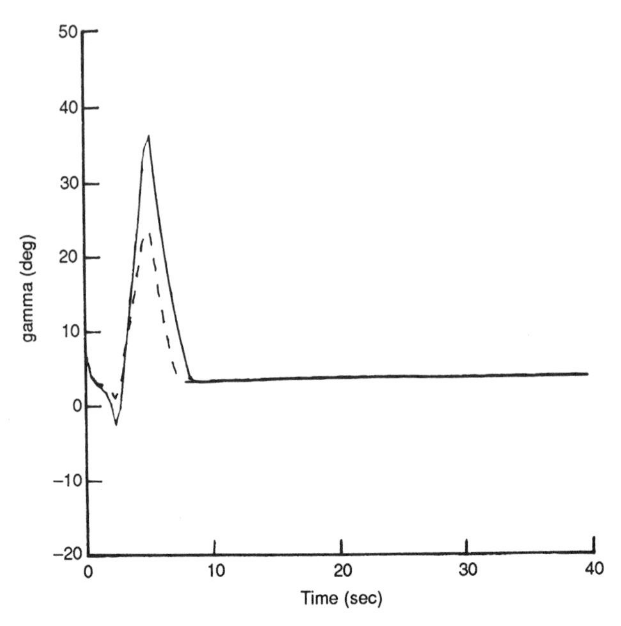


Fig. 63 Proposed Stabilizing Control, $\alpha \leq \alpha_*$: Model 3. Relative Path Inclination $\gamma$ for $V_0 = 100$ (dash), $V_0 = 140$ (solid).

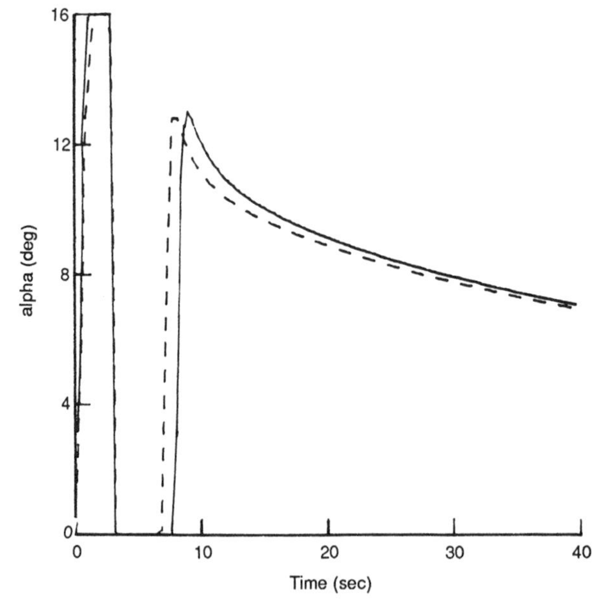


Fig. 64 Proposed Stabilizing Control, $\alpha \leq \alpha_*$: Model 3. Relative Angle of Attack $\alpha$ for $V_0 = 100$ (dash), $V_0 = 140$ (solid).

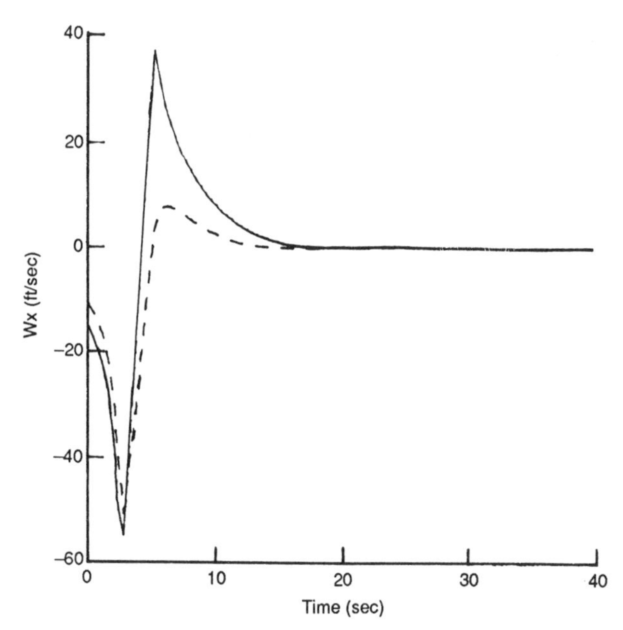


Fig. 65 Proposed Stabilizing Control, $\alpha \leq \alpha_*$ and $|\dot{\alpha}| \leq C$: Model 3. Horizontal Windshear $W_x$ for $V_0 = 100$ (dash), $V_0 = 140$ (solid).

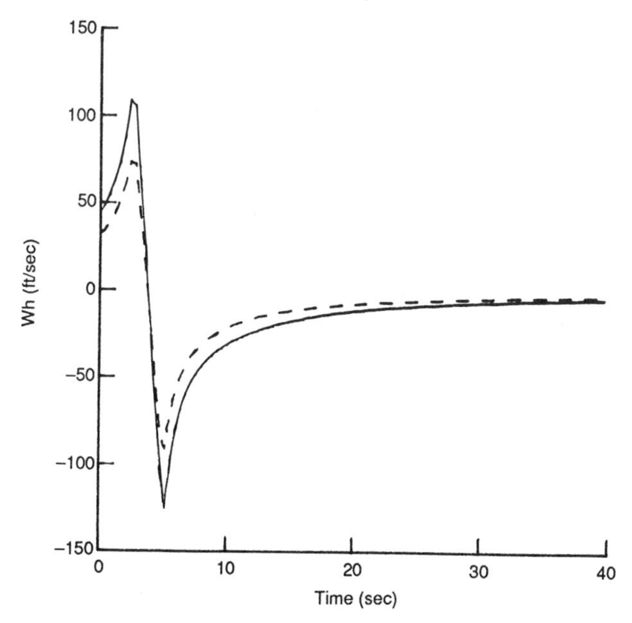


Fig. 66 Proposed Stabilizing Control, $\alpha \leq \alpha_*$ and $|\dot{\alpha}| \leq C$: Model 3. Vertical Windshear $W_h$ for $V_0 = 100$ (dash), $V_0 = 140$ (solid).

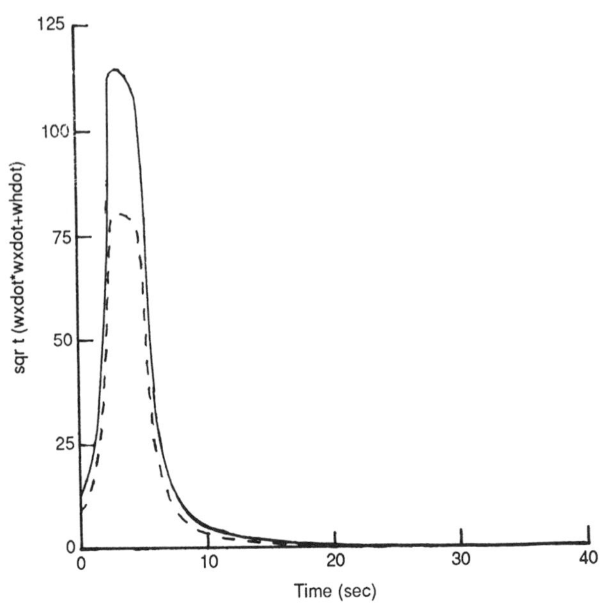


Fig. 67 Proposed Stabilizing Control, $\alpha \leq \alpha_*$ and $|\dot{\alpha}| \leq C$: Model 3. Magnitude of Windshear Acceleration $\sqrt{\dot{W}_x^2 + \dot{W}_h^2}$ for $V_0 = 100$ (dash), $V_0 = 140$ (solid).

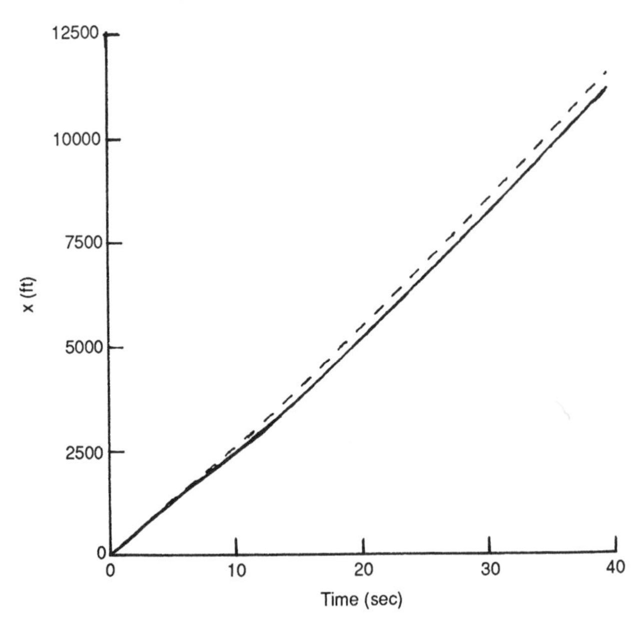


Fig. 68 Proposed Stabilizing Control, $\alpha \leq \alpha_*$ and $|\dot{\alpha}| \leq C$: Model 3. Horizontal Distance $x$ for $V_0 = 100$ (dash), $V_0 = 140$ (solid).

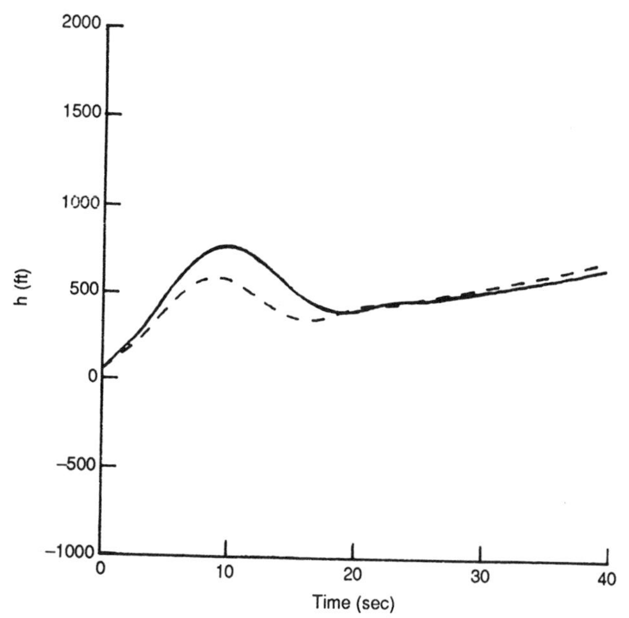


Fig. 69 Proposed Stabilizing Control, $\alpha \leq \alpha_*$ and $|\dot{\alpha}| \leq C$: Model 3. Altitude $h$ for $V_0 = 100$ (dash), $V_0 = 140$ (solid).

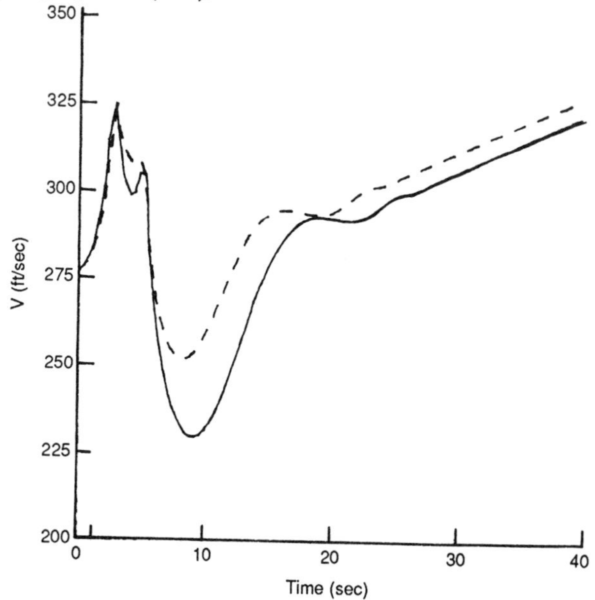


Fig. 70 Proposed Stabilizing Control, $\alpha \leq \alpha_*$ and $|\dot{\alpha}| \leq C$: Model 3. Relative Speed $V$ for $V_0 = 100$ (dash), $V_0 = 140$ (solid).

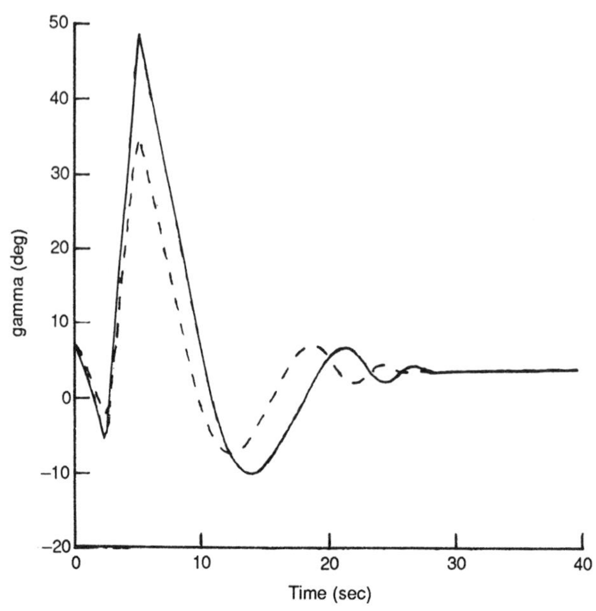


Fig. 71 Proposed Stabilizing Control, $\alpha \leq \alpha_*$ and $|\dot{\alpha}| \leq C$: Model 3. Relative Path Inclination $\gamma$ for $V_0 = 100$ (dash), $V_0 = 140$ (solid).

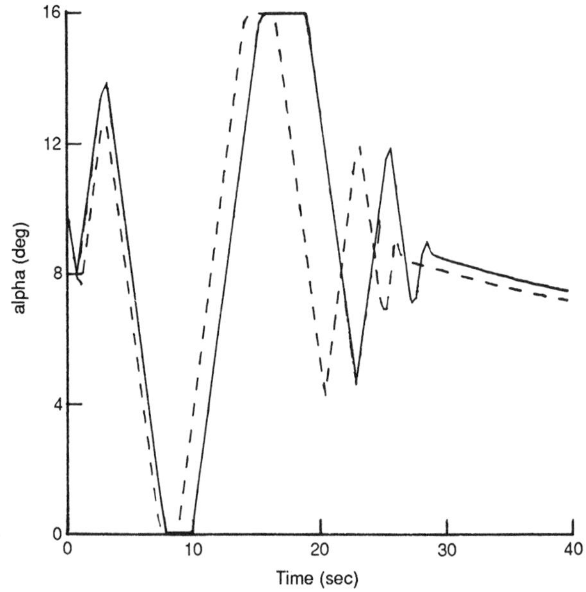


Fig. 72 Proposed Stabilizing Control, $\alpha \leq \alpha_*$ and $|\dot{\alpha}| \leq C$: Model 3. Relative Angle of Attack $\alpha$ for $V_0 = 100$ (dash), $V_0 = 140$ (solid).

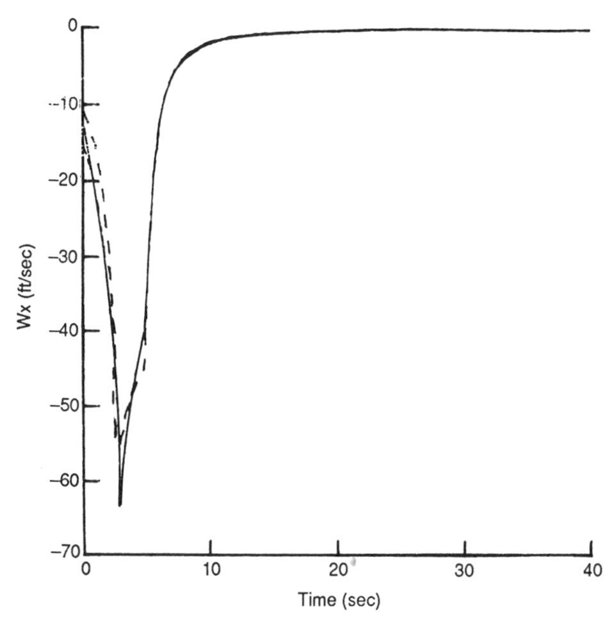


Fig. 73 Simplified Gamma Guidance, [19], $\alpha \leq \alpha_*$ and $|\dot{\alpha}| \leq C$: Model 3. Horizontal Windshear $W_x$ for $V_0 = 100$ (dash), $V_0 = 140$ (solid).

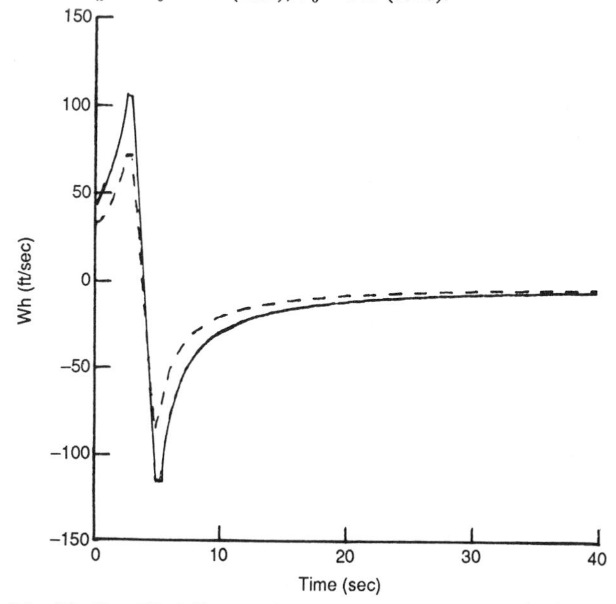


Fig. 74 Simplified Gamma Guidance, [19], $\alpha \leq \alpha_*$ and $|\dot{\alpha}| \leq C$: Model 3. Vertical Windshear $W_h$ for $V_0 = 100$ (dash), $V_0 = 140$ (solid).

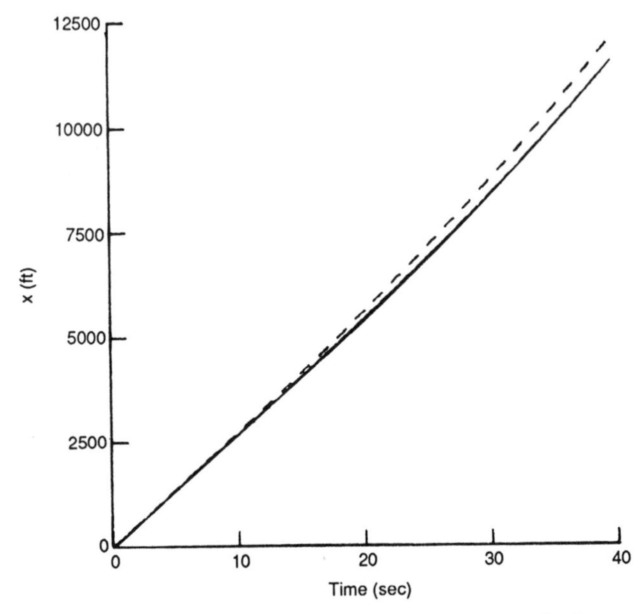


Fig. 75 Simplified Gamma Guidance, [19], $\alpha \leq \alpha_*$ and $|\dot{\alpha}| \leq C$: Model 3. Horizontal Distance $x$ for $V_0 = 100$ (dash), $V_0 = 140$ (solid).

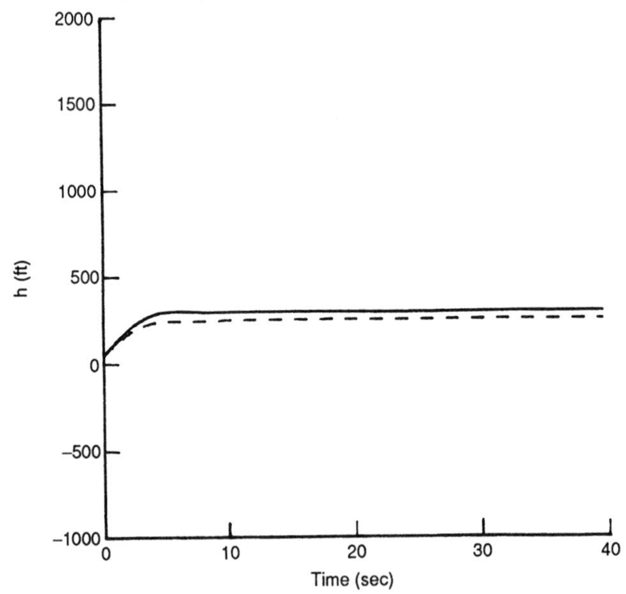


Fig. 76 Simplified Gamma Guidance, [19], $\alpha \leq \alpha_*$ and $|\dot{\alpha}| \leq C$: Model 3. Altitude $h$ for $V_0 = 100$ (dash), $V_0 = 140$ (solid).

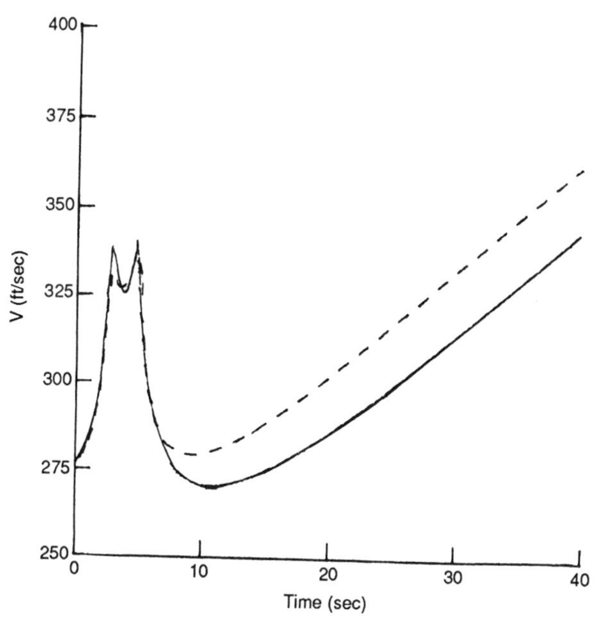


Fig. 77 Simplified Gamma Guidance, [19], $\alpha \le \alpha_*$ and $|\dot{\alpha}| \le C$: Model 3. Relative Speed $V$ for $V_0 = 100$ (dash), $V_0 = 140$ (solid).

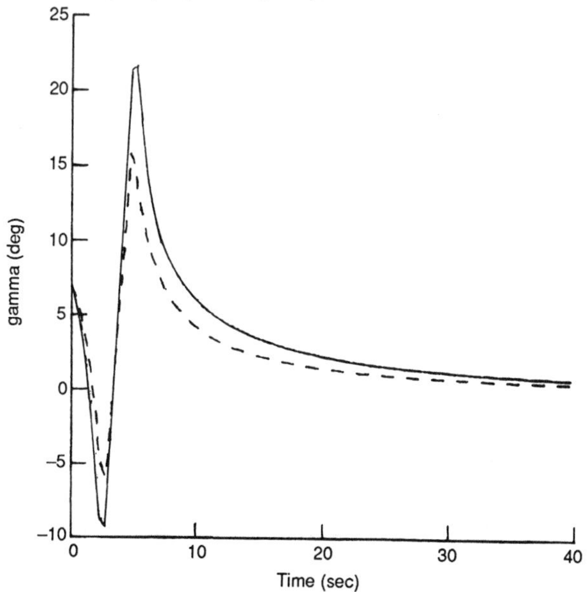


Fig. 78 Simplified Gamma Guidance, [19], $\alpha \le \alpha_*$ and $|\dot{\alpha}| \le C$: Model 3. Relative Path Inclination $\gamma$ for $V_0 = 100$ (dash), $V_0 = 140$ (solid).

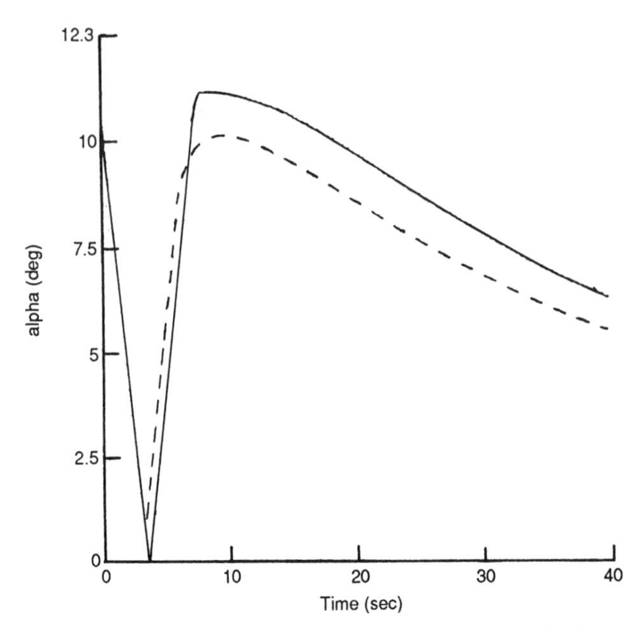


Fig. 79 Simplified Gamma Guidance, [19], $\alpha \leq \alpha_*$ and $|\dot{\alpha}| \leq C$: Model 3. Relative Angle of Attack $\alpha$ for $V_0 = 100$ (dash), $V_0 = 140$ (solid).

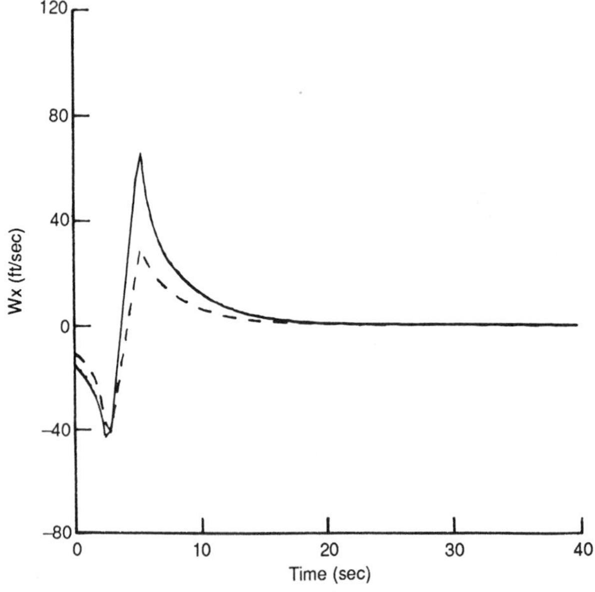


Fig. 80 Nonlinear Feedback Control, [20], $\alpha \leq \alpha_*$ and $|\dot{\alpha}| \leq C$: Model 3. Horizontal Windshear $W_x$ for $V_0 = 100$ (dash), $V_0 = 140$ (solid).

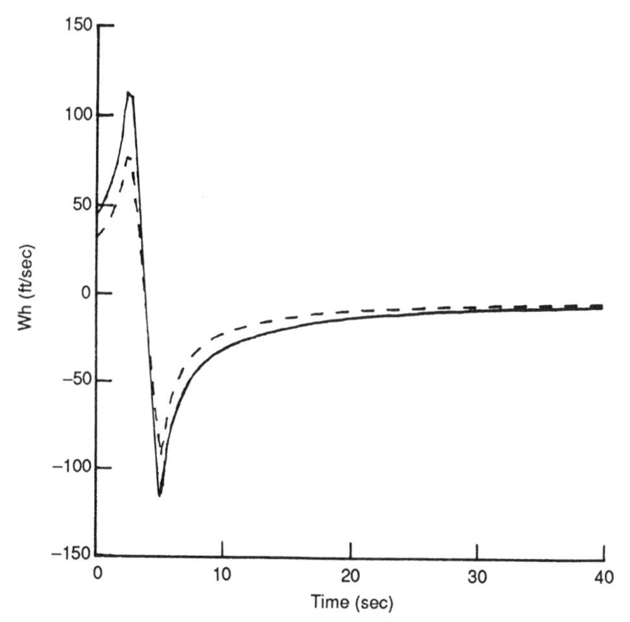


Fig. 81 Nonlinear Feedback Control, [20], $\alpha \leq \alpha_*$ and $|\dot{\alpha}| \leq C$: Model 3. Vertical Windshear $W_h$ for $V_0 = 100$ (dash), $V_0 = 140$ (solid).

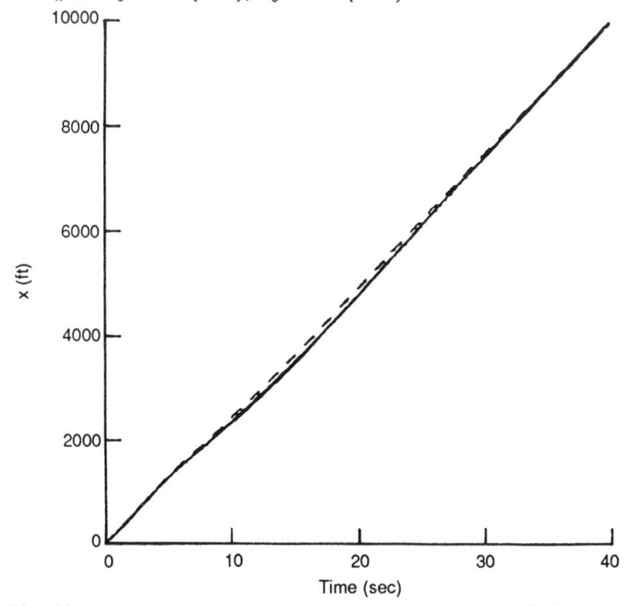


Fig. 82 Nonlinear Feedback Control, [20], $\alpha \leq \alpha_*$ and $|\dot{\alpha}| \leq C$: Model 3. Horizontal Distance $x$ for $V_0 = 100$ (dash), $V_0 = 140$ (solid).

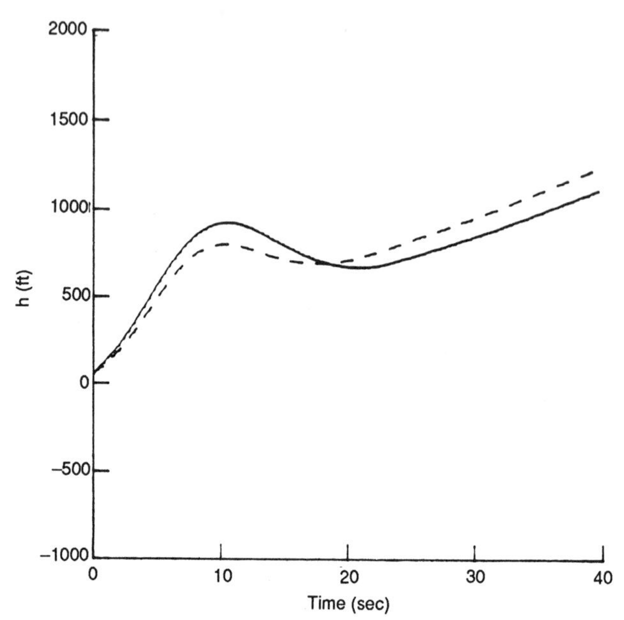


Fig. 83 Nonlinear Feedback Control, [20], $\alpha \leq \alpha_*$ and $|\dot{\alpha}| \leq C$: Model 3. Altitude $h$ for $V_0 = 100$ (dash), $V_0 = 140$ (solid).

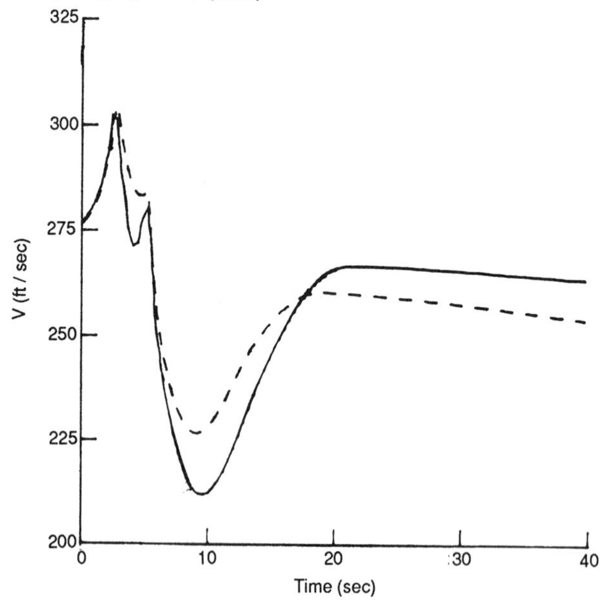


Fig. 84 Nonlinear Feedback Control, [20], $\alpha \leq \alpha_*$ and $|\dot{\alpha}| \leq C$: Model 3. Relative Speed $V$ for $V_0 = 100$ (dash), $V_0 = 140$ (solid).

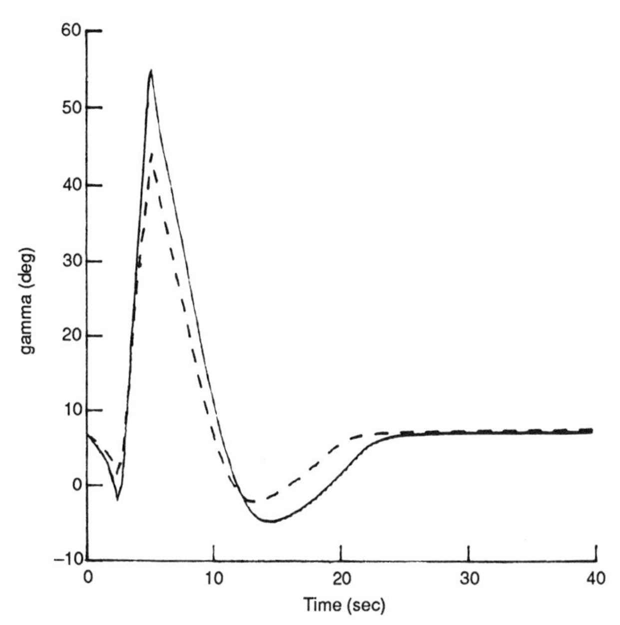


Fig. 85 Nonlinear Feedback Control, [20], $\alpha \leq \alpha_*$ and $|\dot{\alpha}| \leq C$: Model 3. Relative Path Inclination $\gamma$ for $V_0 = 100$ (dash), $V_0 = 140$ (solid).

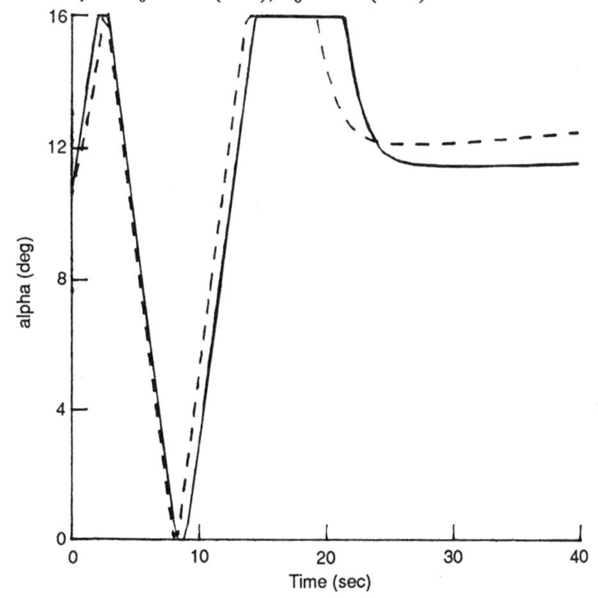


Fig. 86 Nonlinear Feedback Control, [20], $\alpha \leq \alpha_*$ and $|\dot{\alpha}| \leq C$: Model 3. Relative Angle of Attack $\alpha$ for $V_0 = 100$ (dash), $V_0 = 140$ (solid).

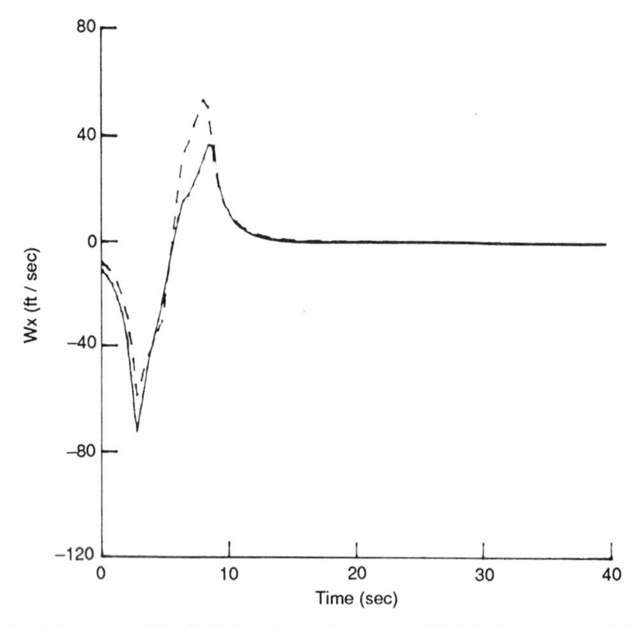


Fig. 87 Proposed Stabilizing Control, $\alpha \leq \alpha_*$: Model 4. Horizontal Windshear $W_x$ for $V_0$ = 100 (dash), $V_0$ = 140 (solid).

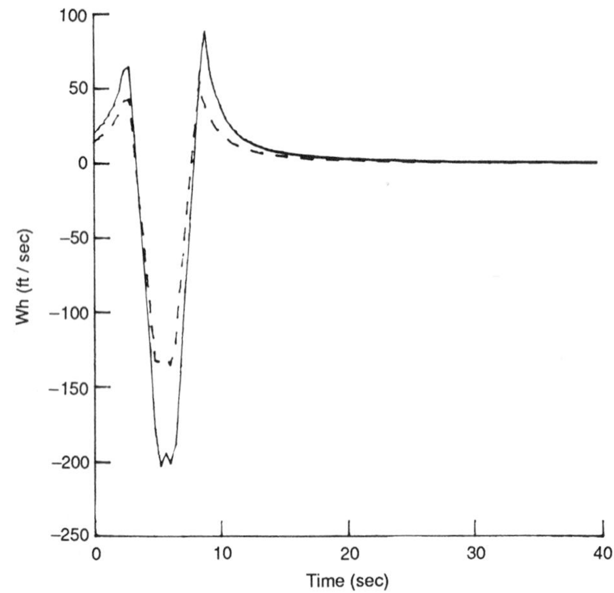


Fig. 88 Proposed Stabilizing Control, $\alpha \leq \alpha_*$: Model 4. Vertical Windshear $W_h$ for $V_0$ = 100 (dash), $V_0$ = 140 (solid).

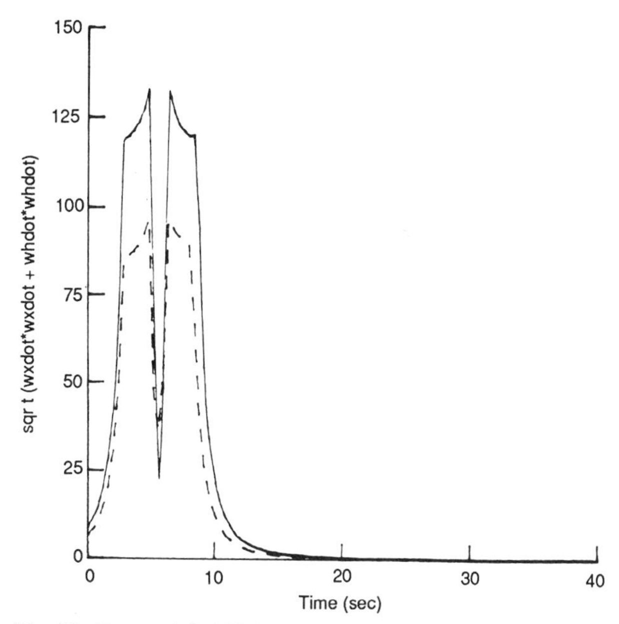


Fig. 89 Proposed Stabilizing Control, $\alpha \leq \alpha_*$: Model 4. Magnitude of Windshear Acceleration $\sqrt{\dot{W}_x^2 + \dot{W}_h^2}$ for $V_0 = 100$ (dash), $V_0 = 140$ (solid).

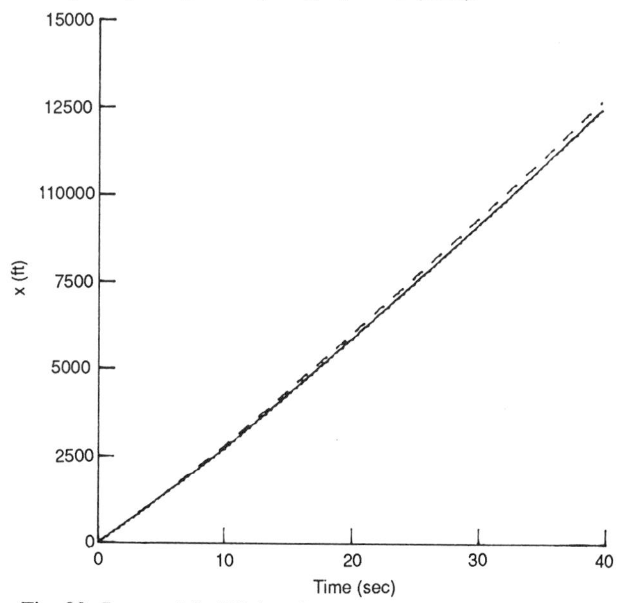


Fig. 90 Proposed Stabilizing Control, $\alpha \leq \alpha_*$: Model 4. Horizontal Distance $x$ for $V_0 = 100$ (dash), $V_0 = 140$ (solid).

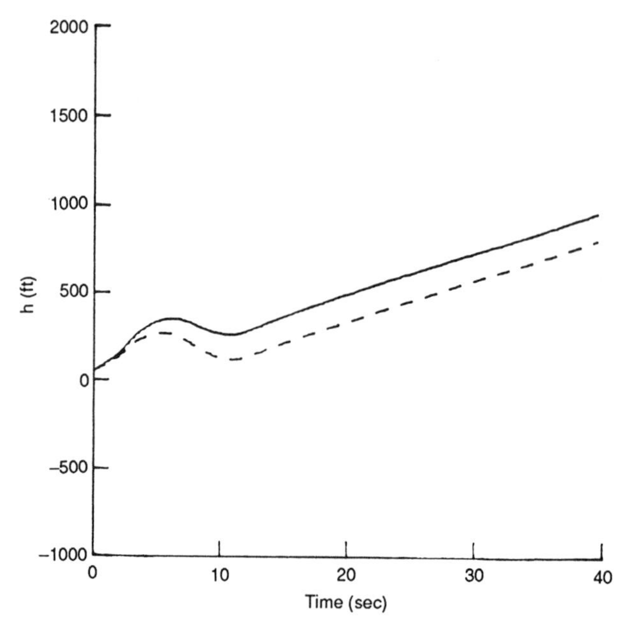


Fig. 91 Proposed Stabilizing Control, $\alpha \leq \alpha_*$: Model 4. Altitude $h$ for $V_0 = 100$ (dash), $V_0 = 140$ (solid).

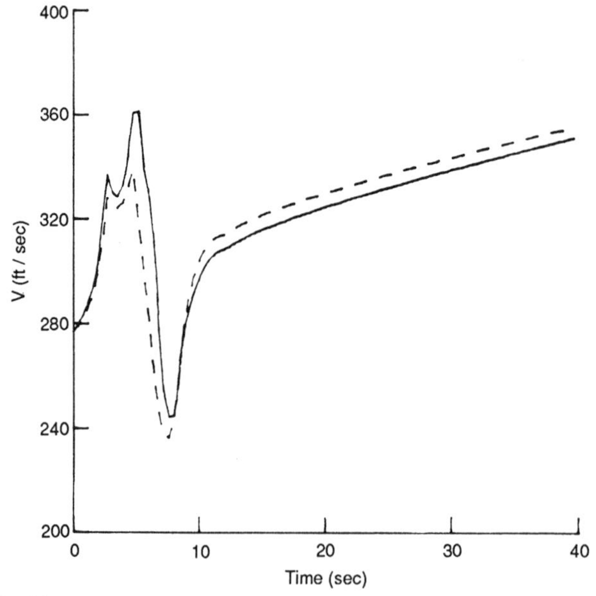


Fig. 92 Proposed Stabilizing Control, $\alpha \leq \alpha_*$: Model 4. Relative Speed $V$ for $V_0 = 100$ (dash), $V_0 = 140$ (solid).

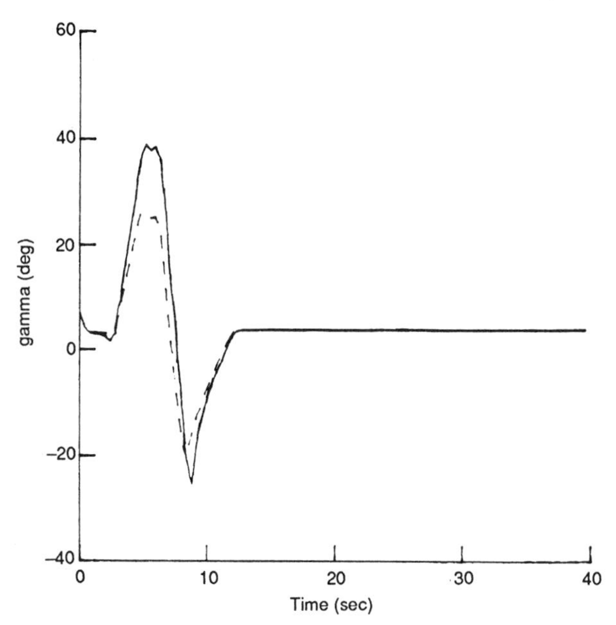


Fig. 93 Proposed Stabilizing Control, $\alpha \leq \alpha_*$: Model 4. Relative Path Inclination $\gamma$ for $V_0 = 100$ (dash), $V_0 = 140$ (solid).

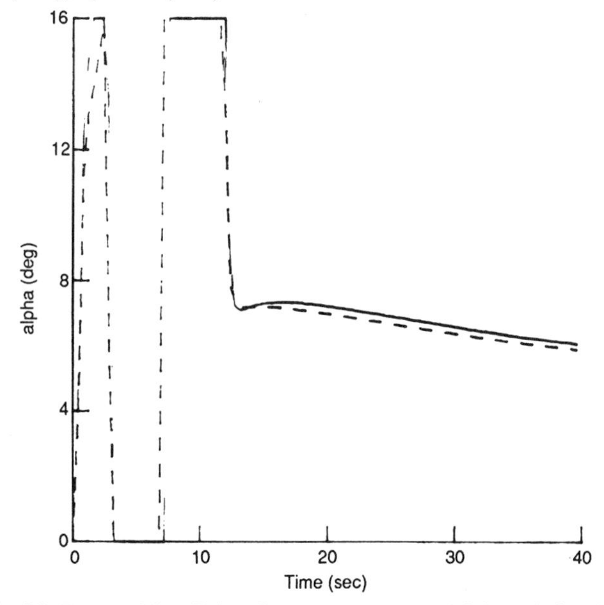


Fig. 94 Proposed Stabilizing Control, $\alpha \leq \alpha_*$: Model 4. Relative Angle of Attack $\alpha$ for $V_0 = 100$ (dash), $V_0 = 140$ (solid).

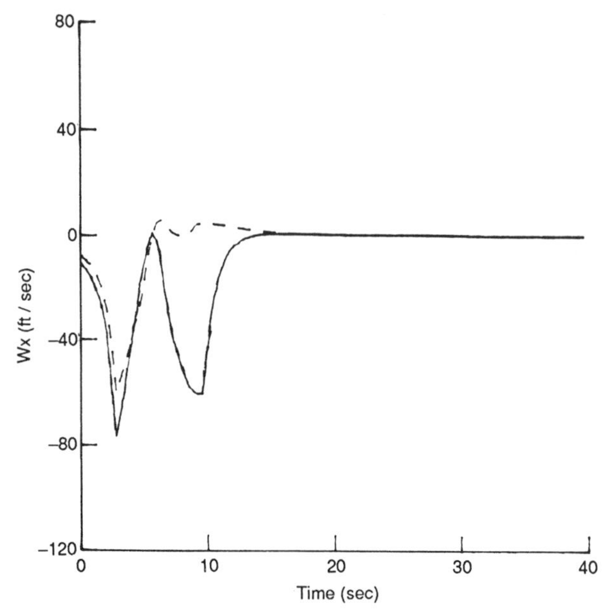


Fig. 95 Proposed Stabilizing Control, $\alpha \leq \alpha_*$ and $|\dot{\alpha}| \leq C$: Model 4. Horizontal Windshear $W_x$ for $V_0 = 100$ (dash), $V_0 = 140$ (solid).

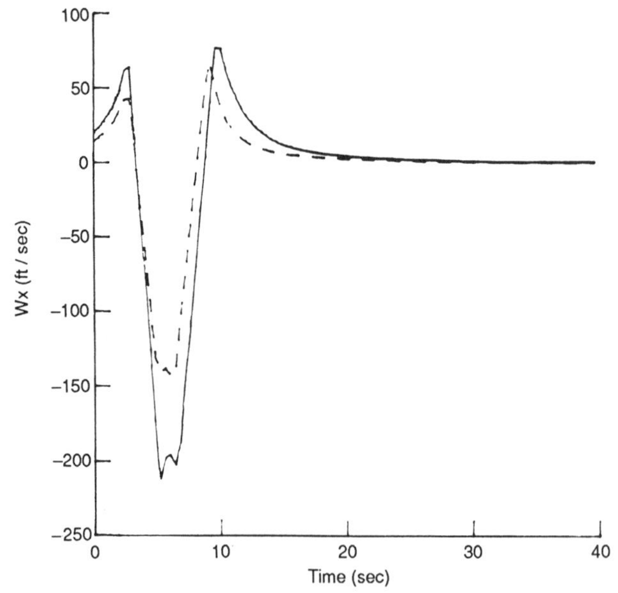


Fig. 96 Proposed Stabilizing Control, $\alpha \leq \alpha_*$ and $|\dot{\alpha}| \leq C$: Model 4. Vertical Windshear $W_h$ for $V_0 = 100$ (dash), $V_0 = 140$ (solid).

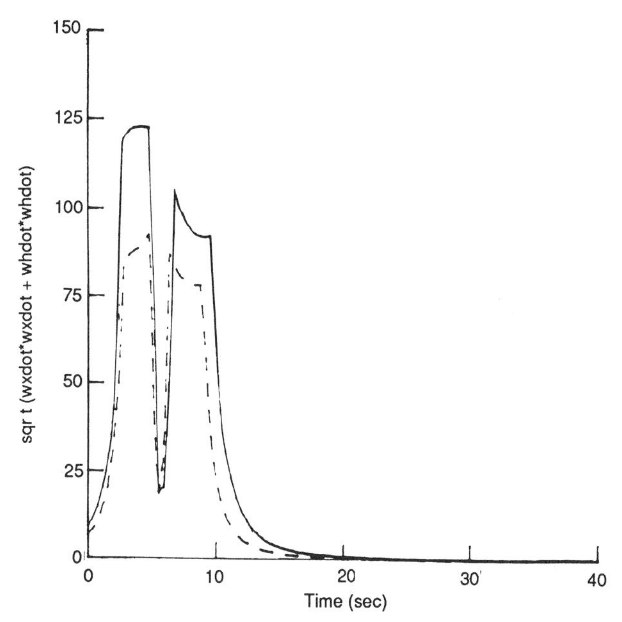


Fig. 97 Proposed Stabilizing Control, $\alpha \le \alpha_*$ and $|\dot{\alpha}| \le C$: Model 4. Magnitude of Windshear Acceleration $\sqrt{\dot{W}_x^2 + \dot{W}_h^2}$ for $V_0 = 100$ (dash), $V_0 = 140$ (solid).

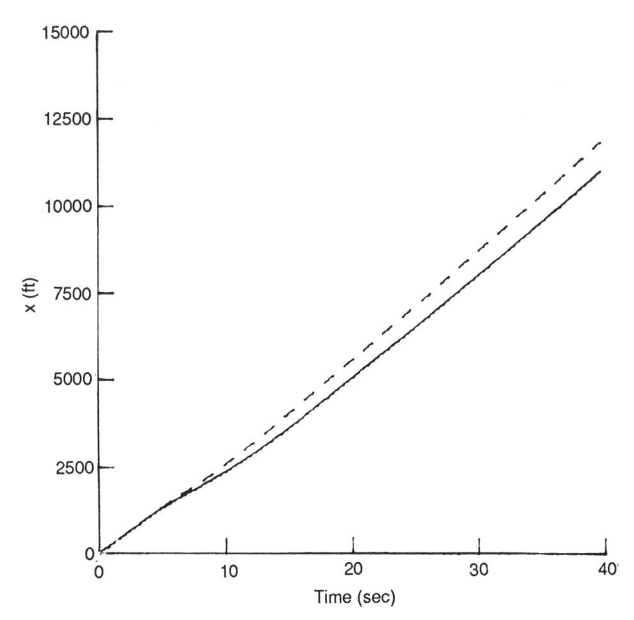


Fig. 98 Proposed Stabilizing Control, $\alpha \le \alpha_*$ and $|\dot{\alpha}| \le C$: Model 4. Horizontal Distance $x$ for $V_0 = 100$ (dash), $V_0 = 140$ (solid).

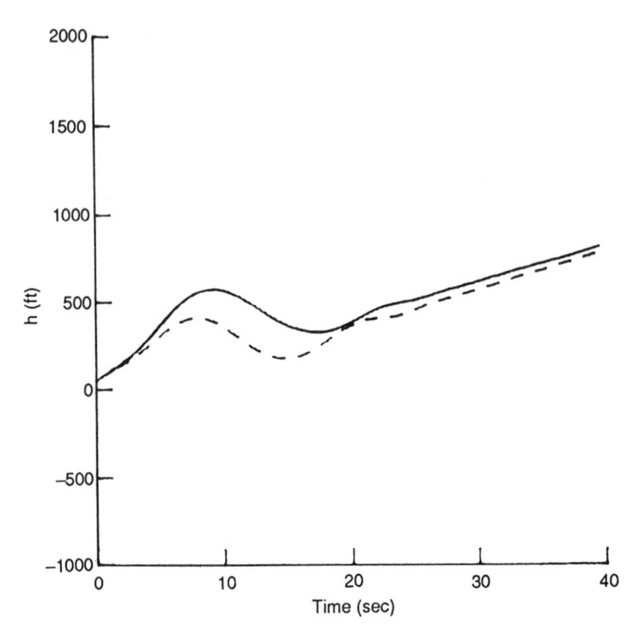


Fig. 99 Proposed Stabilizing Control, $\alpha \leq \alpha_*$ and $|\dot{\alpha}| \leq C$: Model 4. Altitude $h$ for $V_0 = 100$ (dash), $V_0 = 140$ (solid).

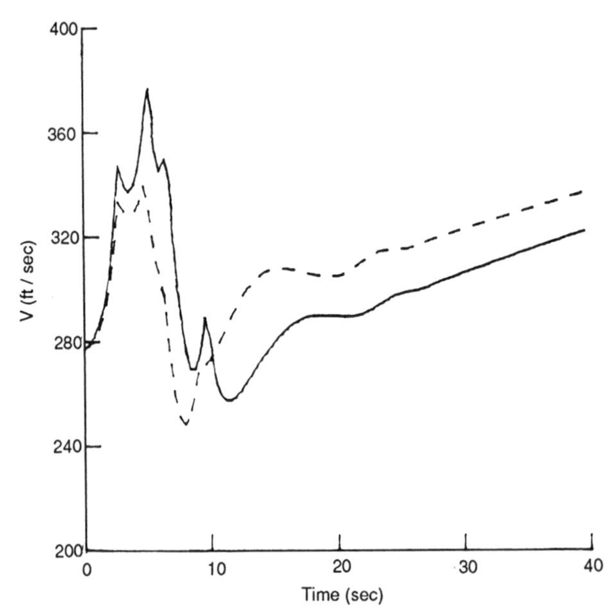


Fig. 100 Proposed Stabilizing Control, $\alpha \leq \alpha_*$ and $|\dot{\alpha}| \leq C$: Model 4. Relative Speed $V$ for $V_0 = 100$ (dash), $V_0 = 140$ (solid).

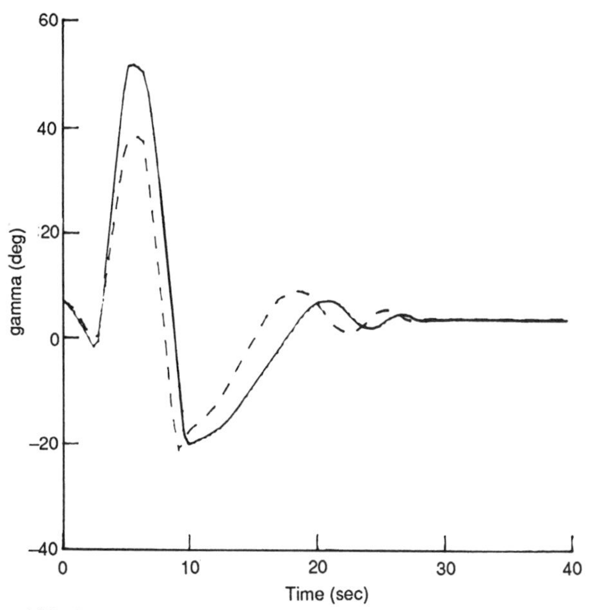


Fig. 101 Proposed Stabilizing Control, $\alpha \leq \alpha_*$ and $|\dot{\alpha}| \leq C$: Model 4. Relative Path Inclination $\gamma$ for $V_0 = 100$ (dash), $V_0 = 140$ (solid).

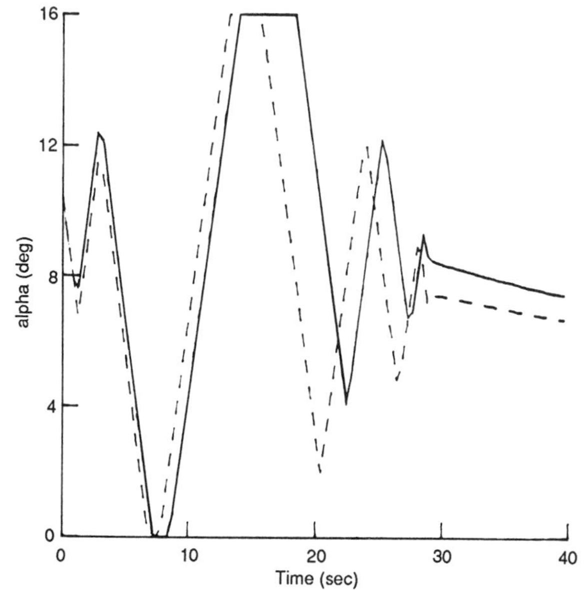


Fig. 102 Proposed Stabilizing Control, $\alpha \leq \alpha_*$ and $|\dot{\alpha}| \leq C$: Model 4. Relative Angle of Attack $\alpha$ for $V_0 = 100$ (dash), $V_0 = 140$ (solid).

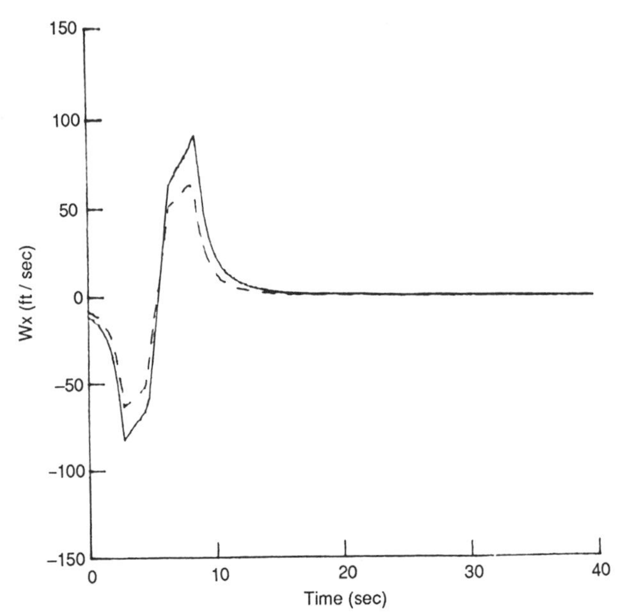


Fig. 103 Simplified Gamma Guidance, [19], $\alpha \leq \alpha_*$ and $|\dot{\alpha}| \leq C$: Model 4. Horizontal Windshear $W_x$ for $V_0 = 100$ (dash), $V_0 = 140$ (solid).

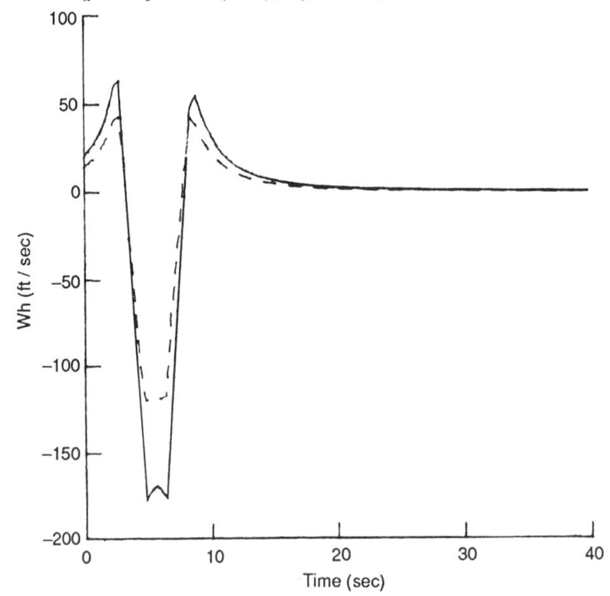


Fig. 104 Simplified Gamma Guidance, [19], $\alpha \leq \alpha_*$ and $|\dot{\alpha}| \leq C$: Model 4. Vertical Windshear $W_h$ for $V_0 = 100$ (dash), $V_0 = 140$ (solid).

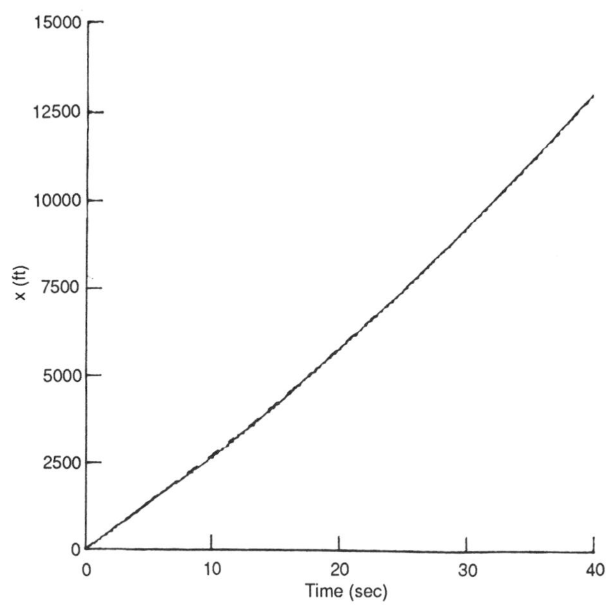


Fig. 105 Simplified Gamma Guidance, [19], $\alpha \le \alpha_*$ and $|\dot{\alpha}| \le C$: Model 4. Horizontal Distance $x$ for $V_0 = 100$ (dash), $V_0 = 140$ (solid).

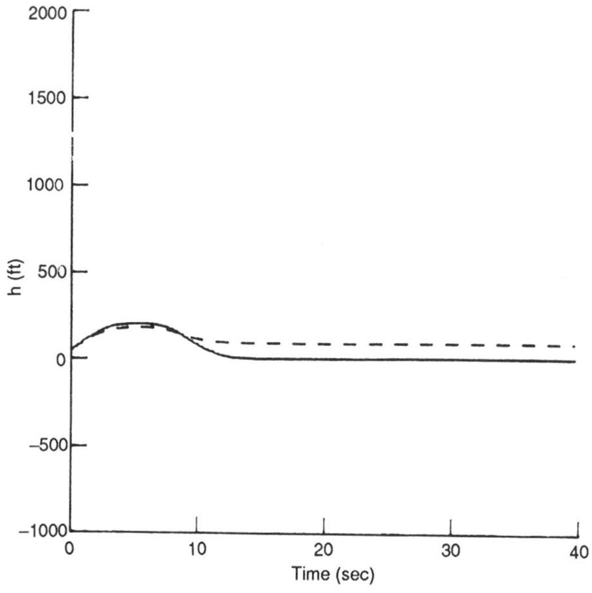


Fig. 106 Simplified Gamma Guidance, [19], $\alpha \le \alpha_*$ and $|\dot{\alpha}| \le C$: Model 4. Altitude $h$ for $V_0 = 100$ (dash), $V_0 = 140$ (solid).

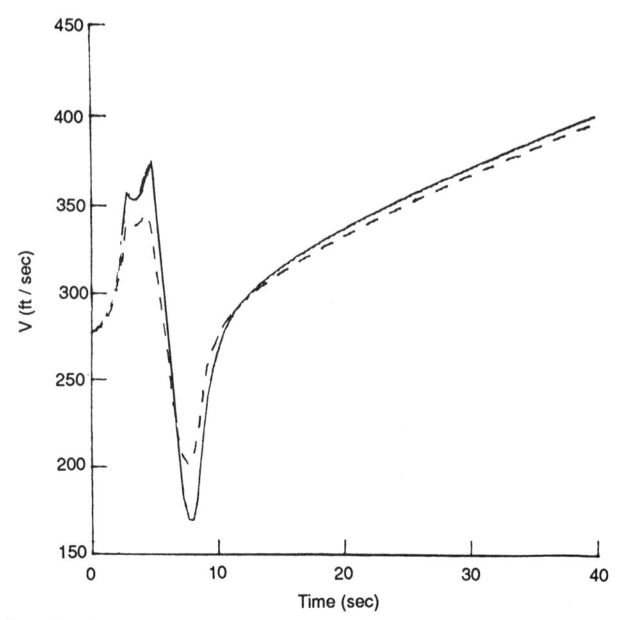


Fig. 107 Simplified Gamma Guidance, [19], $\alpha \leq \alpha_*$ and $|\dot{\alpha}| \leq C$: Model 4. Relative Speed $V$ for $V_0 = 100$ (dash), $V_0 = 140$ (solid).

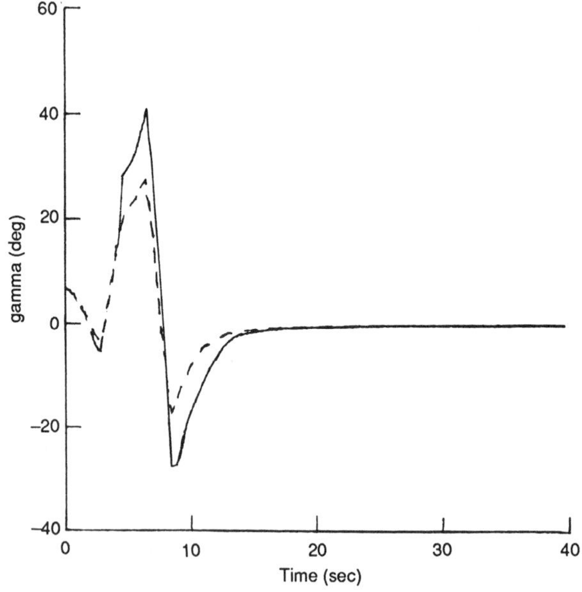


Fig. 108 Simplified Gamma Guidance, [19], $\alpha \leq \alpha_*$ and $|\dot{\alpha}| \leq C$: Model 4. Relative Path Inclination $\gamma$ for $V_0 = 100$ (dash), $V_0 = 140$ (solid).

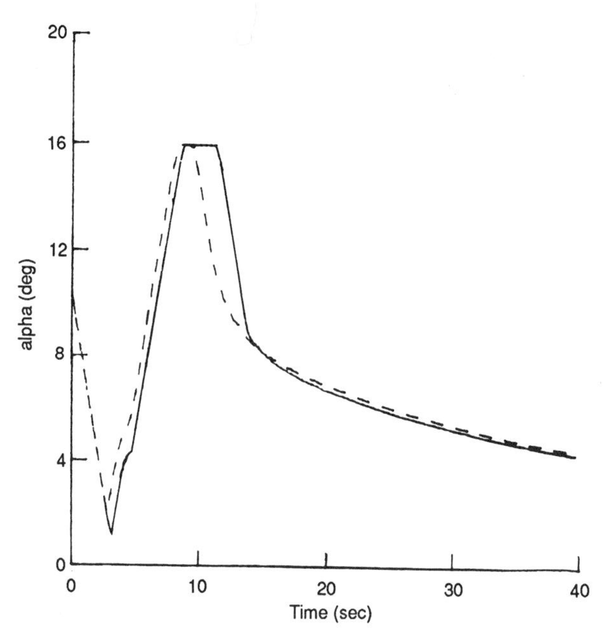


Fig. 109 Simplified Gamma Guidance, [19], $\alpha \leq \alpha_*$ and $|\dot{\alpha}| \leq C$: Model 4. Relative Angle of Attack $\alpha$ for $V_0 = 100$ (dash), $V_0 = 140$ (solid).

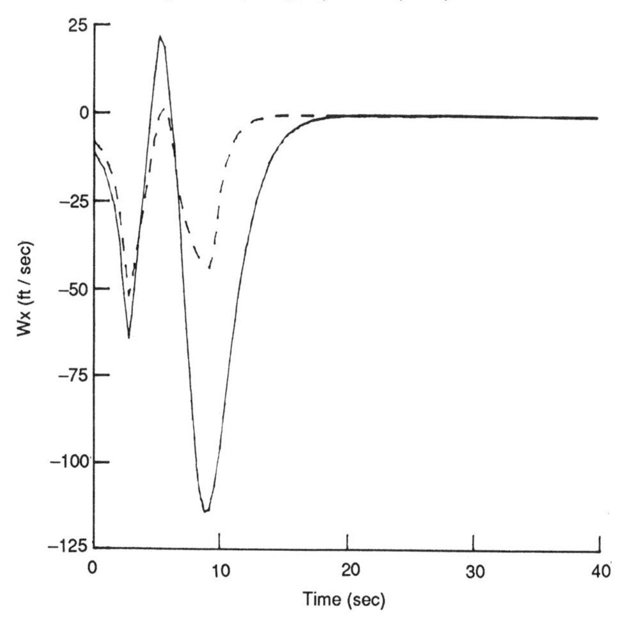


Fig. 110 Nonlinear Feedback Control, [20], $\alpha \leq \alpha_*$ and $|\dot{\alpha}| \leq C$: Model 4. Horizontal Windshear $W_x$ for $V_0 = 100$ (dash), $V_0 = 140$ (solid).

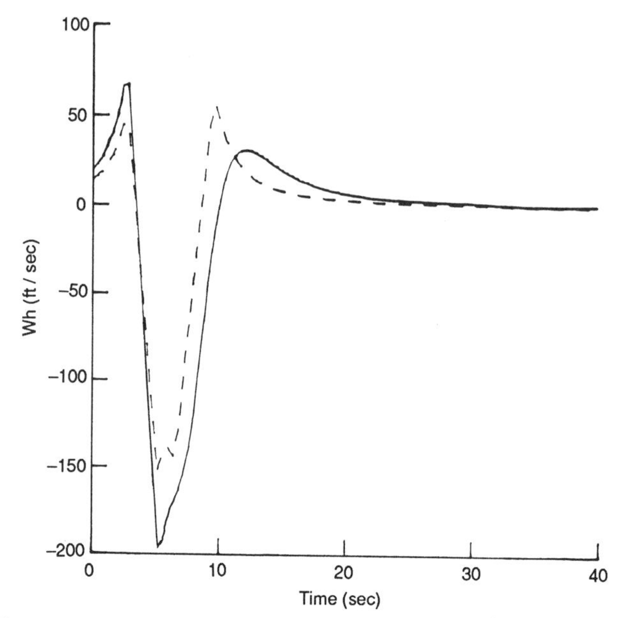


Fig. 111 Nonlinear Feedback Control, [20], $\alpha \leq \alpha_*$ and $|\dot{\alpha}| \leq C$: Model 4. Vertical Windshear $W_h$ for $V_0 = 100$ (dash), $V_0 = 140$ (solid).

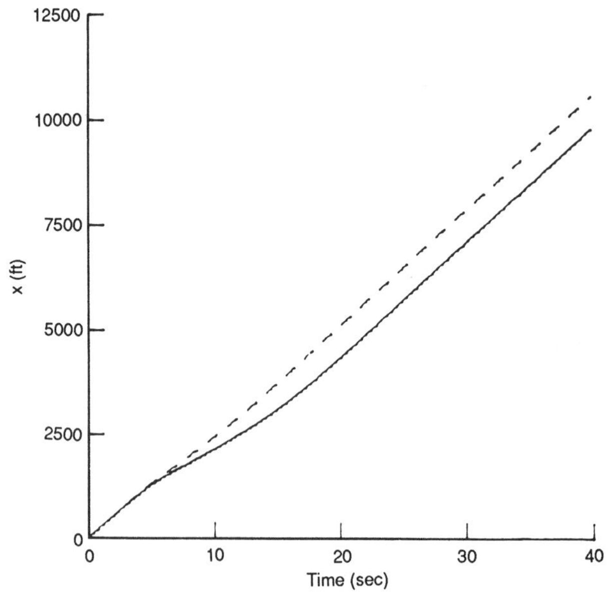


Fig. 112 Nonlinear Feedback Control, [20], $\alpha \leq \alpha_*$ and $|\dot{\alpha}| \leq C$: Model 4. Horizontal Distance $x$ for $V_0 = 100$ (dash), $V_0 = 140$ (solid).

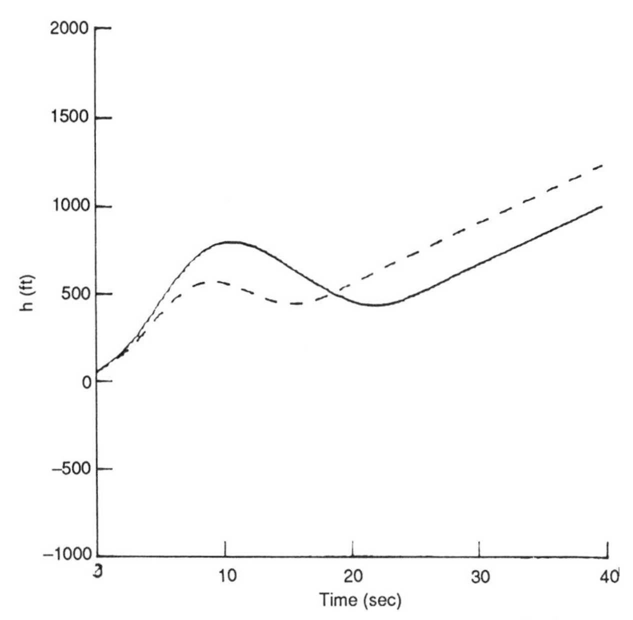


Fig. 113 Nonlinear Feedback Control, [20], $\alpha \leq \alpha_*$ and $|\dot{\alpha}| \leq C$: Model 4. Altitude $h$ for $V_0 = 100$ (dash), $V_0 = 140$ (solid).

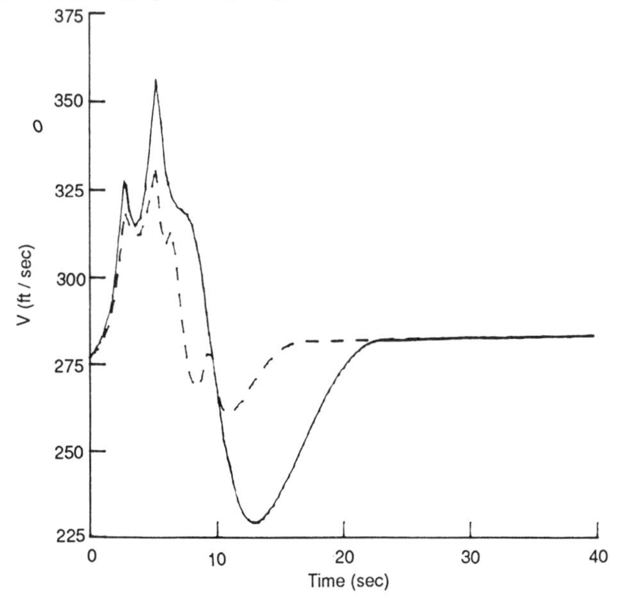


Fig. 114 Nonlinear Feedback Control, [20], $\alpha \leq \alpha_*$ and $|\dot{\alpha}| \leq C$: Model 4. Relative Speed $V$ for $V_0 = 100$ (dash), $V_0 = 140$ (solid).

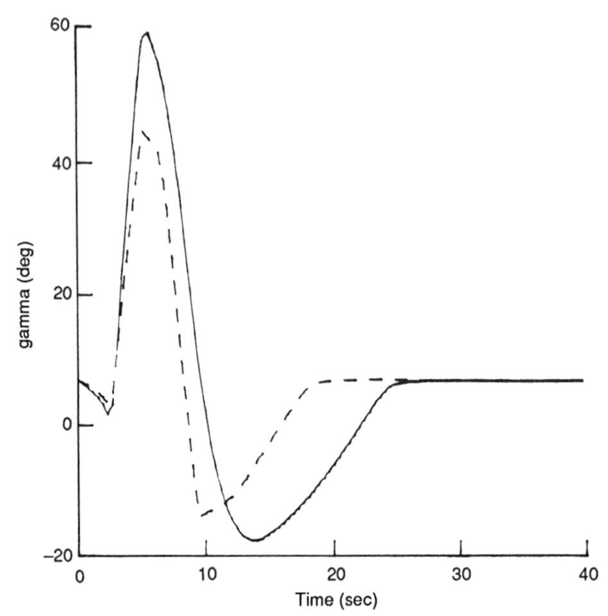


Fig. 115 Nonlinear Feedback Control, [20], $\alpha \leq \alpha_*$ and $|\dot{\alpha}| \leq C$: Model 4. Relative Path Inclination $\gamma$ for $V_0 = 100$ (dash), $V_0 = 140$ (solid).

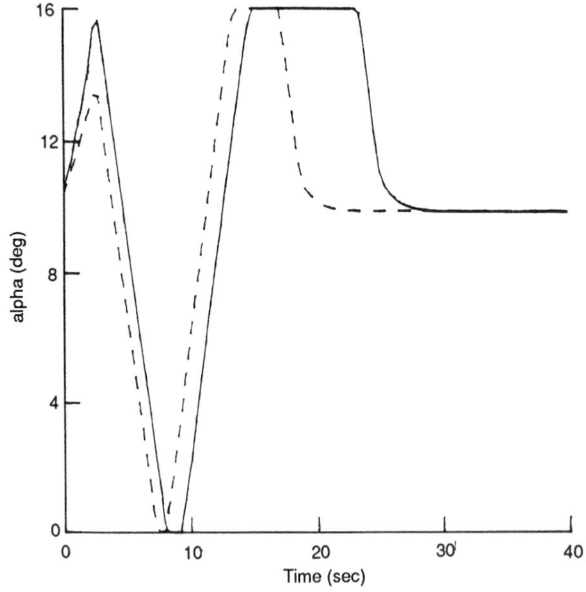


Fig. 116 Nonlinear Feedback Control, [20], $\alpha \leq \alpha_*$ and $|\dot{\alpha}| \leq C$: Model 4. Relative Angle of Attack $\alpha$ for $V_0 = 100$ (dash), $V_0 = 140$ (solid).

## 1. CONCLUSIONS

The investigation reported in this chapter is the first step in a study of the robustness properties (with respect to windshear distribution and intensity) of various aircraft guidance schemes currently espoused.

While it appears that the *stabilizing control*, based on a theory of deterministic control of uncertain systems, is the most robust of the three schemes considered, it is also the least smooth of these schemes, although all three guidance schemes exhibit undesirably large and rapid variations in relative angle of attack.

Current research seeks to address two problems, the problem of increasing robustness as well as that of eliminating impractical variations in the relative angle of attack. Towards these ends we are investigating the efficacy of stabilizing the path inclination $\gamma_e$, rather than the relative path inclination, $\gamma$. Another area of current research concerns the extension of the basic theory of deterministic control of uncertain systems to allow for the *a priori* imposition of bounds on the control and on its rate of change.

Since the study presented here is of a very preliminary nature, we have restricted the discussion to the simpler situation of take-off. The more complicated case of landing during conditions of windshear has been addressed in the literature. Again, the pioneering studies are due to Miele (for instance, see [22-27]). Other investigations of the landing problem can be found in [28] as well as in the Soviet literature (see [29] and references cited therein). It is perhaps noteworthy that the approach taken in the Soviet studies is closely related to that employed here in Section III.B and summarized in Appendix B; it is based on the philosophy of a "game against nature" with uncertainty playing the role of an opposing (potentially destabilizing) player, an approach to control under uncertainty suggested in the 1970's (for instance, see 30,31]).

## APPENDIX A

### GLOBAL PRACTICAL STABILIZABILITY

Our mode of an uncertain system is of the form

$$\dot{y}(t) = F(t,y(t),u(t), \omega ) \tag{A.1}$$

where $t \in \mathbf{R}$ is the time variable, $y(t) \in \mathbf{R}^n$ is the state, and $u(t) \in \mathbf{R}^m$ is the control input. All the uncertainty in the system is represented by the lumped uncertain element $\omega \in \Omega$; it could be an element of $\mathbf{R}^q$ representing unknown constant paramaters and inputs; it could be a function from $\mathbf{R}$ into $\mathbf{R}^q$ representing unknown time varying parameters and inputs; it could also be a function from $\mathbf{R}\times\mathbf{R}^n\times\mathbf{R}^m \to \mathbf{R}^q$ representing nonlinear elements which are difficult to characterize exactly; it could be simply an index. The only information assumed about $\omega$ is the knowledge of a non-empty set $\Omega$ to which it belongs.

Let $G_r$ be the closed ball or radius $r$, that is,

$$G_r = \left\{ y \in \mathbf{R}^n \ : \| y \| \le r. \right\} \tag{A.2}$$

The system (A.1) is said to be *globally practically stabilizable* with respect to the closed ball $G_{d_0}$ iff, given any $\underline{d} > d_0$, there exists a feedback control $p:\mathbf{R}\times\mathbf{R}^n \to \mathbf{R}^m$ for which the following properties hold:

*(i) Existence of Solutions.*

Given $(t_0,y_0) \in \mathbf{R} \times \mathbf{R}^n$, the closed loop system

$$\dot{y}(t) = f(t,y(t),\ p(t,y)\ \omega\ ), \tag{A.3}$$

possesses a solution $y(\cdot) : [t_0,t_1) \to \mathbf{R}^n$, $y(t_0) = y_0$, $t_1 > t_0$.

*(ii) Uniform Boundedness.*

Given $r \in (0, \infty)$, there exists a positive $d(r) < \infty$ such that for all solutions $y(\cdot) : [t_0,t_1) \to \mathbf{R}^n$, $y(t_0) = y_0$, of (A.3).

$\| y_0 \| \leq r \Rightarrow \| y(t) \| \leq d(r)$, for all $t \in [t_0,t_1)$.

*(iii) Extension of Solutions*

Every solution $y(\cdot) : [t_0,t_1) \to \mathbf{R}^n$ of (A.3) can be continued over $[t_0, \infty)$.

*(iv) Uniform Ultimate Boundedness.*

Given any $\bar{d} > \underline{d}$ and any $r \in (0, \infty)$, there is a $T(\bar{d},r) \in [0, \infty)$ such that for every solution $y(\cdot) : [t_0,\infty) \to \mathbf{R}^n$, $y(t_0) = y_0$, of (A.3).

$\| y_0 \| \leq r \Rightarrow y(t) \in G_{\bar{d}}$ for all $t \geq t_0 + T(\bar{d},r)$.

*(v) Uniform Stability*

Given any $\bar{d} > \underline{d}$, there is a positive $\delta(\bar{d})$ such that for every solution $y(\cdot) : [t_0, \infty) \to \mathbf{R}^n$, $y(t_0) = y_0$, of (A.3).

$y_o \in G_{\delta(\bar{d})} \Rightarrow y(t) \in G_{\bar{d}}$ for all $t \geq t_0$.

**APPENDIX B**

STABILIZING CONTROL

The following theorem (see [13] for proof and various definitions) is useful in the construction of practically stabilizing sets of controllers.

*Theorem.* Consider an uncertain system described by (A.1) with $\omega \in \Omega$ and suppose that $\mathbf{P}$ is a collection of feedback control functions $p : \mathbf{R} \times \mathbf{R}^n \to \mathbf{R}^m$. If there exists a candidate Lyapunov function $\mathbf{V} : \mathbf{R} \times \mathbf{R}^n \to \mathbf{R}_+$ and a class K function $\sigma : \mathbf{R}_+ \to \mathbf{R}_+$ such that for each $\varepsilon > 0$ there exists $p^\varepsilon \in \mathbf{P}$ which assures that for all $\omega \in \Omega$

$$\dot{y}(t) = F(t,y(t), p^\varepsilon(t,y(t)), \omega) \tag{B.1}$$

has existence and indefinite extension of solutions and

$$\frac{\partial \mathbf{V}}{\partial t}(t,y) + \frac{\partial \mathbf{V}}{\partial y}(t,y)F(t,y,p^\varepsilon(t,y), \omega) \leq -\sigma(\| y \|) + \varepsilon \tag{B.2}$$

for all $t \in \mathbf{R}, y \in \mathbf{R}^n$, then $\mathbf{P}$ is a practically stabilizing family.

A SPECIFIC CLASS OF UNCERTAIN SYSTEMS

An uncertain system under consideration here is described by (A.1) and satisfies the following assumption.

*Assumption.* There exist a continuous function $B : \mathbf{R} \times \mathbf{R}^n \to \mathbf{R}^{n\times m}$, a candidate Lyapunov function $\mathbf{V} : \mathbf{R} \times \mathbf{R}^n \to \mathbf{R}_+$, a class K function $\sigma : \mathbf{R}_+ \to \mathbf{R}_+$ and continuous functions $\kappa, \xi : \mathbf{R} \times \mathbf{R}^n \to \mathbf{R}_+$ such that

$$F(t,y,u, \omega) = f^s(t,y, \omega) + B(t,y)z(t,y,u, \omega) \tag{B.3}$$

for some functions $f^s$ and $z$ which satisfy:

(1) For each $\omega \in \Omega$ $f^s(\cdot, \omega)$ is continuous and

$$\frac{\partial \mathbf{V}}{\partial t}(t,y) + \frac{\partial \mathbf{V}}{\partial y}(t,y)f^s(t,y, \omega) \leq -\sigma(\| y \|) \tag{B.4}$$

for all $t \in \mathbf{R}, y \in \mathbf{R}^n$.

(2) For each $\omega \in \Omega$, $z(\cdot, \omega)$ is continuous and

$$u^T z(t,y,u, \omega) \geq -\beta_1(t,y, \omega)\| u \| + \beta_2(t,y, \omega)\| u \|^2 \tag{B.5}$$

where $\beta_1(t,y, \omega), \beta_2(t,y, \omega) \geq 0$ and

$$\beta_1(t,y, \omega) \leq \beta_2(t,y, \omega)\, \xi(t,y), \tag{B.6}$$

$$\beta_1(t,y,\omega) \leq \kappa(t,y) \tag{B.7}$$

for all $t \in \mathbf{R}, y \in \mathbf{R}^n, u \in \mathbf{R}^m$ .

*Remark.* If the system has a linear time invariant and stable nominal part, that is

$$f^s = Ay \tag{B.8}$$

where $A \in \mathbf{R}^{nxn}$ and all the eigenvalues of A have negative real parts, then condition (B.4) is satisfied by taking any positive definitive symmetric $Q \in \mathbf{R}^{nxn}$ and letting

$$V(t,y) = \frac{1}{2} y^T P y \qquad \text{for all } (t,y) \in \mathbf{R} \times \mathbf{R}^n \tag{B.9}$$

where $P \in \mathbf{R}^{nxn}$ is the unique positive definite symmetric solution of

$$PA + A^T P + Q = 0. \tag{B.10}$$

PROPOSED CONTROLLERS

Here we present some practically stabilizing controller sets for the systems described by (B.3). Their construction is based on satisfying the hypotheses of the *Theorem* above.

Consider an uncertain system described by (B.3) and let $(B, V, \sigma, \xi, \kappa)$ be any quintuple which assures the satisfaction of the *Assumption* above. Choose any continuous functions $\xi^c, \kappa^c : \mathbf{R} \times \mathbf{R}^n \to \mathbf{R}_+$ *which satisfy*

$$\xi^c(t,y) \geq \xi(t,y), \qquad \kappa^c(t,y) \geq \kappa(t,y), \tag{B.11}$$

*and define*

$$\phi(t,y) \triangleq B(t,y)^T \frac{\partial V}{\partial y}(t,y)^T, \tag{B.12}$$

$$\eta(t,y) \triangleq \kappa^c(t,y)\, \phi(t,y). \tag{B.13}$$

A practically stabilizing family of controllers is the set

$$\mathbf{P} = \left\{ p^\varepsilon : \varepsilon > 0 \right\} \tag{B.14}$$

where $p^\varepsilon : \mathbf{R} \times \mathbf{R}^n \to \mathbf{R}^m$ is any continuous function which satisfies

$$\| \phi(t,y) \| \, p^\varepsilon(t,y) = - \| p^\varepsilon(t,y) \| \, \phi(t,y) \tag{B.15}$$

that is, $p^{\varepsilon}(t,y)$ is opposite in direction to $\phi(t,y)$, and

$$\| \eta(t,y) \| > 0 \Rightarrow \| p^{\varepsilon}(t,y) \| \geq \xi^{c}(t,y)[1 - \| \eta(t,y) \|^{-1} \varepsilon]. \qquad \text{(B.16)}$$

An example of a function satisfying the above requirements is

$$p^{\varepsilon}(t,y) = \begin{cases} -\xi^{c}(t,y)\, \varepsilon^{-1}\, \eta(t,y) & \text{if } \| \eta(t,y) \| \leq \varepsilon \\ -\xi^{c}(t,y)\, \| \eta(t,y) \|^{-1} \eta(t,y) & \text{if } \| \eta(t,y) \| > \varepsilon \end{cases} \qquad \text{(B.17)}$$

Another example is

$$p^{\varepsilon}(t,y) = -\xi^{c}(t,y)\, [\, \| \eta(t,y) \| + \varepsilon\, ]^{-1}\, \eta(t,y). \qquad \text{(B.18)}$$

(See [13,14] for other details).

## ACKNOWLEDGEMENT

This chapter is based on research supported in part by the NSF and AFOSR under Grant ECS - 8602524.

## REFERENCES

1. S.Zhu and B. Etkin, "Fluid Dynamic Model of a Dawnburst," Institute of Aerospace Studies, University of Toronto, UTIAS Report No. 271.

2. A. Miele, T. Wang, C.Y. Tzeng and W.W. Melvin, "Optimization and Guidance of Abort Landing Trajectories in a Windshear," Paper No. AIAA-87-2341, AIAA Guidance, Navigation and Control Conference (1987).

3. W. Frost, "Flight in a Low-Level Wind Shear," NASA CR 3678 (1983).

4. A. Miele, T. Wang, W.W. Melvin and R.L. Bowles, "Maximum Survival Capability of an Aircraft in a Severe Windshear," Journal of Optimization Theory and Applications, Vol. 53, No. 2, (1987).

5. W. Frost and D.W. Camp, "Wind Shear Modeling for Aircraft Hazard Definition," FAA Report No. FAA-RD-77-36, U.S. Department of Transportation, Washington, DC (1977).

6. W. Frost, D.W. Camp and S.T. Wang, "Wind Shear Modeling for Aircraft Hazard Definition," FAA Report No. FAA-RD-78-3, U.S Department of Transportation, Washington, DC (1978).

7. G.G Roetcisoender, W.J. Grantham and E.K. Parks, "The DFW Microburst: Two-Dimensional Multiple Vortex Models for AAL-539," Journal of Aircraft (to appear).

8. F. Caracena, "The Microburst: Common Factor in Recent Aircraft Accidents," Proceedings: Fourth Annual Workshop on Meteorological and Environmental Inputs to Aviation Systems, 1980, NASA CP 2139 / FAA-RD-80-67 (1980).

9. A. Miele, T. Wang and W.W. Melvin, "Guidance Strategies for Near Optimum Take-Off Performance in a Windshear," Journal of Optimizaiton Theory and Applications, Vol. 50, No. 1, (1986).

10. G. Leitmann, "On the Efficacy of Nonlinear Control in Uncertain Linear Systems," Transactions of the ASME, Journal of Dynamic Systems, Measurement and Control, Vol. 102, (1981).

11. M. Corless and G. Leitmann, "Continuous State Feedback Guaranteeing Uniform Ultimate Boundedness for Uncertain Dynamic Systems'" IEEE Transactions on Automatic Control, Vol. AC-26, No. 5, (1981).

12. Y.H Chen and G. Leitmann, "Robustness of Uncertain Systems in the Absence of Matching Assumptions," International Journal of Control, Vol. 45, (1987).

13. M. Corless and G. Leitmann, "Deterministic Control of Uncertain Systems: A Lyapunov Theory Approach," in Deterministic Nonlinear Control of Uncertain Systems: Variable Structure and Lyapunov Control (A. Zinober, ed.), IEE Publishers (to appear).

14. G. Leitmann, "Deterministic Control of Uncertain Systems via a Constructive Use of Lyapunov Stability Theory," SIAM Conference "Control in the 90's", San Francisco (1989).

15. A. Miele, T. Wang and W.W. Melvin, "Optimal Flight Trajectories in the Presence of Windshear, Part I - Take-Off," Paper No. AIAA-85-1843-CP, AIAA Atmoshperic Flight Mechanics Conference (1985).

16. A. Miele, T. Wang and W.W. Melvin, "Optimal Flight Trajectories in the Presence of Windshear, Parts I - IV," Rice University, Aero-Astronautics Report Nos. 191-194 (1985).

17. A. Miele, T. Wang and W.W. Melvin, "Quasi-Steady Flight to Quasi-Steady Flight Transition in a Windshear: Trajectory Optimizaiton and Guidance," Journal of Optimization Theory and Applications, Vol. 54, No. 2, (1987).

18. A. Miele, T. Wang and W.W. Melvin, "Optimization and Acceleration Guidance of Flight Trajectories in a Windshear," Journal of Guidance, Control, and Dynamics, Vol. 10 No. 4, (1987).

19. A. Miele, T. Wang, W.W. Melvin and R.L. Bowles, "Gamma Guidance Schemes for Flight in a Windshear" Journal of Guidance, Control, and Dynamics, Vol. 11 No. 4, (1987).

20. A.E. Bryson, Jr. and Y. Zhao, "Feedback Control for Penetrating a Downburst," Paper No. AIAA-87-2343 (1987).

21. Y.H. Chen and S. Pandey, "Robust Control Strategy for Take-Off Performance in a Windshear," Optimal Control Applications and Methods, Vol. 10, (1989).

22. A. Miele, T. Wang, C.Y. Tzeng and W.W. Melvin, "Optimal Abort Landing Trajectories in the Presence of Windshear," Journal of Optimization Theory and Applications, Vol. 55, No. 2, (1987).

23. A. Miele, T. Wang, H. Wang and W.W. Melvin, "Optimal Penetration Landing Trajectories in the Presence of Windshear," Journal of Optimization Theory and Applications, Vol. 57, No. 1, (1988).

24. A. Miele, T. Wang and W.W. Melvin, "Quasi-Steady Flight to Quasi-Steady Flight Transition for Abort Landing in a Windshear: Trajectory Optimization and Guidance," Journal of Optimization Theory and Applications, Vol. 58, No. 2, (1988).

25. A. Miele, T. Wang, W.W. Melvin and R.L. Bowles, "Acceleration, Gamma, and Theta Guidance Schemes for Abort Landing in a Windshear," Journal of Guidance, Control, and Dynamics, Vol. 12, (1989).

26. A. Miele, T. Wang and W.W. Melvin, "Penetration Landing Guidance Trajectories in the Presence of Windshear," Journal of Guidance, Control, and Dynamics, Vol. 12, (1989).

27. A. Miele, T. Wang, C.Y. Tzeng and W.W. Melvin, "Abort Landing Guidance Trajectories in the Presence of Windshear," Journal of the Franklin Institute, Vol. 326, No. 2, (1989).

28. P.Y. Chu and A.E. Bryson, Jr., "Control of Aircraft Landing Approach in a Windshear," Paper No. AIAA-87-0632, AIAA 25th Aerospace Science Meeting (1987).

29. V.S. Patski and A.I. Subbotin, "Constructive Methods in Differential Game Theory," SIAM Conference "Control in the 90's", San Francisco (1989).

30. G. Leitmann, "Stabilization of Dynamical Systems Under Bounded Input Disturbance and Parameter Uncertianty," in Differential Games and Control Theory II (edited by E.O. Roxin, P.-T. Liu and R.L. Sternberg), Marcel Dekker, New York, 1976.

31. S. Gutman and G. Leitmann, "Stabilizing Feedback Control for Dynamical Systems with Bounded Uncertainty," Proceedings of the IEEE Conference on Decision and Control (1976).

# OVERVIEW OF OPTIMAL TRAJECTORIES FOR FLIGHT IN A WINDSHEAR[1,2]

by

A. Miele[3], T. Wang[4], H. Wang[5], and W. W. Melvin[6]

## 1. INTRODUCTION

Low altitude windshear is a threat to the safety of aircraft in take-off and landing. Over the past 20 years, some 30 aircraft accidents have been attributed

---

[1]This research was supported by NASA Langley Research Center, by Boeing Commercial Airplane Company, by Air Line Pilots Association, and by Texas Advanced Technology Program.

[2]Portions of this paper were presented at the AIAA 27th Aerospace Sciences Meeting, Reno, Nevada, January 9-12, 1989.

[3]Foyt Family Professor of Aerospace Sciences and Mathematical Sciences, Aero-Astronautics Group, Rice University, Houston, Texas.

[4]Senior Research Scientist, Aero-Astronautics Group, Rice University, Houston, Texas.

[5]Graduate Student, Aero-Astronautics Group, Rice University, Houston, Texas.

[6]Captain, Delta Airlines, Atlanta, Georgia; and Chairman, Airworthiness and Performance Committee, Air Line Pilots Association (ALPA), Washington, DC.

to windshear (Refs. 1-3). The most notorious ones are the crash of Eastern Airlines Flight 066 at JFK International Airport (1975), the crash of PANAM Flight 759 at New Orleans International Airport (1982), and the crash of Delta Airlines Flight 191 at Dallas-Fort Worth International Airport (1985). These crashes involves the loss of some 400 people and an insurance settlement in excess of 500 million dollars.

To offset the windshear threat, there are two basic systems: windshear avoidance systems and windshear recovery systems. A windshear avoidance system is designed to alert the pilot to the fact that a windshear encounter might take place; here, the intent is the avoidance of a microburst. A windshear recovery system is designed to guide the pilot in the course of a windshear encounter (Refs. 4-6); here, the intent is to fly smartly across a microburst, if an inadvertent encounter takes place. Obviously, windshear avoidance systems and windshear recovery systems are not mutally exclusive, but complementary to one another.

Concerning windshear recovery systems, considerable work has been done at Rice University over the past five years on both the optimization and the guidance of flight trajectories for take-off, abort landing, and penetration landing (see Refs. 7-15) with particular reference to the B-727 aircraft.

This paper is based on Refs. 12-13. It refers to the optimization of both take-off trajectories and abort landing trajectories and extends to the B-737 and B-747 aircraft the analyses previously made for the B-727 aircraft (see Refs. 7-15). The motivation for the study is threefold: (i) to establish directly whether the properties of the optimal take-off trajectories and the optimal abort landing trajectories for the B-727 aircraft extend to the B-737 and B-747 aircraft; (ii) to establish indirectly whether the take-off guidance schemes and abort landing guidance schemes developed for the B-727 aircraft qualitatively extend to the B-737 and B-747 aircraft; and (iii) since the windshear behavior of these aircraft is governed to a large degree by the thrust-to-weight ratio, to establish the effect of the thrust-to-weight ratio on survival capability in take-off and abort landing.

## 2. NOTATIONS

Throughout the paper, the following notations are employed:

$D$ = drag force, lb;

$g$ = acceleration of gravity, ft $sec^{-2}$;

$h$ = altitude, ft;

$L$ = lift force, lb;

$m$ = mass, lb $ft^{-1}$ $sec^{2}$;

$S$ = reference surface area, $ft^2$;

$t$ = running time, sec;

$T$ = thrust force, lb;

$V$ = relative velocity, ft $sec^{-1}$;

$W$ = mg = weight, lb;

$W_h$ = h-component of wind velocity, ft $sec^{-1}$;

$W_x$ = x-component of wind velocity, ft $sec^{-1}$;

$x$ = horizontal distance, ft;

$\alpha$ = angle of attack (wing), rad;

$\beta$ = engine power setting;

$\gamma$ = relative path inclination, rad;

$\gamma_e$ = absolute path inclination, rad;

$\delta$ = thrust inclination, rad;

$\delta_F$ = flap deflection, rad;

$\theta$ = pitch attitude angle (wing), rad;

$\lambda$ = wind intensity parameter;

$\tau$ = final time, sec.

## 3. SYSTEM DESCRIPTION

In this paper, we make use of the relative wind-axes system in connection with the following assumptions: (a) the aircraft is a particle of constant mass; (b) flight takes place in a vertical plane; (c) Newton's law is valid in an Earth-fixed system; and (d) the wind flow field is steady.

With the above premises, the equations of motion include the kinematical equations

$$\dot{x} = V\cos\gamma + W_x, \tag{1a}$$

$$\dot{h} = V\sin\gamma + W_h, \tag{1b}$$

and the dynamical equations

$$\dot{V} = (T/m)\cos(\alpha + \delta) - D/m - g\sin\gamma$$

$$- (\dot{W}_x\cos\gamma + \dot{W}_h\sin\gamma), \tag{2a}$$

$$\dot{\gamma} = (T/mV)\sin(\alpha + \delta) + L/mV - (g/V)\cos\gamma$$

$$+ (1/V)(\dot{W}_x\sin\gamma - \dot{W}_h\cos\gamma). \tag{2b}$$

Because of assumption (d), the total derivatives of the wind velocity components and the corresponding partial derivatives satisfy the relations

$$\dot{W}_x = (\partial W_x/\partial x)(V\cos\gamma + W_x) + (\partial W_x/\partial h)(V\sin\gamma + W_h), \tag{3a}$$

$$\dot{W}_h = (\partial W_h/\partial x)(V\cos\gamma + W_x) + (\partial W_h/\partial h)(V\sin\gamma + W_h). \tag{3b}$$

These equations must be supplemented by the functional relations

$$T = T(h,V,\beta), \tag{4a}$$

$$D = D(h,V,\alpha), \qquad L = L(h,V,\alpha), \tag{4b}$$

$$W_x = W_x(x,h), \qquad W_h = W_h(x,h), \tag{4c}$$

and by the analytical relations

$$\theta = \alpha + \gamma, \tag{5a}$$

$$\gamma_e = \arctan[(V\sin\gamma + W_h)/(V\cos\gamma + W_x)]. \tag{5b}$$

The differential system (1)-(4) involves four state variables $[x(t), h(t), V(t), \gamma(t)]$ and two control variables $[\alpha(t), \beta(t)]$. However, the number of control variables reduces to one (the angle of attack), if the power setting is specified in advance. The quantities (5) can be computed a posteriori, once the values of the state and the control are known.

3.1. Inequality Constraints

The angle of attack $\alpha$ and its time derivative $\dot{\alpha}$ are subject to the inequalities

$$\alpha \leq \alpha_*, \tag{6a}$$

$$-\dot{\alpha}_* \leq \dot{\alpha} \leq \dot{\alpha}_*, \tag{6b}$$

where $\alpha_*$ is a prescribed upper bound and $\dot{\alpha}_*$ is a prescribed, positive constant. These inequalities are enforced indirectly via transformation techniques, which convert the inequality constraints into equality constraints. Note that, after the transformation, $\alpha(t)$ becomes a state variable (Refs. 12-13).

The power setting $\beta$ and its time derivative $\dot{\beta}$ are subject to the inequalities

$$\beta_* \leq \beta \leq 1, \tag{7a}$$

$$-\dot{\beta}_* \leq \dot{\beta} \leq \dot{\beta}_*, \tag{7b}$$

where $\beta_*$ is a prescribed lower bound and $\dot{\beta}_*$ is a prescribed, positive constant. These inequalities are automatically satisfied, owing to the following considerations. For the take-off problem, maximum power setting is employed; hence,

$$\beta = 1, \qquad 0 \leq t \leq \tau. \tag{8}$$

For the abort landing problem, the power setting is increased at a constant time rate $\dot{\beta}_0$, consistent with (7b), from the initial value $\beta_0$ to the maximum value; afterward, the power setting is held constant; hence,

$$\beta = \beta_0 + \dot{\beta}_0 t, \qquad 0 \leq t \leq \sigma, \tag{9a}$$

$$\beta = 1, \qquad \sigma \leq t \leq \tau, \tag{9b}$$

where $\sigma = (1 - \beta_0)/\dot{\beta}_0$. In the examples of this paper, $\dot{\beta}_0 = 0.2/\text{sec}$.

3.2. Wind Model

The wind model employed in this paper involves the combination of shear (transition from headwind to tailwind) and downdraft (Ref. 16). Analytically, it is represented by the relations

$$W_x = \lambda A(x), \tag{10a}$$

$$W_h = \lambda(h/h_*)B(x), \tag{10b}$$

with

$$\Delta W_x = \lambda \Delta W_{x*}, \tag{10c}$$

$$\Delta W_h = \lambda(\Delta W_{x*}/2)h/h_*. \tag{10d}$$

Here, the parameter $\lambda$ characterizes the intensity of the windshear/downdraft combination; the function $A(x)$ represents the profile of the horizontal wind versus the horizontal distance (Table 1); and the function $B(x)$ represents the profile of the vertical wind versus the horizontal distance (Table 1). Also, $\Delta W_x$ is the horizontal wind velocity difference (maximum tailwind minus maximum headwind); $\Delta W_h$ is the vertical wind velocity difference (maximum updraft minus maximum downdraft); $\Delta W_{x*} = 100$ fps is a reference value for the horizontal wind velocity difference; and $h_* = 1000$ ft is a reference value for the altitude.

Decreasing values of $\lambda$ (hence, decreasing values of $\Delta W_x$) correspond to milder windshears; conversely, increasing values of $\lambda$ (hence, increasing values of $\Delta W_x$) correspond to more severe windshears. Therefore, by changing the value of $\lambda$, one can generate shear/downdraft combinations ranging from extremely mild to extremely severe.

To sum up, the windshear model (10) has the following properties: (a) it represents the transition from a uniform headwind to a uniform tailwind, with nearly constant shear in the core of the downburst; (b) the downdraft achieves maximum negative value at the center of the downburst; (c) the downdraft vanishes at $h = 0$; and (d) the wind velocity components nearly satisfy the continuity equation and the irrotationality condition in the core of the downburst.

## 4. OPTIMAL TAKE-OFF TRAJECTORIES

We refer to the take-off problem and to the differential system (1)-(3). We assume that: the aircraft is airborne; the thrust, the drag, and the lift are given by Eqs. (4); in particular, the wind flow field is given by Eqs. (10), with $\lambda$ specified; the power setting $\beta(t)$ is given by Eq. (8); the angle of attack $\alpha(t)$ is subject to Ineqs. (6).

The initial values $x_0$, $h_0$, $V_0$ are specified; the remaining initial values $\gamma_0$, $\alpha_0$ are computed using the assumption of quasi-steady flight prior to the windshear onset. At the specified final time $\tau$, gamma recovery is required, hence $\gamma_\tau = \gamma_0$.

With the above understanding, we formulate the following optimization problem.

### 4.1. Minimax Problem

Minimize the peak value of the modulus of the difference between the absolute path inclination and a reference value, assumed constant. In this problem, the performance index is given by

$$I = \max_t |\gamma_e - \gamma_{eR}|, \qquad 0 \leq t \leq \tau, \tag{11a}$$

where

$$\gamma_e = \arctan[(V\sin\gamma + W_h)/(V\cos\gamma + W_x)], \tag{11b}$$

$$\gamma_{eR} = \gamma_{e0}. \tag{11c}$$

This is a minimax problem or Chebyshev problem of optimal control. It can be reformulated as a Bolza problem of optimal control, in which one minimizes the integral performance index

$$J = \int_0^\tau (\gamma_e - \gamma_{eR})^q dt, \tag{12}$$

for large values of the positive, even exponent q (for instance, q = 6).

### 4.2. Sequential Gradient-Restoration Algorithm

The transformed problem is a Bolza problem of optimal control, which can be solved using the family of sequential gradient-restoration algorithms for optimal control problems (SGRA) in either the primal formulation (PSGRA) or the dual formulation (DSGRA). In this work, the dual formulation is used. The algorithmic details can be found in Refs. 17-21; they are omitted here, for the sake of brevity.

In computing optimal take-off trajectories for the B-727, B-737, and B-747 aircraft, DSGRA was programmed in FORTRAN 77. The numerical results were obtained in double-precision arithmetic. The computations were performed using the NAS-AS-9000 computer of Rice University.

The following values of the wind intensity parameter were considered:

$$\lambda = 0.6,\ 0.8,\ 1.0,\ 1.2. \tag{13a}$$

These values correspond to the following wind velocity differences:

$$\Delta W_x = 100\ \lambda = 60,\ 80,\ 100,\ 120 \text{ fps}. \tag{13b}$$

4.3. Aircraft Data

The aircraft under consideration are the following: the Boeing B-727-200, powered by three Pratt and Whittney JT8D-17 engines; the Boeing B-737-300, powered by two General Electric CFM56-3B1 engines; and the Boeing B-747-200, powered by four Pratt and Whittney JT9D-7R4G2 engines. It is assumed that the runway is located at sea-level altitude and that the ambient temperature is 100 deg F.

Table 1. Wind functions A(x) and B(x).

| x (ft) | A(x) (fps) | B(x) (fps) | x (ft) | A(x) (fps) | B(x) (fps) |
|---|---|---|---|---|---|
| 0 | -50.00 | 0.00 | 2300 | 0.00 | -50.00 |
| 200 | -49.58 | -0.23 | 2400 | 2.50 | -50.00 |
| 400 | -47.08 | -1.80 | 2600 | 7.50 | -50.00 |
| 600 | -42.50 | -6.08 | 2800 | 12.50 | -49.78 |
| 800 | -37.50 | -14.20 | 3000 | 17.50 | -48.20 |
| 1000 | -32.50 | -25.00 | 3200 | 22.50 | -43.92 |
| 1200 | -27.50 | -35.80 | 3400 | 27.50 | -35.80 |
| 1400 | -22.50 | -43.92 | 3600 | 32.50 | -25.00 |
| 1600 | -17.50 | -48.20 | 3800 | 37.50 | -14.20 |
| 1800 | -12.50 | -49.78 | 4000 | 42.50 | -6.08 |
| 2000 | -7.50 | -50.00 | 4200 | 47.08 | -1.80 |
| 2200 | -2.50 | -50.00 | 4400 | 49.58 | -0.23 |
| 2300 | 0.00 | -50.00 | 4600 | 50.00 | 0.00 |

Table 2. Take-off problem, aircraft data.

| | B-727-200 | B-737-300 | B-747-200 | Units |
|---|---|---|---|---|
| W | 190,000 | 130,000 | 833,000 | lb |
| S | 1,560 | 980 | 5,500 | $ft^2$ |
| $\delta_F$ | 15.0 | 5.0 | 20.0 | deg |
| $\alpha_*$ | 16.0 | 18.2 | 16.4 | deg |
| $\dot{\alpha}_*$ | 3.0 | 3.0 | 3.0 | deg/sec |
| $V_*$ | 143.0 | 134.0 | 158.4 | knots |
| $V_2$ | 164.0 | 157.0 | 187.0 | knots |
| $V_0$ | 174.0 | 167.0 | 197.0 | knots |

Table 3. Take-off problem, boundary conditions.

| | B-727-200 | B-737-300 | B-747-200 | Units |
|---|---|---|---|---|
| $x_0$ | 0.0 | 0.0 | 0.0 | ft |
| $h_0$ | 50.0 | 50.0 | 50.0 | ft |
| $V_0$ | 293.7 | 281.9 | 332.5 | fps |
| $\gamma_0$ | 6.330 | 8.200 | 5.046 | deg |
| $\alpha_0$ | 9.859 | 11.009 | 9.425 | deg |
| $\tau$ | 40.0 | 40.0 | 40.0 | sec |
| $\gamma_\tau$ | 6.330 | 8.200 | 5.046 | deg |

For the above aircraft, Table 2 shows the following data: the weight W, the wing surface S, the flap deflection $\delta_F$, the stick-shaker angle of attack $\alpha_*$, the limiting angle of attack rate $\dot{\alpha}_*$, the stick-shaker velocity for quasi-steady level flight $V_*$, the FAA certification velocity $V_2$, and the initial velocity $V_0$. Note that $V_0 = V_2 + 10$ knots.

Also for the above aircraft, Table 3 shows the boundary conditions for the optimization problem, specifically: (i) the initial values of the distance $x_0$, the altitude $h_0$, the velocity $V_0$, the relative path inclination $\gamma_0$, and the angle of attack $\alpha_0$; and (ii) the final values of the time $\tau$ and the relative path inclination $\gamma_\tau$. Note that the final time $\tau = 40$ sec is more than twice the duration of the windshear encounter (18 sec for the B-727, 18 sec for the B-737, and 15 sec for the B-747).

4.4. Numerical Results

For the B-727, B-737, and B-747 aircraft, optimal trajectories were computed for the boundary conditions of Table 3 and the windshear intensities (13). The results are given in Figs. 1-3, which show the following quantities as functions of the time t: the horizontal wind $W_x$; the vertical wind $W_h$; the altitude h; the relative velocity V; the angle of attack $\alpha$; and

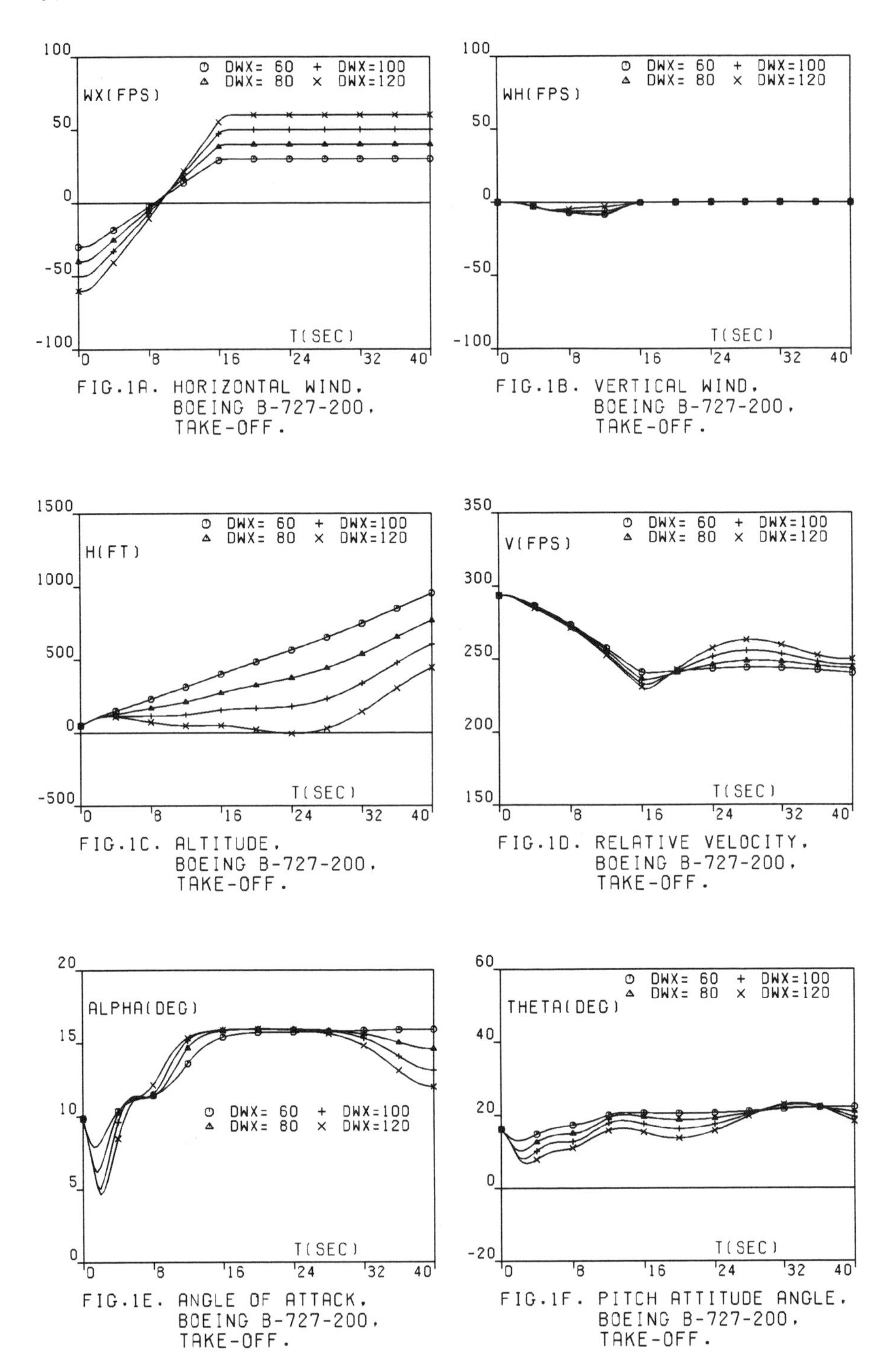


FIG.1A. HORIZONTAL WIND. BOEING B-727-200. TAKE-OFF.

FIG.1B. VERTICAL WIND. BOEING B-727-200. TAKE-OFF.

FIG.1C. ALTITUDE. BOEING B-727-200. TAKE-OFF.

FIG.1D. RELATIVE VELOCITY. BOEING B-727-200. TAKE-OFF.

FIG.1E. ANGLE OF ATTACK. BOEING B-727-200. TAKE-OFF.

FIG.1F. PITCH ATTITUDE ANGLE. BOEING B-727-200. TAKE-OFF.

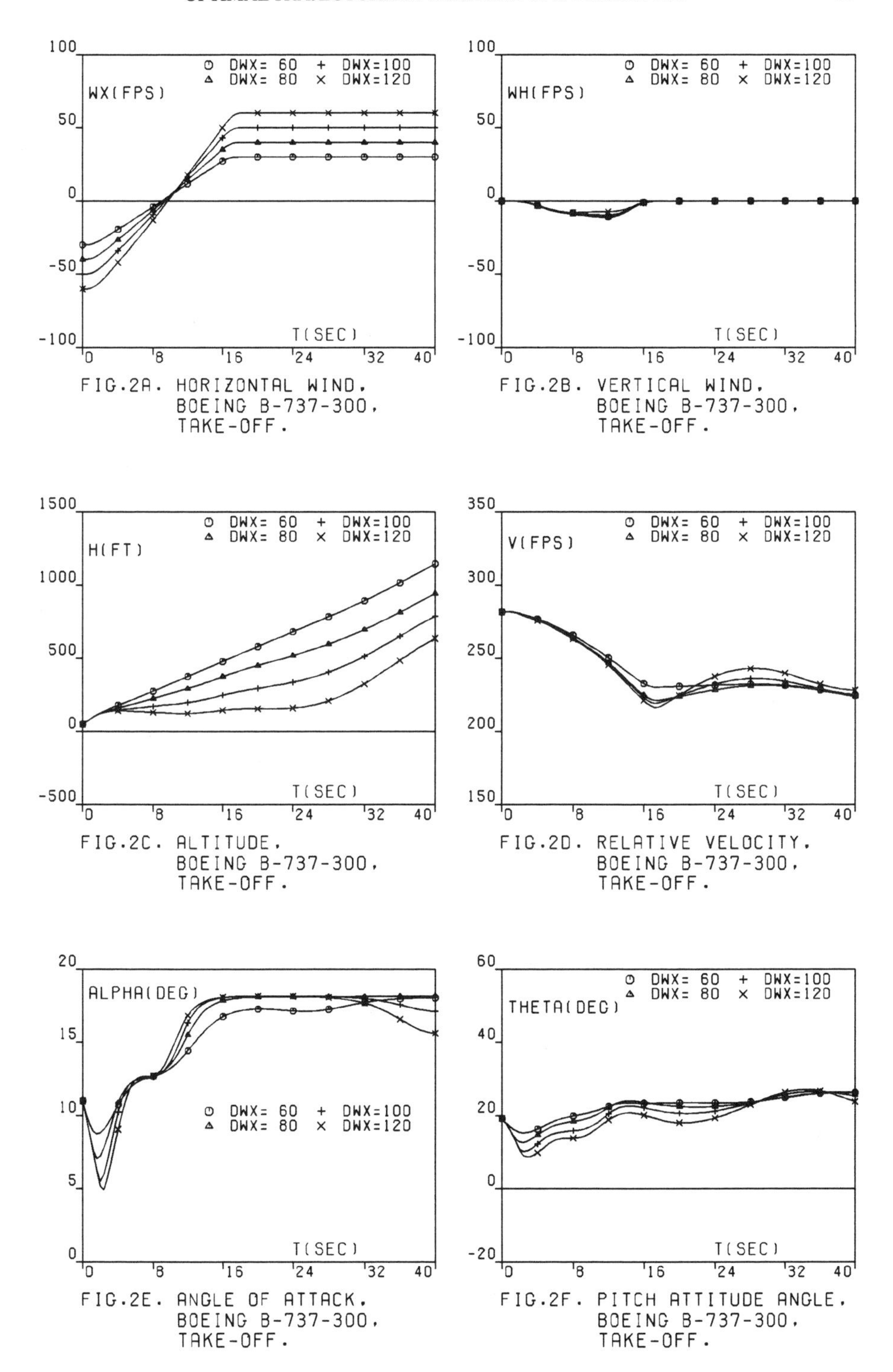


FIG.2A. HORIZONTAL WIND. BOEING B-737-300. TAKE-OFF.

FIG.2B. VERTICAL WIND. BOEING B-737-300. TAKE-OFF.

FIG.2C. ALTITUDE. BOEING B-737-300. TAKE-OFF.

FIG.2D. RELATIVE VELOCITY. BOEING B-737-300. TAKE-OFF.

FIG.2E. ANGLE OF ATTACK. BOEING B-737-300. TAKE-OFF.

FIG.2F. PITCH ATTITUDE ANGLE. BOEING B-737-300. TAKE-OFF.

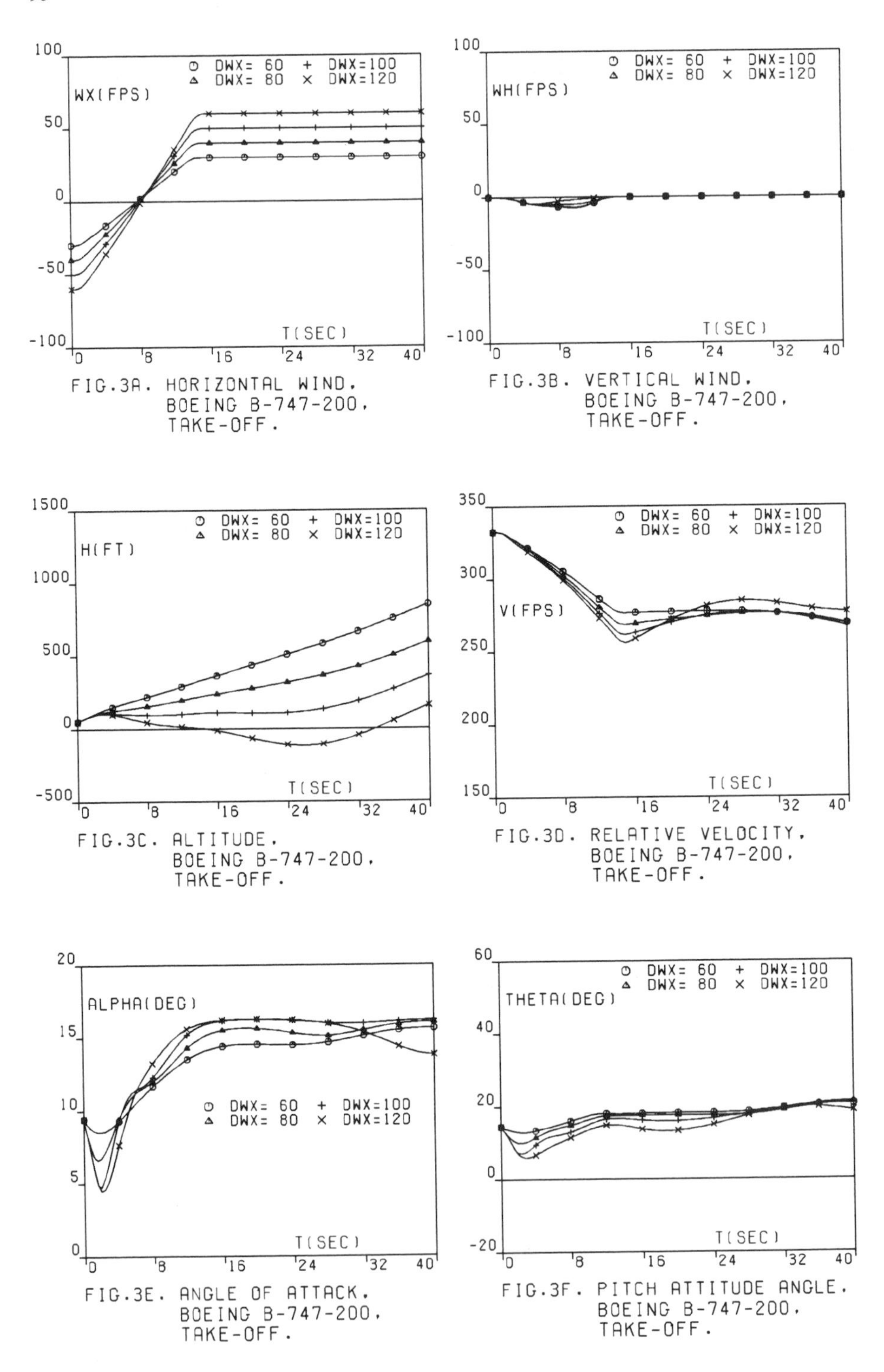


FIG.3A. HORIZONTAL WIND. BOEING B-747-200. TAKE-OFF.

FIG.3B. VERTICAL WIND. BOEING B-747-200. TAKE-OFF.

FIG.3C. ALTITUDE. BOEING B-747-200. TAKE-OFF.

FIG.3D. RELATIVE VELOCITY. BOEING B-747-200. TAKE-OFF.

FIG.3E. ANGLE OF ATTACK. BOEING B-747-200. TAKE-OFF.

FIG.3F. PITCH ATTITUDE ANGLE. BOEING B-747-200. TAKE-OFF.

the pitch attitude angle $\theta$.

After inspecting Figs. 1-3, the following comments appear to be valid for the B-727, B-737, and B-747 aircraft:

(i) the relative velocity decreases in the shear region and increases in the aftershear region; minimum velocity is achieved at the end of the shear and is nearly independent of the intensity of the shear;

(ii) the angle of attack exhibits an initial decrease, followed by a gradual, sustained increase; the peak value of the angle of attack is achieved near the end of the shear;

(iii) the pitch attitude angle exhibits an initial decrease, followed by a gradual, wavelike increase; the larger values of the pitch attitude angle are achieved near the end of the shear;

(iv) in the shear region, the average value of the path inclination decreases as the windshear intensity increases; it is positive for weak-to-moderate windshears, nearly zero for strong-to-severe windshears, and negative for extremely severe windshears;

(v) for weak-to-moderate windshears, the altitude profile is characterized by a continuous climb; for strong-to-severe windshears, the altitude profile is characterized by an initial climb, followed by nearly horizontal flight, followed by renewed climbing after the aircraft has passed through the shear region; for extremely severe windshears, the altitude profile is characterized by an initial climb, followed by descending flight until the aircraft finally hits the ground.

### 4.5. Guidance Implications

The results on optimal trajectories can be useful in the construction of near-optimal guidance schemes if one exploits the correlation existing between various physical quantities and the shear/downdraft factor, which is defined by

$$F = \dot{W}_x/g - W_h/V. \tag{14}$$

This factor attempts to combine the effects of the shear and the downdraft into a single entity and is a measure of the intensity of the shear/downdraft combination (for details, see Refs. 9, 10, 12). The following comments are pertinent:

(a) in the shear region, the average value of the absolute path inclination is monotonically related to the average value of the shear/downdraft factor; this observation is the basis of the gamma guidance scheme (Ref. 10);

(b) in the shear region, the average value of the relative acceleration is monotonically related to the average value of the shear/downdraft factor; this observation is the basis of the acceleration guidance scheme (Ref. 9);

(c) because the optimal trajectories of the B-727, B-737, and B-747 aircraft exhibit the same qualitative behavior, it appears that the near-optimal guidance schemes developed for the B-727 aircraft can be extended to the B-737 and B-747 aircraft, albeit with some quantitative modification.

4.6. Trajectory Comparison

It is of interest to compare the optimal trajectories of the B-727, B-737, and B-747 aircraft. The comparison is shown in Fig. 4 in terms of the altitude profile h(t) and the thrust-to-weight ratio function $T/W = f(V/V_0)$.

Inspection of Figs. 4A-4D shows that, for the same windshear intensity, the function h(t) of the B-737 lies above that of the B-727, which in turn lies above that of the B-747; in other words, for the take-off problem, the windshear behavior of the B-737 is better than that of the B-727, which in turn is better than that of the B-747. The major reason for this result is the following: owing to FAA certification requirements for flight with one engine shut-off, the thrust-to-weight ratio of the B-737 is higher than that

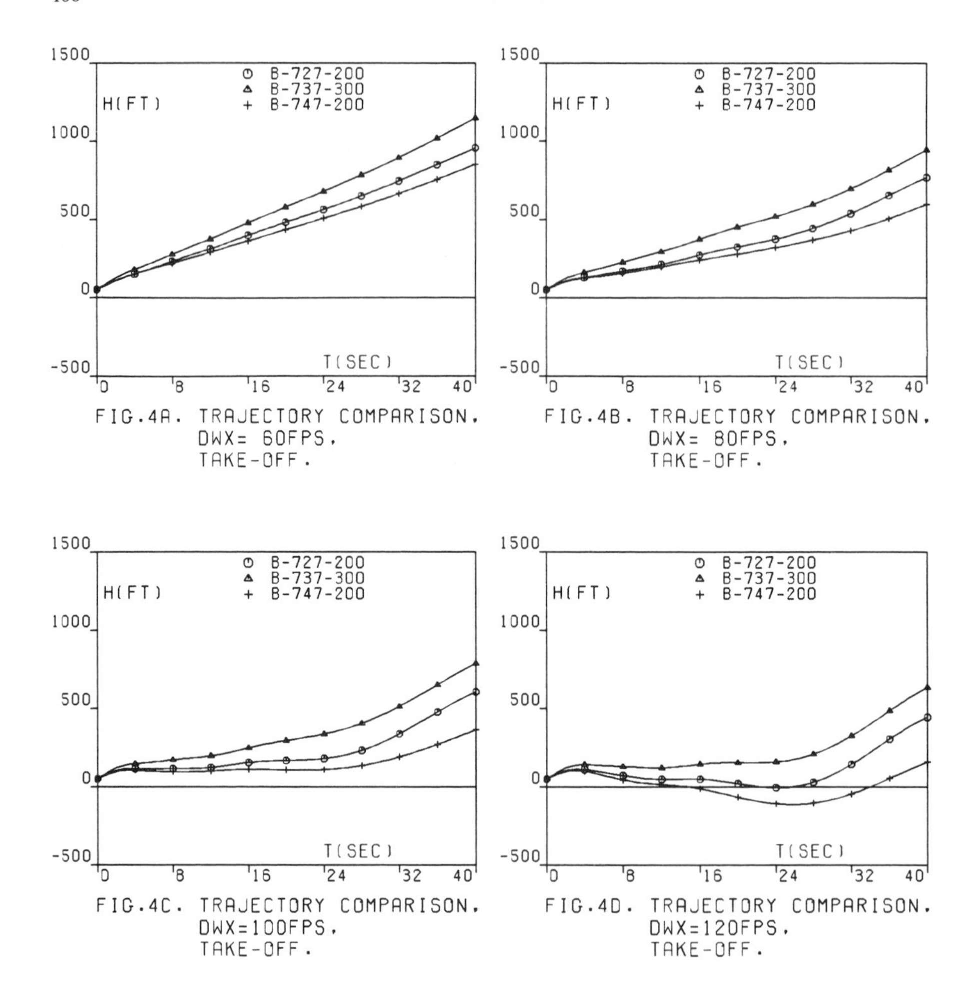


FIG.4A. TRAJECTORY COMPARISON, DWX= 60FPS, TAKE-OFF.

FIG.4B. TRAJECTORY COMPARISON, DWX= 80FPS, TAKE-OFF.

FIG.4C. TRAJECTORY COMPARISON, DWX=100FPS, TAKE-OFF.

FIG.4D. TRAJECTORY COMPARISON, DWX=120FPS, TAKE-OFF.

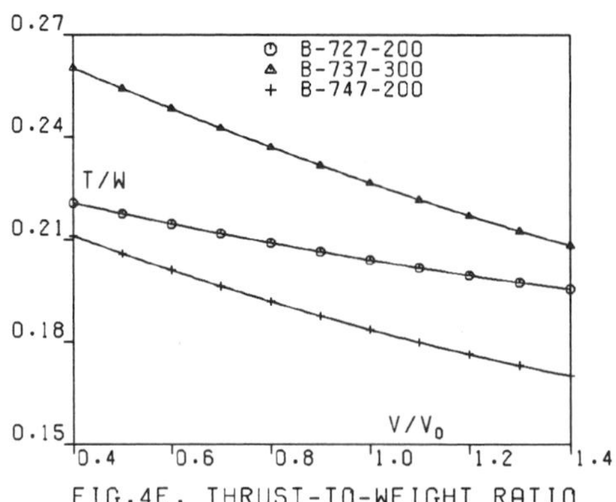


FIG.4E. THRUST-TO-WEIGHT RATIO COMPARISON, BETA=1.0, TAKE-OFF.

of the B-727, which in turn is higher than that of the B-747; see Fig. 4E, which shows the thrust-to-weight ratio T/W as a function of the relative velocity $V/V_0$ for the three aircraft.

## 4.7. Survival Capability

Here, we analyze the survival capability of an aircraft in a severe windshear. Indicative of this survival capability is the windshear/downdraft combination which results in the minimum altitude being equal to the ground altitude.

To analyze this important problem, we recall the one-parameter family of windshear models (10), in which the parameter $\lambda$ characterizes the intensity of the windshear/downdraft combination. By increasing the value of $\lambda$, more intense windshear/downdraft combinations are generated until a critical value $\lambda_c$ is found (hence, a critical value $\Delta W_{xc}$ is found), such that $h_{min} = 0$ for a given trajectory type.

More precisely, we consider the following trajectories (Refs. 9,10,12): the optimal trajectory (OT); the constant pitch trajectory (CPT); and the maximum angle of attack trajectory (MAAT). The results are shown in Table 4, which supplies the critical wind velocity difference $\Delta W_{xc}$, and in Table 5, which supplies the windshear efficiency ratio WER, defined to be

$$WER = (\lambda_c)_{PT}/(\lambda_c)_{OT} = (\Delta W_{xc})_{PT}/(\Delta W_{xc})_{OT}; \qquad (15)$$

here, the subscript PT denotes a particular trajectory and the subscript OT denotes the optimal trajectory.

From Tables 4-5, the following conclusions can be inferred:

(i) for the OT, the survival capability of the B-737 is higher than that of the B-727, which in turn is higher than that of the B-747;

(ii) for the CPT, the survival capability of the B-737 is higher than that of the B-727, which in turn is higher than that of the B-747;

(iii) for the MAAT, the survival capability of the B-737 is higher than that of the B-727, which in turn is higer than that of the B-747;

(iv) the qualitative explanation of the above results lies primarily in the fact that the thrust-to-weight ratio of the B-737 is larger than that of the B-727, which in turn is larger than that of the B-747;

(v) in terms of survival capability, if one defines the windshear efficiency of the OT to be 100%, that of the CPT is 86% to 90% and that of the MAAT is 42% to 48%, depending on the type of aircraft.

## 5. OPTIMAL ABORT LANDING TRAJECTORIES

We refer to the abort landing problem and to the differential system (1)-(3). We assume that: the thrust, the drag, and the lift are given by Eqs. (4);

in particular, the wind flow field is given by Eqs. (10), with $\lambda$ specified; the power setting $\beta(t)$ is given by Eqs. (9); the angle of attack $\alpha(t)$ is subject to Ineqs. (6).

The initial values $x_0$, $h_0$, $V_0$, $\gamma_{e0} = -3.0$ deg are specified; the remaining initial values $\gamma_0$, $\alpha_0$, $\beta_0$ are computed using the assumption of quasi-steady flight prior to the windshear onset. At the specified final time $\tau$, gamma recovery is required , hence $\gamma_\tau = \gamma_*$, where $\gamma_*$ is the relative path inclination for quasi-steady steepest climb.

With the above understanding, we formulate the following optimization problem.

### 5.1. Minimax Problem

Minimize the peak value of the altitude drop; that is, minimize the peak value of the difference between a constant reference altitude and the instantaneous altitude. In this problem, the performance index is given by

$$I = \max_t (h_R - h), \qquad 0 \leq t \leq \tau. \qquad (16)$$

This is a minimax problem or Chebyshev problem of optimal control. It can be reformulated as a Bolza problem of optimal control, in which one minimizes the integral performance index

Table 4. Take-off problem, survival capability $\Delta W_{xc}$(fps), $h_0 = 50$ ft.

| Aircraft | W (lb) | OT | CPT | MAAT |
|---|---|---|---|---|
| B-727-200 | 190,000 | 119.3 | 103.2 | 57.5 |
| B-737-300 | 130,000 | 139.3 | 125.8 | 58.3 |
| B-747-200 | 833,000 | 110.0 | 97.0 | 51.3 |

Table 5. Take-off problem, windshear efficiency ratio WER, $h_0 = 50$ ft.

| Aircraft | W (lb) | OT | CPT | MAAT |
|---|---|---|---|---|
| B-727-200 | 190,000 | 1.000 | 0.865 | 0.482 |
| B-737-300 | 130,000 | 1.000 | 0.903 | 0.419 |
| B-747-200 | 833,000 | 1.000 | 0.882 | 0.466 |

Table 6. Abort landing problem, aircraft data.

| | B-727-200 | B-737-300 | B-747-200 | Units |
|---|---|---|---|---|
| $W$ | 150,000 | 114,000 | 630,000 | lb |
| $S$ | 1,560 | 980 | 5,500 | $ft^2$ |
| $\delta_F$ | 30.0 | 40.0 | 25.0 | deg |
| $\alpha_*$ | 17.2 | 16.0 | 15.2 | deg |
| $\dot{\alpha}_*$ | 3.0 | 3.0 | 3.0 | deg/sec |
| $\beta_*$ | 0.2 | 0.2 | 0.2 | — |
| $\dot{\beta}_*$ | 0.3 | 0.3 | 0.3 | 1/sec |
| $V_*$ | 107.3 | 106.9 | 133.5 | knots |
| $V_{ref}$ | 137.0 | 138.0 | 165.0 | knots |
| $V_0$ | 147.0 | 148.0 | 175.0 | knots |

$$J = \int_0^\tau (h_R - h)^q dt, \tag{17}$$

for large values of the positive, even exponent $q$ (for instance, $q = 6$).

## 5.2. Sequential Gradient-Restoration Algorithm

The transformed problem is a Bolza problem of optimal control, which can be solved using the family of sequential gradient-restoration algorithms for optimal control problems (SGRA) in either the primal formulation (PSGRA) or the dual formulation (DSGRA). In this work, the dual formulation is used. The algorithmic details can be found in Refs. 17-21; they are omitted here, for the sake of brevity.

In computing optimal abort landing trajectories for the B-727, B-737, and B-747 aircraft, DSGRA was programmed in FORTRAN 77. The numerical results were obtained in double-precision arithmetic. The computations were performed using the NAS-AS-9000 computer of Rice University.

The following values of the wind intensity parameter were considered:

$$\lambda = 1.0, 1.2, 1.4, 1.6. \tag{18a}$$

These values correspond to the following wind velocity differences:

$$\Delta W_x = 100\ \lambda = 100,\ 120,\ 140,\ 160\ \text{fps}. \qquad (18b)$$

5.3. Aircraft Data

The aircraft under consideration are the following: the Boeing B-727-200, powered by three Pratt and Whittney JT8D-17 engines; the Boeing B-737-300, powered by two General Electric CFM56-3B1 engines; and the Boeing B-747-200, powered by four Pratt and Whittney JT9D-7R4G2 engines. It is assumed that the runway is located at sea-level altitude and that the ambient temperature is 100 deg F.

For the above aircraft, Table 6 shows the following data: the weight W, the wing surface S, the flap deflection $\delta_F$, the stick-shaker angle of attack $\alpha_*$, the limiting angle of attack rate $\dot{\alpha}_*$, the power setting lower bound $\beta_*$, the limiting power setting rate $\dot{\beta}_*$, the stick-shaker velocity for quasi-steady level flight $V_*$, the FAA certification velocity $V_{ref}$, and the initial velocity $V_0$. Note that $V_0 = V_{ref} +$ 10 knots.

Also for the above aircraft, Table 7 shows the boundary conditions for the optimization problem, specifically: (i) the initial value of the distance $x_0$, the altitude $h_0$, the velocity $V_0$, and the absolute path inclination $\gamma_{e0}$; and (ii) the final values of the time $\tau$ and the relative path inclination $\gamma_\tau$. Note that the

Table 7. Abort landing problem, boundary conditions.

| | B-727-200 | B-737-300 | B-747-200 | Units |
|---|---|---|---|---|
| $x_0$ | 0.0 | 0.0 | 0.0 | ft |
| $h_0$ | 600.0 | 600.0 | 600.0 | ft |
| $V_0$ | 248.1 | 249.8 | 295.4 | fps |
| $\gamma_{e0}$ | -3.000 | -3.000 | -3.000 | deg |
| $\tau$ | 40.0 | 40.0 | 40.0 | sec |
| $\gamma_\tau$ | 7.435 | 7.233 | 7.656 | deg |

Table 8. Abort landing problem, survival capability $\Delta W_{xc}$(fps), $h_0$ = 600 ft.

| Aircraft | W (lb) | OT | CPT | MAAT |
|---|---|---|---|---|
| B-727-200 | 150,000 | 191.5 | 144.1 | 81.8 |
| B-737-300 | 114,000 | 189.6 | 141.9 | 82.1 |
| B-747-200 | 630,000 | 186.8 | 140.7 | 91.5 |

Table 9. Abort landing problem, windshear efficiency ratio WER, $h_0$ = 600 ft.

| Aircraft | W (lb) | OT | CPT | MAAT |
|---|---|---|---|---|
| B-727-200 | 150,000 | 1.000 | 0.752 | 0.427 |
| B-737-300 | 114,000 | 1.000 | 0.748 | 0.433 |
| B-747-200 | 630,000 | 1.000 | 0.753 | 0.490 |

final time $\tau = 40$ sec is about twice the duration of the windshear encounter (22 sec for the B-727, 22 sec for the B-737, and 20 sec for the B-747).

5.4. Numerical Results

For the B-727, B-737, and B-747 aircraft, optimal trajectories were computed for the boundary conditions of Table 7 and the windshear intensities (18). The results are given in Figs. 5-7, which show the following quantities as functions of the time t: the horizontal wind $W_x$; the vertical wind $W_h$; the altitude h; the relative velocity V; the angle of attack $\alpha$; and the pitch attitude angle $\theta$.

After inspecting Figs. 5-7, the following comments appear to be valid for the B-727, B-737, and B-747 aircraft in severe windshears:

(i) the relative velocity decreases in the shear region and increases in the aftershear region; minimum velocity is achieved at the end of the shear and is nearly independent of the intensity of the shear;

(ii) the angle of attack exhibits an initial decrease, followed by a gradual, sustained increase; the peak value of the angle of attack is achieved near the end of the shear;

(iii) the pitch attitude angle exhibits an initial decrease, followed by a gradual, wavelike increase; the

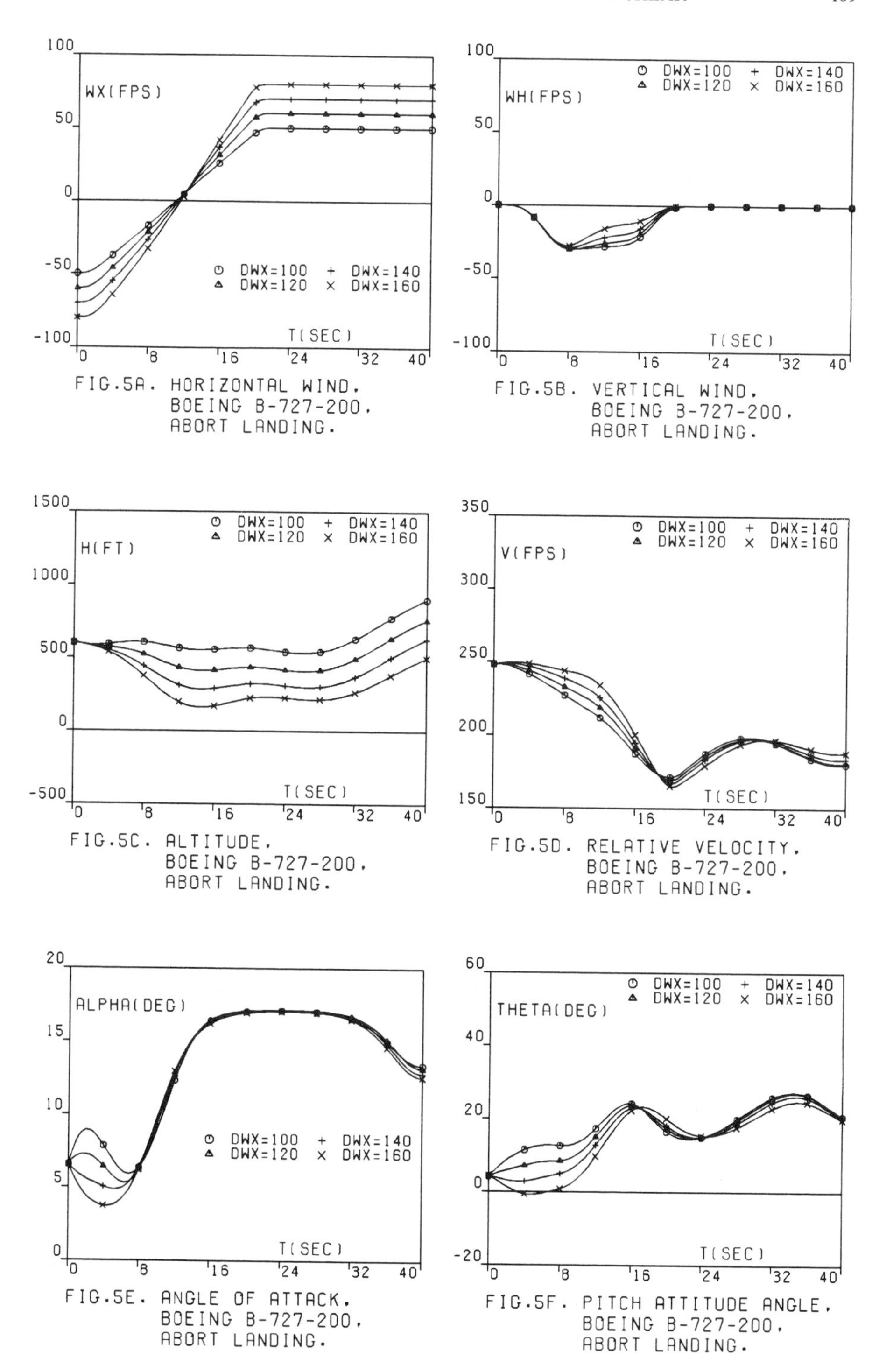


FIG.5A. HORIZONTAL WIND. BOEING B-727-200. ABORT LANDING.

FIG.5B. VERTICAL WIND. BOEING B-727-200. ABORT LANDING.

FIG.5C. ALTITUDE. BOEING B-727-200. ABORT LANDING.

FIG.5D. RELATIVE VELOCITY. BOEING B-727-200. ABORT LANDING.

FIG.5E. ANGLE OF ATTACK. BOEING B-727-200. ABORT LANDING.

FIG.5F. PITCH ATTITUDE ANGLE. BOEING B-727-200. ABORT LANDING.

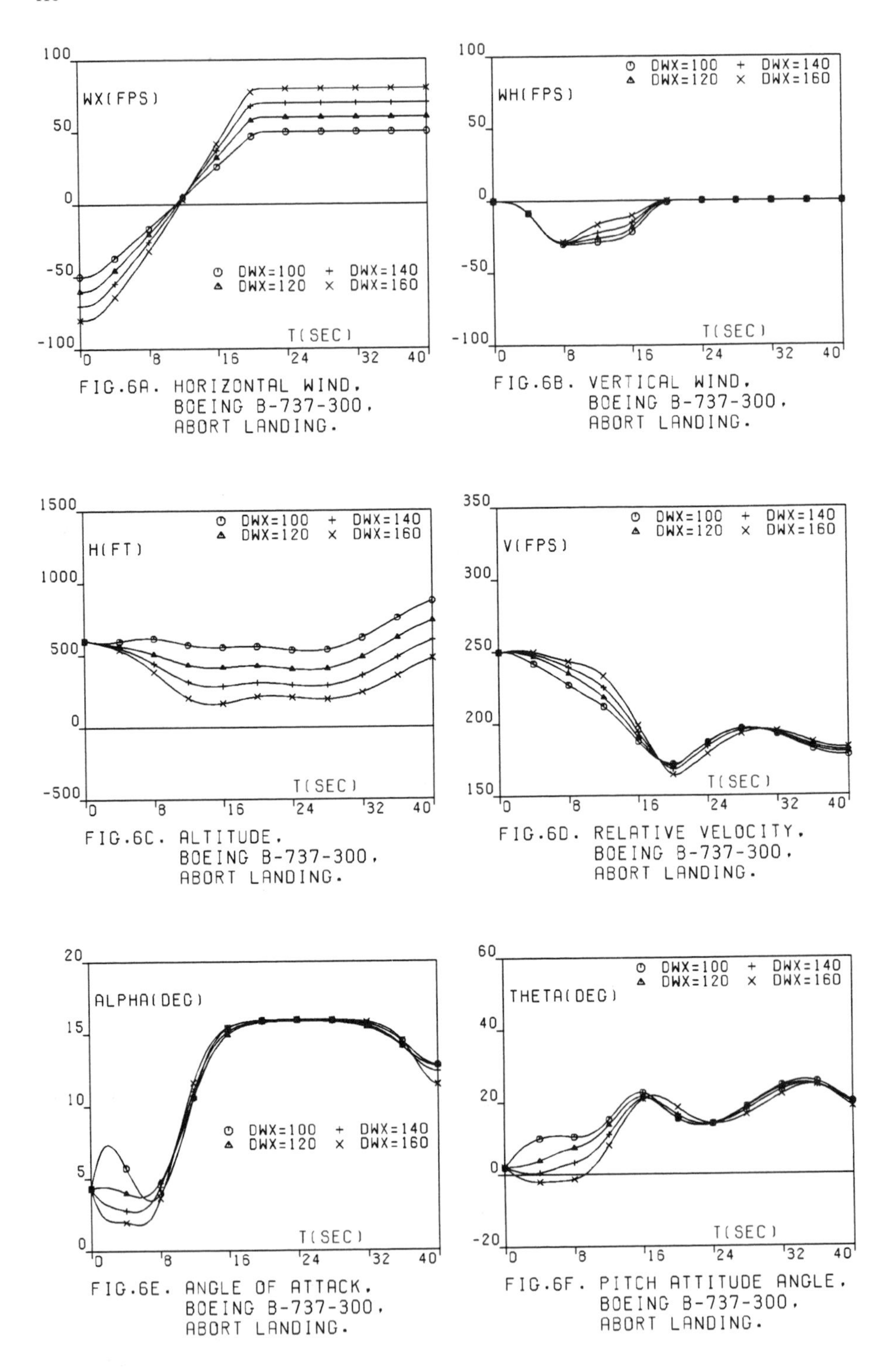


FIG.6A. HORIZONTAL WIND. BOEING B-737-300. ABORT LANDING.

FIG.6B. VERTICAL WIND. BOEING B-737-300. ABORT LANDING.

FIG.6C. ALTITUDE. BOEING B-737-300. ABORT LANDING.

FIG.6D. RELATIVE VELOCITY. BOEING B-737-300. ABORT LANDING.

FIG.6E. ANGLE OF ATTACK. BOEING B-737-300. ABORT LANDING.

FIG.6F. PITCH ATTITUDE ANGLE. BOEING B-737-300. ABORT LANDING.

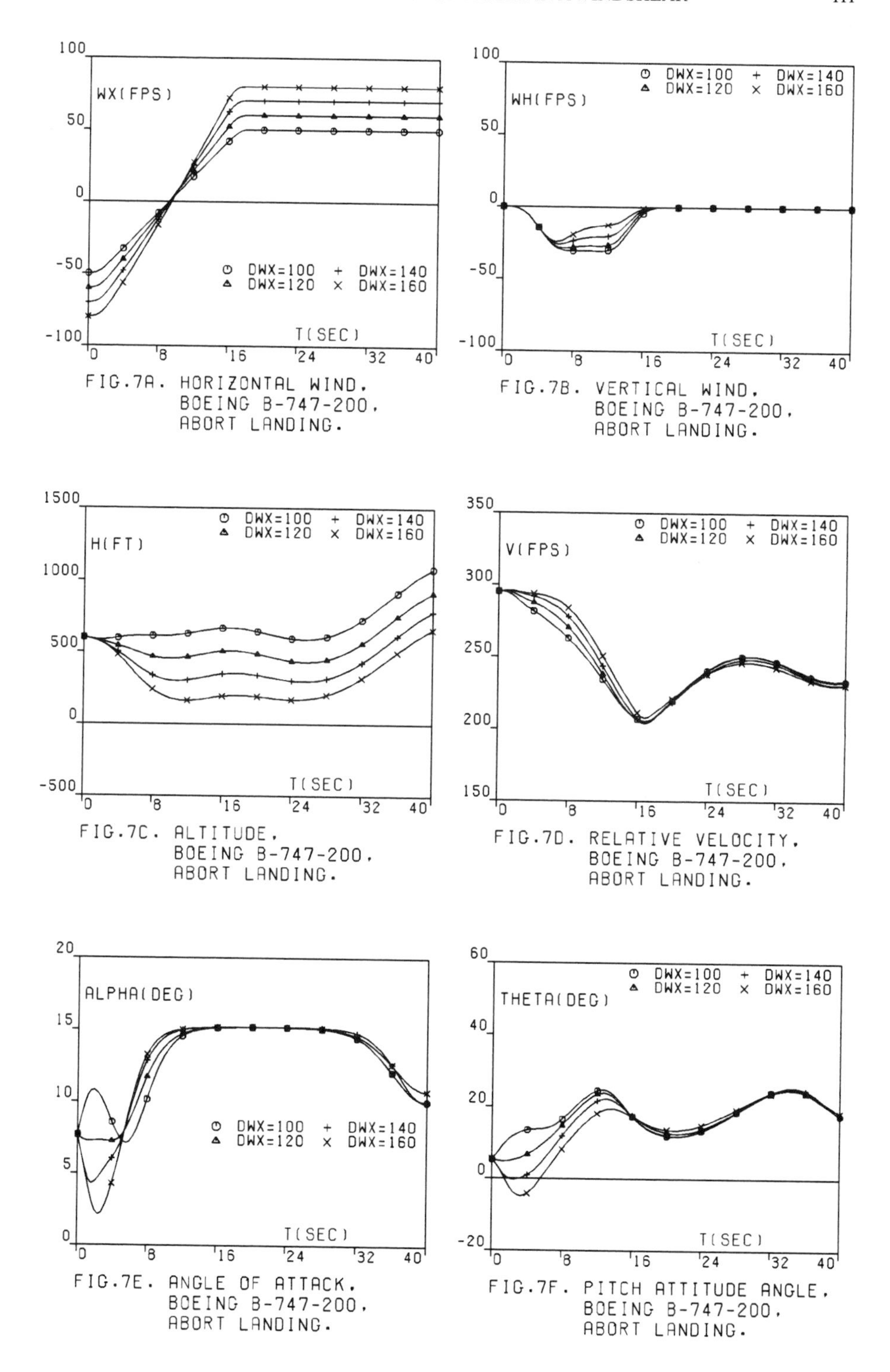


FIG.7A. HORIZONTAL WIND. BOEING B-747-200. ABORT LANDING.

FIG.7B. VERTICAL WIND. BOEING B-747-200. ABORT LANDING.

FIG.7C. ALTITUDE. BOEING B-747-200. ABORT LANDING.

FIG.7D. RELATIVE VELOCITY. BOEING B-747-200. ABORT LANDING.

FIG.7E. ANGLE OF ATTACK. BOEING B-747-200. ABORT LANDING.

FIG.7F. PITCH ATTITUDE ANGLE. BOEING B-747-200. ABORT LANDING.

larger values of the pitch attitude angle are achieved near the end of the shear;

(iv) the optimal trajectory includes three branches: a descending flight branch, followed by a nearly horizontal flight branch, followed by an ascending flight branch after the aircraft has passed through the shear region;

(v) the peak altitude drop depends on the windshear intensity and the initial altitude; it increases as the windshear intensity increases and the initial altitude increases (Ref. 13).

### 5.5. Guidance Implications

The results on optimal trajectories can be useful in the construction of near-optimal guidance schemes if one exploits the correlation existing between various physical quantities and the shear/downdraft factor (14). The following comments are pertinent:

(a) based on the altitude profile of the optimal trajectories, an abort landing guidance scheme requires two switches: the first switch occurs in the shear region and transfers the aircraft from descending flight to level flight; the second switch occurs in the aftershear region and transfers the aircraft from level flight to ascending flight;

(b) in the descending flight branch (shear region),

the average value of the absolute path inclination is monotonically related to the average value of the shear/downdraft factor; this observation is the basis of the gamma guidance scheme (Refs. 13-14);

(c) in the level flight branch (shear region), the average value of the relative acceleration is monotonically related to the average value of the shear/downdraft factor; this observation is the basis of the acceleration guidance scheme (Refs. 13-14);

(d) because the optimal trajectories of the B-727, B-737, and B-747 aircraft exhibit the same qualitative behavior, it appears that the near-optimal guidance schemes developed for the B-727 aircraft can be extended to the B-737 and B-747 aircraft, albeit with some quantitative modification.

5.6. Trajectory Comparison

It is of interest to compare the optimal trajectories of the B-727, B-737, and B-747 aircraft. The comparison is shown in Fig. 8 in terms of the altitude profile h(t) and the thrust-to-weight ratio function $T/W = f(V/V_0)$.

Inspection of Figs. 8A-8D shows that, for the same windshear intensity, the function h(t) of the B-737 is close to that of the B-727 and that of the B-747; in other words, for the abort landing problem, the

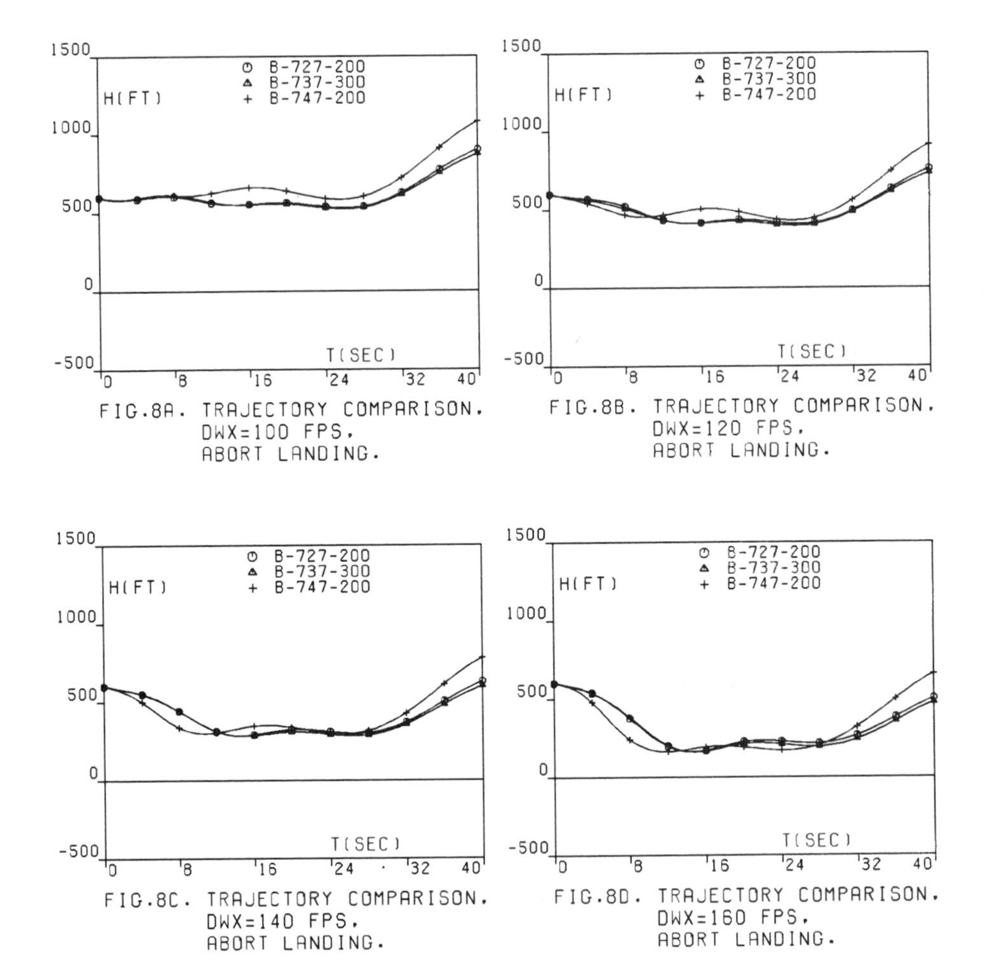


FIG.8A. TRAJECTORY COMPARISON. DWX=100 FPS. ABORT LANDING.

FIG.8B. TRAJECTORY COMPARISON. DWX=120 FPS. ABORT LANDING.

FIG.8C. TRAJECTORY COMPARISON. DWX=140 FPS. ABORT LANDING.

FIG.8D. TRAJECTORY COMPARISON. DWX=160 FPS. ABORT LANDING.

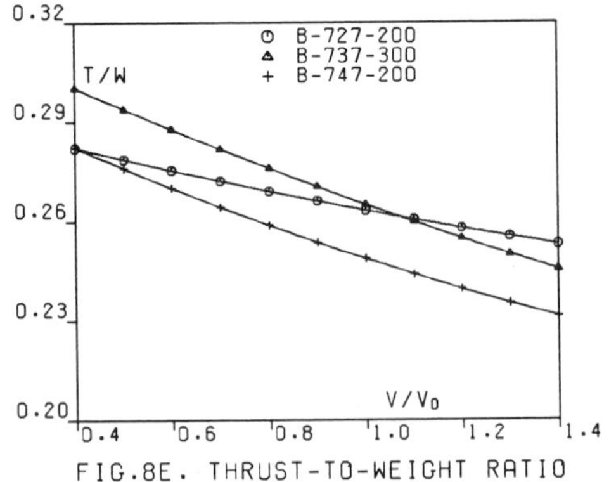


FIG.8E. THRUST-TO-WEIGHT RATIO COMPARISON, BETA=1.0. ABORT LANDING.

windshear behavior of the B-727, the B-737, and the B-747 is about the same. The major reasons for this results are the following: (i) for the abort landing problem, the thrust-to-weight ratios of the B-727 and the B-737 are nearly the same at $V = V_0$ (Fig. 8E); in addition, the initial velocities are nearly the same; and (ii) for the B-747, the thrust-to-weight ratio at $V = V_0$ is slightly smaller (Fig. 8E); but this is compensated by the fact that, owing to its higher wing loading W/S, the B-747 has a higher initial velocity.

### 5.7. Survival Capability

Here, we analyze the survival capability of an aircraft in a severe windshear. Indicative of this survival capability is the windshear/downdraft combination which results in the minimum altitude being equal to the ground altitude.

To analyze this important problem, we recall the one-parameter family of windshear models (10), in which the parameter $\lambda$ characterizes the intensity of the windshear/downdraft combination. By increasing the value of $\lambda$, more intense windshear/downdraft combinations are generated until a critical value $\lambda_c$ is found (hence, a critical value $\Delta W_{xc}$ is found), such that $h_{min} = 0$ for a given trajectory type.

More precisely, we consider the following

trajectories (Refs. 13-14): the optimal trajectory (OT); the constant pitch trajectory (CPT); and the maximum angle of attack trajectory (MAAT). The results are shown in Table 8, which supplies the critical wind velocity difference $\Delta W_{xc}$, and in Table 9, which supplies the windshear efficiency ratio WER, defined by Eq. (15).

From Tables 8-9, the following conclusions can be inferred:

(i) for each trajectory type, the survival capability of the B-727 is nearly the same as that of the B-737 and that of the B-747;

(ii) the qualitative explanation of the above result lies in the fact that the thrust-to-weight ratios of the B-727 and the B-737 are nearly the same at $V = V_0$; in addition, the initial velocities are nearly the same; while the thrust-to-weight ratio of the B-747 is slightly smaller, this is compensated by the fact that, owing to its higher wing loading W/S, the B-747 has a higher initial velocity;

(iii) in terms of survival capability, if one defines the windshear efficiency of the OT to be 100%, that of the CPT is about 75%, and that of the MAAT is between 43% and 49%, depending on the type of aircraft.

## 6. CONCLUSIONS

This paper is concerned with the optimal flight

trajectories of three different aircraft (B-727, B-737, and B-747) in the presence of windshear. The take-off problem and the abort landing problem are considered with reference to flight in a vertical plane.

Both the take-off problem and the abort landing problem are characterized by windshear inertia forces, whose dynamical effect is equivalent to a drag increase or a thrust decrease. In extremely severe windshears, the windshear inertia force can be as large as the drag of the aircraft and sometimes as large as the thrust of the engines. Under these conditions, the controllability of the aircraft is diminished, and this results occasionally in a crash.

6.1. Take-Off Problem

Optimal trajectories are computed by minimizing the peak value of the deviation of the absolute path inclination from a reference value. They exhibit the following properties:

(i) the relative velocity decreases in the shear region and increases in the aftershear region; minimum velocity is achieved at the end of the shear;

(ii) the angle of attack exhibits an initial decrease, followed by a gradual, sustained increase; the peak value of the angle of attack is achieved near the end of the shear;

(iii) for weak-to-moderate windshears, the altitude profile is characterized by a continuous climb; for strong-to-severe windshears, the altitude profile is characterized by an initial climb, followed by nearly horizontal flight, followed by renewed climbing after the aircraft has passed through the shear region; for extremely severe windshears, the altitude profile is characterized by an initial climb, followed by descending flight until the aircraft finally hits the ground;

(iv) since the windshear inertia force is proportional to the aircraft weight, the thrust-to-weight ratio is an important index for survival capability in a severe windshear; for the take-off problem, the thrust-to-weight ratio of the B-737 (two engines) is larger than that of the B-727 (three engines), which in turn is larger than that of the B-747 (four engines); this explains why, for the take-off problem, the survival capability of the B-737 in a severe windshear is superior to that of the B-727, which in turn is superior to that of the B-747;

(v) because the optimal trajectories of the B-727, B-737, and B-747 aircraft exhibit the same qualitative behavior, it appears that the near-optimal guidance schemes developed for the B-727 aircraft can be extended to the B-737 and B-747 aircraft, albeit with some quantitative modification.

6.2. Abort Landing Problem

Optimal trajectories are computed by minimizing the peak value of the altitude drop. With reference to strong-to-severe windshears, they exhibit the following properties:

(i) the relative velocity decreases in the shear region and increases in the aftershear region; minimum velocity is achieved at the end of the shear;

(ii) the angle of attack exhibits an initial decrease, followed by a gradual, sustained increase; the peak value of the angle of attack is achieved near the end of the shear;

(iii) the altitude profile of the optimal trajectory includes three branches: a descending flight branch, followed by a nearly horizontal flight branch, followed by an ascending flight branch after the aircraft has passed through the shear region;

(iv) since the windshear inertia force is proportional to the aircraft weight, the thrust-to-weight ratio is an important index for survival capability in a severe windshear; for the abort landing problem, the thrust-to-weight ratio of the B-737 (two engines) is close to that of the B-727 (three engines) and that of the B-747 (four engines); this explains why, for the abort landing problem, the survival capability of the B-737 in a severe windshear is close to that of the B-727 and that of the B-747;

(v) because the optimal trajectories of the B-727, B-737, and B-747 aircraft exhibit the same qualitative behavior, it appears that the near-optimal guidance schemes developed for the B-727 aircraft can be extended to the B-737 and B-747 aircraft, albeit with some quantitative modification.

REFERENCES

1. FUJITA, T. T., "The Downburst", Department of Geophysical Sciences, University of Chicago, Chicago, Illinois, 1985.

2. ANONYMOUS, N. N., "Aircraft Accident Report: Pan American World Airways, Clipper 759, Boeing B-727-235, N4737, New Orleans International Airport, Kenner, Louisiana, July 9, 1982", Report No. NTSB-AAR-8302, National Transportation Safety Board, Washington, DC, 1983.

3. ANONYMOUS, N. N., "Aircraft Accident Report: Delta Air Lines, Lockheed L-1011-3851, N726DA, Dallas-Fort Worth International Airport, Texas, August 2, 1985", Report No. NTSB-AAR-8605, National Transportation Safety Board, Washington, DC, 1986.

4. PSIAKI, M. L., and STENGEL, R. F., "Optimal Flight Paths through Microburst Wind Profiles", Journal of Aircraft, Vol. 23, No. 8, pp. 629-635, 1986.

5. ANONYMOUS, N. N., "Flight Path Control in Windshear", Boeing Airliner, pp. 1-12, January-March 1985.

6. ANONYMOUS, N. N., "Windshear Training Aid",Vols. 1 and 2, Federal Aviation Administration, Washington, DC, 1987.

7. MIELE, A., WANG, T., and MELVIN, W. W., "Optimal Take-Off Trajectories in the Presence of Windshear", Journal of Optimization Theory and Applications, Vol. 49, No. 1, pp. 1-45, 1986.

8. MIELE, A., WANG, T., TZENG, C. Y., and MELVIN, W. W., "Optimal Abort Landing Trajectories in the Presence of Windshear", Journal of Optimization Theory and Applications, Vol. 55, No. 2, pp. 165-202, 1987.

9. MIELE, A., WANG, T., and MELVIN, W. W., "Optimization and Acceleration Guidance of Flight Trajectories in a Windshear", Journal of Guidance, Control, and Dynamics, Vol. 10, No. 4, pp. 368-377, 1987.

10. MIELE, A., WANG, T., MELVIN, W. W., and BOWLES, R. L., "Gamma Guidance Schemes for Flight in a Windshear", Journal of Guidance, Control, and

Dynamics, Vol. 11, No. 4, pp. 320-327, 1988.

11. MIELE, A., WANG, T., and MELVIN, W. W., "Optimal Penetration Landing Trajectories in the Presence of Windshear", Journal of Optimization Theory and Applications, Vol. 57, No. 1, pp. 1-40, 1988.

12. MIELE, A., and WANG, H., "Optimal Take-Off Trajectories for B-727, B-737, and B-747 Aircraft in the Presence of Windshear", Aero-Astronautics Report No. 229, Rice University, 1988.

13. MIELE, A., and WANG, H., "Optimal Abort Landing Trajectories for B-727, B-737, and B-747 Aircraft in the Presence of Windshear", Aero-Astronautics Report No. 230, Rice University, 1988.

14. MIELE, A., WANG, T., MELVIN, W. W., and BOWLES, R. L., "Acceleration, Gamma, and Theta Guidance for Abort Landing in a Windshear", Journal of Guidance, Control, and Dynamics, Vol. 12, 1989.

15. MIELE, A., WANG, T., and MELVIN, W. W., "Penetration Landing Guidance Trajectories in the Presence of Windshear", Journal of Guidance, Control, and Dynamics, Vol. 12, 1989.

16. IVAN, M., "A Ring-Vortex Downburst Model for Flight Simulation", Journal of Aircraft, Vol. 23, No. 3, pp. 232-236, 1986.

17. MIELE, A., and WANG, T., "Primal-Dual Properties

of Sequential Gradient-Restoration Algorithms for Optimal Control Problems, Part 1, Basic Problem", Integral Methods in Science and Engineering, Edited by F. R. Payne et al, Hemisphere Publishing Corporation, Washington, DC, pp. 577-607, 1986.

18. MIELE, A., and WANG, T., "Primal-Dual Properties of Sequential Gradient-Restoration Algorithms for Optimal Control Problems, Part 2, General Problem", Journal of Mathematical Analysis and Applications, Vol. 119, Nos. 1-2, pp. 21-54, 1986.

19. MIELE, A., PRITCHARD, R. E., and DAMOULAKIS, J. N., "Sequential Gradient-Restoration Algorithm for Optimal Control Problems", Journal of Optimization Theory and Applications, Vol. 5, No. 4, pp. 235-282, 1970.

20. MIELE, A., "Recent Advances in Gradient Algorithms for Optimal Control Problems", Journal of Optimization Theory and Applications, Vol. 17, Nos. 5-6, pp. 361-430, 1975.

21. MIELE, A., "Gradient Algorithms for the Optimization of Dynamic Systems", Control and Dynamic Systems, Advances in Theory and Application, Edited by C. T. Leondes, Academic Press, New York, New York, Vol. 16, pp. 1-52, 1980.

# CONTROLLERS FOR UNCERTAIN MECHANICAL SYSTEMS WITH ROBUSTNESS IN THE PRESENCE OF UNMODELLED FLEXIBILITIES†

MARTIN CORLESS
School of Aeronautics and Astronautics
Purdue University
West Lafayette, Indiana 47907
USA

## ABSTRACT

We consider a class of uncertain mechanical systems containing flexible elements and subject to memoryless output-feedback controllers. The damping and stiffness properties of some of the flexible elements are parameterized linearly in $\mu^{-1}$ and $\mu^{-2}$, respectively, where $\mu > 0$ and these components become more rigid as $\mu$ approaches zero. We propose a class of "stabilizing" controllers for a system model in which the above components are rigid. Subject to a "linear growth condition," the controllers also stabilize the model in which the components are flexible, provided $\mu > 0$ is sufficiently small.

## 1. INTRODUCTION

The effect of the flexibility of mechanical elements is becoming more significant in engineering applications, e.g., light high-speed robotic manipulators and flexible space structures. We consider here the problem of obtaining memoryless, stabilizing, feedback controllers for a class of uncertain mechanical systems with flexible elements. These elements are not rigid and can deform. The uncertainties are characterized deterministically rather than stochastically.

† Based on research supported by the U.S. National Science Foundation under grant MSM-87-06927.

An example of a system with a deterministic uncertainty is one which contains an uncertain disturbance input or an uncertain parameter about which the only information available is an upper bound on its magnitude.

In general, if one models some of the flexible elements as rigid components, a simpler model results and controller design is simplified. However, one should then assure that the stability properties of the feedback-controlled system are robust in the presence of the previously unmodelled flexibilities.

In this paper, we present "stabilizing" controllers whose designs are based on a model of the mechanical system in which some of the flexible elements are modelled as rigid components. These controllers also have the following robustness property. Consider a model of the system in which the above components are treated as flexible components whose damping and stiffness properties are parameterized linearly in $\mu^{-1}$ and $\mu^{-2}$, respectively, where $\mu > 0$ and these components become more rigid as $\mu$ approaches zero. Then the controllers also "stabilize" this model, provided $\mu$ is sufficiently small.

Controller design is based on the constructive use of Lyapunov functions; see, e.g., [1-4, 8-14, 17-19].

## 2. A CLASS OF MECHANICAL SYSTEMS WITH FLEXIBLE COMPONENTS

Consider a mechanical system which at each instant of *time* $t \in \mathbb{R}$ is subject to a *control input* $u(t) \in \mathbb{R}^m$. Suppose the system contains certain flexible components (hereafter called the *neglected components)* whose flexibilities are neglected in the design of a feedback controller generating $u(t)$, i.e., they are modelled as rigid components for controller design.

Letting $q(t) \in \mathbb{R}^N$ denote a vector of *generalized coordinates* which describe the configuration of the mechanical system at t, we assume that, when modelled as rigid bodies, the neglected components give rise to a linear constraint

$$\Theta q = 0 \tag{2.1}$$

where $\Theta \in \mathbb{R}^{L \times N}$ has rank $L < N$; see the examples in [6, 7]. Also, we suppose that there are no other possible kinematical constraints on the system.

We model all uncertainty in the system by a *lumped uncertain element* $\omega$. The only information assumed available on $\omega$ is the knowledge of a non-empty set $\Omega$ to which it belongs.

Letting

$$\dot{q}(t) \triangleq \frac{dq}{dt}(t)\,,$$

we suppose that the kinetic energy of the system is equal to[1]

$$\frac{1}{2}\dot{q}^T M(\omega)\dot{q}$$

where the *system mass matrix* $M(\omega) \in \mathbb{R}^{N \times N}$ is symmetric and positive definite.

Modelling the neglected components a la [5], the motion of the system can be described by

$$M(\omega)\ddot{q} = \chi(t, q, \dot{q}, u, \omega) + \Theta^T\lambda \tag{2.2}$$

where $\Theta^T\lambda$ represents the sum of the generalized forces exerted by the neglected components and $\chi$ represents the sum of all the other generalized forces. We assume that for each $\omega \in \Omega$, $\chi(\cdot, \omega)$: $\mathbb{R} \times \mathbb{R}^N \times \mathbb{R}^N \times \mathbb{R}^m \to \mathbb{R}^m$ is continuous. We suppose that the *measurement vector* $z(t) \in \mathbb{R}^l$ available for feedback control is given by

$$z = D(t, q, \dot{q}, \omega) \tag{2.3}$$

where $D(\cdot, \omega)$ is continuous.

## 2.1. "Rigid" Model

Consider first the situation in which the neglected components are modelled as rigid elements. Then (2.1) holds and the system can be described by

---

(1) Sometimes we omit arguments.

$$M(\omega)\ddot{q} = \chi(t, q, \dot{q}, u, \omega) + \Theta^T \lambda \tag{2.4a}$$

$$\Theta q = 0 \tag{2.4b}$$

$$z = D(t, q, \dot{q}, \omega) \tag{2.4c}$$

To obtain a system description without coordinate constraints, choose any matrix $U \in \mathbb{R}^{N\times\overline{N}}$ of rank $\overline{N} \triangleq N - L$ which satisfies

$$\Theta U = 0 \ , \tag{2.5}$$

i.e., the columns of U span the null space of $\Theta$. Such a matrix exists since $\text{rank}(\Theta) = L$; hence the dimension of the null space of $\Theta$ equals $\overline{N}$. Now let $\Phi \in \mathbb{R}^{\overline{N}\times N}$ and $V \in \mathbb{R}^{N\times L}$ be any matrices which satisfy

$$\begin{bmatrix} \Phi \\ \Theta \end{bmatrix}^{-1} = [U\ V] \ ; \tag{2.6}$$

e.g., consider

$$\Phi \triangleq (U^T U)^{-1} U^T \ , \ V \triangleq \Theta^T (\Theta\Theta^T)^{-1} \ . \tag{2.7}$$

Defining new coordinate vectors,

$$\phi \triangleq \Phi q \ , \ \theta \triangleq \Theta q \ , \tag{2.8}$$

one has

$$q = U\phi + V\theta \tag{2.9}$$

and premultiplying (2.4a) by $U^T$, system (2.4) can be described by

$$\overline{M}(\omega)\ddot{\phi} = \overline{\chi}(t, \phi, \dot{\phi}, u, \omega) \tag{2.10a}$$

$$z = \overline{D}(t, \phi, \dot{\phi}, \omega) \tag{2.10b}$$

with

$$\theta = 0 \tag{2.11}$$

and

$$\bar{M}(\omega) \triangleq U^T M(\omega) U \tag{2.12a}$$

$$\bar{\chi}(t, \phi, \dot{\phi}, u, \omega) \triangleq U^T \chi(t, U\phi, U\dot{\phi}, u, \omega) \tag{2.12b}$$

$$\bar{D}(t, \phi, \dot{\phi}, \omega) = D(t, U\phi, U^T\dot{\phi}, \omega) \tag{2.12c}$$

Although the model described by (2.10) may contain other flexible components we shall, for convenience, refer to it as the *"rigid" model.*

**Remark 2.1.** An alternative way to obtain an appropriate matrix U is to obtain first a matrix $\Phi$ such that $\begin{bmatrix} \Phi \\ \Theta \end{bmatrix}$ is invertible. Then $U \in \mathbb{R}^{N \times \bar{N}}$ is defined via (2.6). Satisfaction of (2.6) guarantees satisfaction of (2.5).

## 2.2. Flexible Model

Suppose now the neglected components are considered flexible, i.e., they are not rigid and can deform; hence constraint (2.1) no longer holds. Following [5] and assuming the components to be linear, their effect on the system can be represented by letting

$$\lambda = -C\Theta\dot{q} - K\Theta q \tag{2.13}$$

in (2.2). The matrix $K \in \mathbb{R}^{L \times L}$, which is assumed symmetric and positive definite, represents the stiffness properties of the components and $C \in \mathbb{R}^{L \times L}$, which is assumed positive definite, represents the damping properties of the components. For robustness considerations we shall let

$$K = \mu^{-2} K^o, \quad C = \mu^{-1} C^o, \tag{2.14}$$

where $\mu > 0$, and consider behavior for sufficiently small $\mu$. Substituting (2.13), (2.14) into (2.2)-(2.3), the system is now described

$$M(\omega)\ddot{q} = \chi(t, q, \dot{q}, u, \omega) - \mu^{-1}\Theta^T C^o \Theta\dot{q} - \mu^{-2}\Theta^T K^o \Theta q, \tag{2.15a}$$

$$z = D(t, q, \dot{q}, \omega). \tag{2.15b}$$

We shall refer to (2.15) as the *flexible model.*

The following assumption puts some regularity conditions on the functions $\chi$ and D.

**Assumption A1.**[(2)] For each $\omega \in \Omega$, there is a real number $k \geq 0$ such that the following inequalities hold for all $t \in \mathbb{R}$, $q, \dot{q} \in \mathbb{R}^N$, and $u \in \mathbb{R}^m$.

$$||\chi||, ||\frac{\partial \chi}{\partial t}|| \leq k(1 + ||q|| + ||\dot{q}|| + ||u||) , \tag{2.16a}$$

$$||D||, ||\frac{\partial D}{\partial t}|| \leq k(1 + ||q|| + ||\dot{q}||) , \tag{2.16b}$$

$$||\frac{\partial \chi}{\partial q}||, ||\frac{\partial \chi}{\partial \dot{q}}||, ||\frac{\partial \chi}{\partial u}||, ||\frac{\partial D}{\partial q}||, ||\frac{\partial D}{\partial \dot{q}}|| \leq k . \tag{2.16c}$$

Note that the above assumption is readily satisfied by a linear system whose time-varying coefficients are bounded and have bounded derivatives.

## 3. PROBLEM STATEMENT

We shall consider the control u(t) to be given by a *memoryless feedback controller* $p: \mathbb{R}\times\mathbb{R}^l \to \mathbb{R}^m$ operating on z(t), i.e.,

$$u(t) = p(t, z(t)) . \tag{3.1}$$

Roughly speaking, the problem we wish to consider is as follows. Utilizing *only the information available on the "rigid" model,* obtain a feedback controller p whose utilization assures that

(i) the feedback-controlled "rigid" model is "stable" about zero, and

(ii) the feedback-controlled flexible model is "stable" about zero, provided $\mu > 0$ is sufficiently small.

Ideally "stable" means global uniform asymptotic stability. However, for systems with uncertain disturbance inputs, asymptotic stability may not be achievable, so, we content ourselves with "stable" behavior which is close to

---

(2) If a derivative appears in a condition, this implicitly assumes that the derivative exists.

asymptotic stability.

To obtain a more precise problem statement, we introduce state vectors

$$\xi \triangleq \begin{bmatrix} q \\ \dot{q} \end{bmatrix}, \quad x \triangleq \begin{bmatrix} \phi \\ \dot{\phi} \end{bmatrix} = \Phi \zeta \ . \tag{3.2}$$

The "rigid" model is described by

$$\dot{x} = \bar{F}(t, x, u, \omega) , \tag{3.3a}$$

$$z = \bar{d}(t, x, \omega) , \tag{3.3b}$$

where

$$\bar{F}(t, x, u, \omega) \triangleq \begin{bmatrix} \dot{\phi} \\ \bar{M}(\omega)^{-1}\bar{\chi}(t, \phi, \dot{\phi}, u, \omega) \end{bmatrix}, \tag{3.4a}$$

$$\bar{d}(t, x, \omega) \triangleq \bar{D}(t, \phi, \dot{\phi}, \omega) ; \tag{3.4b}$$

the flexible model is described by

$$\dot{\xi} = F(t, \xi, u, \mu, \omega) , \tag{3.5a}$$

$$z = d(t, \xi, \omega) , \tag{3.5b}$$

where

$$F(t, \xi, u, \mu, \omega) \triangleq \begin{bmatrix} \dot{q} \\ M(\omega)^{-1}[\chi(t, q, \dot{q}, u, \omega) - \mu^{-1}\Theta^T C^o \Theta \dot{q} - \mu^{-2}\Theta^T K^o \Theta q] \end{bmatrix}, \tag{3.6a}$$

$$d(t, \xi, \omega) \triangleq D(t, q, \dot{q}, \omega) . \tag{3.6b}$$

The feedback-controlled "rigid" model is described by

$$\dot{x} = \bar{F}(t, x, p(t, \bar{d}(t, x, \omega)), \omega) \tag{3.7}$$

and the feedback-controlled flexible model is described by

$$\dot{\xi} = F(t, \xi, p(t, d(t, \xi, \omega)), \mu, \omega) . \tag{3.8}$$

The problem is as follows. Using only the information available on the "rigid" model, obtain a function $p: \mathbb{R} \times \mathbb{R}^l \to \mathbb{R}^m$ which assures that

(i) system (3.7) asymptotically tracks[3] 0 to within a bounded set, and

(ii) system (3.8) asymptotically tracks 0 to within a bounded set, provided $\mu$ is sufficiently small.

## 4. PROPOSED CONTROLLERS

The following assumption yields "stabilizing" controllers for the "rigid" model.

**Assumption A2.** There exists a continuous function $p: \mathbb{R}\times\mathbb{R}^{l} \to \mathbb{R}^{m}$ such that for some symmetric, positive definite matrices[4] $P, Q \in \mathbb{R}^{n\times n}$ and non-negative numbers[5] a, b,

$$x^T P\bar{F}(t, x, p(t, \bar{d}(t, x, \omega)), \omega) \le -\|x\|_Q^2 + a\|x\|_Q + b \tag{4.1}$$

for all $t \in \mathbb{R}$, $x \in \mathbb{R}^n$, and $\omega \in \Omega$.

Roughly speaking, the following theorem states that any function p which assures satisfaction of A2 is a "stabilizing" controller for the "rigid" model.

**Theorem 4.1.** Consider an uncertain "rigid" model described by (2.10) or (3.3), satisfying Assumption A2, and subject to feedback control given by (3.1) where p assures A2. Then, the feedback-controlled "rigid" model, (3.7), asymptotically tracks 0 to within the set

$$\mathcal{B}_o \overset{\Delta}{=} \{x \in \mathbb{R}^n \mid \|x\|_P \le d_o\} \tag{4.2}$$

where[6]

---

(3) Appendix A contains a definition.

(4) $n \overset{\Delta}{=} 2\bar{N} = 2(N - L)$

(5) If $Q \in \mathbb{R}^{n\times n}$ is symmetric and positive-definite and $x \in \mathbb{R}^n$, $\|x\|_Q \overset{\Delta}{=} (x^T Q x)^{1/2}$

(6) If all the eigenvalues of $M \in \mathbb{R}^{n\times n}$ are real, $\lambda_{max(min)}(M)$ is the maximum (minimum) eigenvalue of M.

$$d_o \triangleq [\lambda_{max}(Q^{-1}P)]^{1/2}[a/2 + (a^2/4 + b)^{1/2}] . \tag{4.3}$$

**Proof.** The proof proceeds by considering the function $U_o: \mathbb{R}^n \to \mathbb{R}$, given by

$$U_o(x) = x^T P x , \tag{4.4}$$

as a candidate Lyapunov function for (3.7). Utilizing (4.1), it follows that along any solution of (3.7),

$$\frac{dU_o(x(t))}{dt} \le -2\,\|x(t)\|_Q^2 + 2a\|x(t)\|_Q + 2b .$$

Thus, $\dfrac{dU_o(x(t))}{dt} < 0$ for all t such that

$$\|x(t)\|_Q > a/2 + (a^2/4 + b)^{1/2} ;$$

hence $\dfrac{dU_o(x(t))}{dt} < 0$ for all t satisfying

$$U_o(x(t)) > d_o^2 .$$

Standard arguments in Lyapunov theory complete the proof; see, e.g., [11].

**Remark 4.1.** If Assumption A2 is satisfied with

$$a = b = 0 ,$$

then the corresponding controller yields a feedback-controlled "rigid" model which is globally uniformly asymptotically stable about 0.

In order to obtain controllers which are also stabilizing for the flexible model, the following assumption is introduced.

**Assumption A3.** Assumption A2 is assured with a function p which, for some non-negative number k, satisfies

$$\|p(t, z)\|, \quad \|\frac{\partial p}{\partial t}(t, z)\| \leq k(1+\|z\|), \tag{4.5a}$$

$$\|\frac{\partial p}{\partial z}(t, z)\| \leq k \tag{4.5b}$$

for all $t \in \mathbb{R}$, $z \in \mathbb{R}^l$.

A proposed feedback controller is any function p which assures satisfaction of Assumptions A2 and A3.

## 5. ROBUSTNESS IN THE PRESENCE OF UNMODELLED FLEXIBILITIES

The following result assures us that a controller whose design is based on satisfying the requirements of Assumptions A2 and A3 for the "rigid" model will also "stabilize" the flexible model, provided Assumption A1 is satisfied and $\mu$ is sufficiently small.

**Theorem 5.1.** Consider an uncertain flexible model described by (2.15) or (3.5) where $M(\omega)$, $K^o$ are symmetric and

$$M(\omega),\ C^o,\ K^o > 0\,. \tag{5.1}$$

Suppose A1 is satisfied and the corresponding "rigid" model satisfies A2 and A3 with a controller p. Then for each $\omega \in \Omega$ there exists $\mu^* > 0$ such that the following hold.

(i) For each $\mu \in (0, \mu^*)$, there is a bounded set $C_\mu \subset \mathbb{R}^{2N}$ such that the feedback-controlled flexible model (3.8) asymptotically tracks 0 to within $C_\mu$.

(ii) If

$$B_\mu \triangleq \{x \in \mathbb{R}^{2\overline{N}} : x = \Phi\zeta,\ \zeta \in C_\mu\} \tag{5.2}$$

then $B_\mu$ approaches $B_0$ as $\mu$ approaches 0 in the following sense.

$$\lim_{\mu \to 0} d(B_0, B_\mu) = 0 \tag{5.3a}$$

where

$$d(B_0, B_\mu) \triangleq \sup_{x \in B_\mu} \inf_{x_0 \in B_0} \|x - x_0\| \ . \tag{5.3b}$$

**Proof.** A detailed proof is contained in Appendix C. We outline that proof here. First one introduces state vectors

$$x \triangleq \begin{bmatrix} \phi \\ \dot{\phi} \end{bmatrix}, \qquad y \triangleq \begin{bmatrix} \mu^{-2}\theta \\ \mu^{-1}\dot{\theta} \end{bmatrix} \tag{5.4}$$

where $\phi$ and $\theta$ are as described in Section 2.1. The corresponding state-space representation of the feedback-controlled flexible model has the following form

$$\dot{x} = f(t, x, y, \mu, \omega) \tag{5.5a}$$

$$\mu\dot{y} = g(t, x, y, \mu, \omega) \tag{5.5b}$$

i.e., it is a singularly perturbed system with singular perturbation parameter $\mu$; see Appendix B. Also, the state-space representation (3.7) of the feedback-controlled "rigid" model is the reduced-order system (i.e., the system obtained by letting $\mu = 0$ in (5.5)) associated with (5.5). The proof then demonstrates that (5.5) satisfies the hypotheses of Theorem B.1. Application of Theorem B.1 and noting that

$$x = \Phi\xi \ , \qquad y = \begin{bmatrix} \mu^{-2} & 0 \\ 0 & \mu^{-1} \end{bmatrix} \Theta\xi$$

yields the desired result.

**Remarks 5.1.**

(i) The above theorem states that, if $\mu > 0$ is sufficiently small, then, in a qualitative sense, the stability properties of the feedback-controlled flexible model are the same as those of the feedback-controlled "rigid" model. Also, in a quantitative sense, the behavior of state x in the flexible model approaches that of the "rigid" model as $\mu$ approaches 0.

(ii) Note that $\mu^*$ can depend on $\omega$.

## 6. EXAMPLES OF PROPOSED CONTROLLERS

In this section, we consider a specific class of uncertain mechanical systems whose "rigid" models satisfy Assumptions A2, A3. For these systems, we exhibit "stabilizing" controllers which are robust in the presence of unmodelled flexibilities. Two main characterizations of the "rigid" models treated here are that the number of independent scalar control inputs is the same as the number of coordinates and the complete state is available for feedback.

### 6.1. A Specific Class of Uncertain "Rigid" Models

Consider an uncertain mechanical system whose "rigid" model (2.10) is described by

$$\overline{M}(\omega)\ddot{\phi} = \overline{U}(t, \phi, \dot{\phi}, \omega) + Wu \tag{6.1a}$$

with measurement vector

$$z = \begin{bmatrix} \phi \\ \dot{\phi} \end{bmatrix} \tag{6.1b}$$

where $t \in \mathbb{R}$, $\phi \in \mathbb{R}^{\overline{N}}$, and $u \in \mathbb{R}^{\overline{N}}$; the uncertain element $\omega$ belongs to a known set $\Omega$; $\overline{M}(\omega) \in \mathbb{R}^{\overline{N}\times\overline{N}}$ is symmetric; $W \in \mathbb{R}^{\overline{N}\times\overline{N}}$; and, for each $\omega \in \Omega$, the function $\overline{U}(\cdot, \omega)$: $\mathbb{R}\times\mathbb{R}^{\overline{N}}\times\mathbb{R}^{\overline{N}} \to \mathbb{R}^{\overline{N}}$ is continuous.

The following assumption is satisfied.

**Assumption 6.1.**

(a) W is nonsingular.

(b) There exist real numbers $\underline{\beta}, \overline{\beta} > 0$ such that for all $\omega \in \Omega$

$$\lambda_{min}[\overline{M}(\omega)] \geq \underline{\beta}\,, \tag{6.2a}$$

$$\lambda_{max}[\overline{M}(\omega)] \leq \overline{\beta}\,, \tag{6.2b}$$

and

$$||\overline{U}(t, \phi, \dot{\phi}, \omega)|| \leq \overline{\beta}[1 + ||\phi|| + ||\dot{\phi}||] \tag{6.2c}$$

for all $t \in \mathbb{R}$, $\phi \in \mathbb{R}^{\overline{N}}$, and $\dot{\phi} \in \mathbb{R}^{\overline{N}}$.

To demonstrate that Assumption 6.1 implies A2-A3, we present some controllers which assure satisfaction of A2-A3.

## 6.2. Examples of Proposed Controllers

Choosing any nonsingular matrix $T \in \mathbb{R}^{\bar{N} \times \bar{N}}$ and defining

$$x \triangleq \begin{bmatrix} \phi \\ \dot{\phi} \end{bmatrix}$$

the "rigid" model can be described by (3.3a)-(3.3b) with

$$\bar{F}(t, x, u, \omega) = Ax + B[h(t, x, \omega) + G(\omega)T^T Wu] \; , \tag{6.3a}$$

$$\bar{d}(t, x, \omega) = x \; , \tag{6.3b}$$

where

$$A \triangleq \begin{bmatrix} 0 & I \\ 0 & 0 \end{bmatrix}, \quad B \triangleq \begin{bmatrix} 0 \\ T \end{bmatrix}, \tag{6.3c}$$

$$h(t, x, \omega) \triangleq T^{-1}\bar{M}(\omega)^{-1}\bar{U}(t, \phi, \dot{\phi}, \omega) \; , \tag{6.3d}$$

$$G(\omega) \triangleq [T^T \bar{M}(\omega) T]^{-1} \; . \tag{6.3e}$$

A proposed controller is any function $p\colon\ \mathbb{R} \times \mathbb{R}^{\bar{N}} \to \mathbb{R}^{\bar{N}}$ of the form

$$p(t, z) = (T^T W)^{-1}[p^o(t, z) + p^\varepsilon(z)] \tag{6.4}$$

where $p^\varepsilon$ is specified below and $p^o$ is any function satisfying requirements (4.5) of A3; $p^o$ is chosen to reduce the magnitude of the uncertain term

$$e(t, x, \omega) \triangleq h(t, x, \omega) + G(\omega)p^o(t, x) \; . \tag{6.5}$$

### 6.2.1. Construction of $p^\varepsilon$.

First choose any positive definite symmetric matrix $Q \in \mathbb{R}^{n \times n}$, $n \triangleq 2\bar{N}$, and any positive real number $\sigma$ and solve the Riccati equation

$$PA + A^T P - 2\sigma PBB^T P + 2Q = 0 \tag{6.6}$$

for a positive definite symmetric $P \in \mathbb{R}^{n \times n}$; since (A, B) is controllable such a solution exists.

Choose any non-negative numbers $\gamma$, $\rho$, $\kappa$ which, for all $\omega \in \Omega$, satisfy

$$\gamma \geq \lambda(\omega)[\bar{\sigma} + \frac{1}{4}\beta_1(\omega)^2]\,, \tag{6.7a}$$

$$\rho \geq \lambda(\omega)\beta_o(\omega)\,, \tag{6.7b}$$

$$\kappa \geq \beta_o(\omega)\,, \tag{6.7c}$$

where $\lambda(\omega)$, $\beta_o(\omega)$, $\beta_1(\omega)$, $\bar{\sigma}$, are chosen to satisfy

$$\lambda_{max}[T^T\overline{M}(\omega)T] \leq \lambda(\omega)\,, \tag{6.7d}$$

$$||e(t, x, \omega)|| \leq \beta_o(\omega) + \beta_1(\omega)||x||_Q\,, \tag{6.7e}$$

$$\left.\begin{array}{l} \bar{\sigma} \geq \sigma \quad \text{if} \quad \beta_1(\omega) \equiv 0 \\ \bar{\sigma} > \sigma \quad \text{if} \quad \beta_1(\omega) \not\equiv 0 \end{array}\right\}. \tag{6.7f}$$

Part (b) of Assumption 6.1 guarantees the existence of the above bounds.

Now let $s: \mathbb{R}^{\bar{N}} \to \mathbb{R}^{\bar{N}}$ be any differentiable function with bounded derivative which satisfies

$$||\eta||\, s(\eta) = ||s(\eta)||\, \eta\,, \tag{6.8a}$$

$$||\eta|| \geq 1 \Rightarrow ||s(\eta)|| \geq 1 - ||\eta||^{-1}\,, \tag{6.8b}$$

for $\eta \in \mathbb{R}^{\bar{N}}$.

Then, for any $\varepsilon > 0$,

$$p^{\varepsilon}(z) \triangleq -\gamma B^T P z - \rho s(\varepsilon^{-1}\kappa B^T P z)\,. \tag{6.9}$$

As an example of a function satisfying the above requirements on s, consider

$$s(\eta) = (||\eta|| + 1)^{-1}\eta\,. \tag{6.10}$$

After some calculations (see [3, 11] for similar calculations) one can show that

$$x^T P \bar{F}(t, x, p(t, x), \omega) \leq -\,||x||^2_{\underline{Q}} + \varepsilon \tag{6.11}$$

where $\underline{Q} \triangleq \delta Q$ and

$$\delta \triangleq \begin{cases} \inf_{\omega \in \Omega} \; [1 + \frac{1}{4}(\bar{\sigma} - \sigma)^{-1} \beta_1(\omega)^2]^{-1} & \text{if} \quad \bar{\sigma} > \sigma \\ 1 & \text{if} \quad \bar{\sigma} = \sigma ; \end{cases}$$

hence p assures satisfaction of Assumption A2. Assumption A3 can be readily verified.

Note that, as a consequence of (6.11) and Theorem 4.1, the above controllers assure asymptotic tracking of 0 to within the set

$$B_\varepsilon \triangleq \{x \in \mathbb{R}^n \mid \|x\|_P \le d_\varepsilon\}$$

$$d_\varepsilon \triangleq [\lambda_{max}(Q^{-1}P)\delta^{-1}\varepsilon]^{1/2} \quad .$$

Hence, one can obtain tracking to within any desired degree of accuracy by choosing $\varepsilon$ sufficiently small.

## APPENDIX A

Consider any system described by

$$\dot{z} = Z(t, z) \tag{A.1}$$

where $t \in \mathbb{R}$, $z \in \mathbb{R}^q$, $Z: \mathbb{R} \times \mathbb{R}^q \to \mathbb{R}^q$ and let $B$ be a set containing $0 \in \mathbb{R}^q$.

**Definition A.1.** System (A.1) *asymptotically B-tracks* 0 or *asymptotically tracks* 0 *to within B* iff it has the following properties.

(i) *Existence of solutions.* Given any $t_o \in \mathbb{R}$, $z_o \in \mathbb{R}^q$, there exists a solution $z(\cdot)$ of (A.1) with $z(t_o) = z_o$.

(ii) *Indefinite extension of solutions.* Every solution $z(\cdot)$: $[t_o, t_1) \to \mathbb{R}^q$ of (A.1) has an extension over $[t_o, \infty)$.

(iii) *Global uniform boundedness.* Given any bound $r \in \mathbb{R}_+$, there exists a bound $d(r) \in \mathbb{R}_+$ such that for any $t_o \in \mathbb{R}$ and any solution $z(\cdot)$ of (A.1),

$$\|z(t_o)\| \le r \;\; => \;\; \|z(t)\| \le d(r) \quad \forall\, t \ge t_o \, .$$

(iv) *Global uniform attractivity of $B$.* Given any bound $r \in \mathbb{R}_+$ and any neighborhood $B_\varepsilon$ of $B$, there exists $T(r, B_\varepsilon) \in \mathbb{R}_+$ such that for any $t_o \in \mathbb{R}$ and any solution $z(\cdot)$ of (A.1),

$$\|z(t_o)\| \leq r \implies z(t) \in B_\varepsilon \qquad \forall\, t \geq t_o + T(r, B_\varepsilon) .$$

## APPENDIX B: SINGULARLY PERTURBED SYSTEMS

In this appendix, we introduce some terminology and a theorem, Theorem B.1, for singularly perturbed systems; see [15, 16]. Theorem B.1 is used in the proof of Theorem 5.1.

Consider a singularly perturbed system described by

$$\dot{x} = f(t, x, y, \mu) , \tag{B.1a}$$

$$\mu \dot{y} = g(t, x, y, \mu) . \tag{B.1b}$$

where $t \in \mathbb{R}$ is time; $x \in \mathbb{R}^n$ and $y \in \mathbb{R}^l$ describe the state of the system; and $\mu \in (0, \infty)$ is the *singular perturbation parameter.*

The *reduced-order system* associated with (B.1) is obtained by letting $\mu = 0$, i.e.,

$$\dot{x} = f(t, x, y, 0) , \tag{B.2a}$$

$$0 = g(t, x, y, 0) . \tag{B.2b}$$

**Assumption B.1.** There exists a unique function $h: \mathbb{R} \times \mathbb{R}^n \to \mathbb{R}^l$ such that

$$g(t, x, h(t, x), 0) \equiv 0 . \tag{B.3}$$

As a consequence of Assumption B.1, (B.2) is equivalent to

$$\dot{x} = \bar{f}(t, x) \tag{B.4}$$

$$y = h(t, x) \tag{B.5}$$

where

$$\bar{f}(t, x) \triangleq f(t, x, h(t, x)) \quad . \tag{B.6}$$

**Assumption B.2.** The function $\bar{f}$ is continuous and there exist positive-definite symmetric matrices P, $Q \in \mathbb{R}^{n \times n}$ and scalars a, $b \geq 0$ such that

$$x^T P \bar{f}(t, x) \leq -\|x\|_Q^2 + a\|x\|_Q + b \tag{B.7}$$

for all $t \in \mathbb{R}$ and $x \in \mathbb{R}^n$.

The above assumption assures that the reduced-order system (B.4) asymptotically tracks 0 to within the set

$$B_0 \triangleq \{x \in \mathbb{R}^n: \|x\|_P \leq d_o\} \tag{B.8}$$

where

$$d_o \triangleq [\lambda_{max}(Q^{-1}P)]^{1/2}[a/2 + (a^2/4 + b)^{1/2}] \quad . \tag{B.9}$$

The proof of this fact is the same as that of Theorem 4.1; see also Remark 4.1.

For each fixed $t^* \in \mathbb{R}$ and $x^* \in \mathbb{R}^n$, the *boundary-layer system* associated with (B.1) is defined by

$$\dot{\eta}(\tau) = g(t^*, x^*, \eta(\tau), 0) \quad . \tag{B.10}$$

Assumption B.1 assures that (B.10) has a unique equilibrium point $h(t^*, x^*)$. The following assumption implies that (B.10) is globally asymptotically stable and the stability properties are uniform with respect to $t^*$ and $x^*$.

**Assumption B.3.** There exist positive-definite symmetric matrices R, $S \in \mathbb{R}^{l \times l}$ such that

$$[\eta - h(t, x)]^T R g(t, x, \eta, 0) \leq -\|\eta - h(t, x)\|_S^2 \tag{B.11}$$

for all $t \in \mathbb{R}$, $x \in \mathbb{R}^n$, and $\eta \in \mathbb{R}^l$.

In order to guarantee that, for $\mu > 0$ sufficiently small, the full order system (B.1) has the same stability properties as that of the reduced-order system (B.4), we require that the functions f, g, and h satisfy the following regularity conditions.

**Assumption B.4.** The functions f and g are continuous and there exists a scalar $c \geq 0$ and continuous functions $o_1, o_2: \mathbb{R}_+ \to \mathbb{R}_+$ such that for all $t \in \mathbb{R}$, $x \in \mathbb{R}^n$, and $y \in \mathbb{R}^l$

(i)

$$||f(t, x, y, 0) - f(t, x, h(t, x), 0)|| \leq c||y - h(t, x)|| \ , \tag{B.12a}$$

(ii)

$$||\bar{f}(t, x)||, \ \ ||h(t, x)|| \ , \ ||\frac{\partial h}{\partial t}(t, x)|| \leq c(1 + ||x||) \ , \tag{B.12b}$$

$$||\frac{\partial h}{\partial x}(t, x)|| \leq c \ , \tag{B.12c}$$

(iii) and for all $\mu \in \mathbb{R}_+$,

$$||f(t, x, y, \mu) - f(t, x, y, 0)|| \leq o_1(\mu)[1 + ||x|| + ||y||] \ , \tag{B.12d}$$

$$||g(t, x, y, \mu) - g(t, x, y, 0)|| \leq \mu o_2(\mu)[1 + ||x|| + ||y||] \ , \tag{B.12e}$$

with

$$o_1(\mu) = 0 \ . \tag{B.12f}$$

We have now the following result concerning the stability properties of (B.1).

**Theorem B.1.** Suppose a system described by (B.1) satisfies Assumptions B.1-B.4. Then there exists $\mu^* > 0$ such that the following hold.

(i) For each $\mu \in (0, \mu^*)$, there is a bounded set $C_\mu \subset \mathbb{R}^n \times \mathbb{R}^l$ such that (B.1) asymptotically tracks 0 to within $C_\mu$.

(ii) If

$$B_\mu \triangleq \{x \in \mathbb{R}^n: (x, y) \in C_\mu \text{ for some } y \in \mathbb{R}^l\}, \tag{B.13}$$

then $B_\mu$ approaches $B_0$ (defined by (B.8), (B.9)) as $\mu$ approaches 0 in the following sense.

$$\lim_{\mu \to 0} d(B_0, B_\mu) = 0$$

where

$$d(B_0, B_\mu) \triangleq \sup_{x \in B_\mu} \inf_{x_0 \in B_0} \|x - x_0\| \quad .$$

**Proof.** Introducing a new state

$$e(t) \triangleq y(t) - h(t, x(t)) \quad , \tag{B.14}$$

a system description equivalent to (B.1) is given by

$$\dot{x} = \tilde{f}(t, x, e, \mu) \tag{B.15a}$$

$$\mu \dot{e} = \tilde{g}(t, x, e, \mu) \tag{B.15b}$$

where

$$\tilde{f}(t, x, e, \mu) \triangleq f(t, x, h(t, x) + e, \mu) \tag{B.15c}$$

$$\tilde{g}(t, x, e, \mu) \triangleq g(t, x, h(t, x) + e, \mu) - \mu \frac{\partial h}{\partial t}(t, x) - \mu \frac{\partial h}{\partial x}(t, x)\tilde{f}(t, x, e, \mu) \tag{B.15d}$$

We will show first that (B.15) asymptotically tracks 0 to within a bounded neighborhood.

Note that, as a consequence of Assumptions B.1 - B.4, we have

$$\tilde{f}(t, x, 0, 0) = \bar{f}(t, x) \tag{B.16a}$$

$$\|\tilde{f}(t, x, e, 0) - \bar{f}(t, x)\| \le c\|e\| \tag{B.16b}$$

$$e^T R \tilde{g}(t, x, e, 0) \le -\|e\|_S^2 \tag{B.16c}$$

and

$$\|\tilde{f}(t, x, e, \mu) - \tilde{f}(t, x, e, 0)\| \le \tilde{o}_1(\mu)[1 + \|x\| + \|e\|] \tag{B.16d}$$

$$\|\tilde{g}(t, x, e, \mu) - \tilde{g}(t, x, e, 0)\| \le \mu \tilde{o}_2(\mu)[1 + \|x\| + \|e\|] \tag{B.16e}$$

where

$$\tilde{o}_1(\mu) = o_1(\mu)(1 + c) \tag{B.16f}$$

$$\tilde{o}_2(\mu) = [o_2(\mu) + c o_1(\mu) + c](1 + c) \quad ; \tag{B.16g}$$

hence,

$$\tilde{o}_1(\mu) = 0 \quad . \tag{B.16h}$$

As a candidate Lyapunov function for (B.15) we consider $U_\mu : \mathbb{R}^n \times \mathbb{R}^l \to \mathbb{R}_+$ defined by

$$U_\mu(x, e) \triangleq x^T P x + \mu^{1/2} e^T R e \quad . \tag{B.17}$$

Along any solution of (B.15), we have

$$\frac{d}{dt} U_\mu(x(t), e(t)) = L_\mu(t, x(t), e(t)) \tag{B.18}$$

where

$$L_\mu(t, x, e) \triangleq 2x^T P \tilde{f}(t, x, e, \mu) + 2\mu^{-1/2} e^T R \tilde{g}(t, x, e, \mu) \quad . \tag{B.19}$$

Utilizing (B.7) and (B.16), it is shown in Calculation B.1 (after this proof) that for all $\mu > 0$, there exists $\lambda_\mu \in \mathbb{R}$ and $\tilde{a}_\mu \in \mathbb{R}_+$ such that for all $t \in \mathbb{R}$, $x \in \mathbb{R}^n$, $y \in \mathbb{R}^l$

$$L_\mu(t, x, e) \le -2\lambda_\mu W_\mu(x, e) + 2\tilde{a}_\mu W_\mu(x, e)^{1/2} + 2b \tag{B.20}$$

where

$$W_\mu(x, e) \triangleq x^T Q x + \mu^{1/2} k e^T S e \tag{B.21a}$$

with

$$k \triangleq \lambda_{min}(P^{-1}Q)/\lambda_{min}(R^{-1}S) \quad ; \tag{B.21b}$$

in addition,

$$\lim_{\mu \to 0} \lambda_\mu = 1 \ , \qquad \lim_{\mu \to 0} \tilde{a}_\mu = a \ . \tag{B.22}$$

It follows from (B.22) that there exists $\mu^* > 0$ such that $\lambda_\mu > 0$ for all $\mu \in (0, \mu^*)$. Note also that

$$W_\mu(x, e) \ge \lambda_{min}(P^{-1}Q)\, U_\mu(x, e) \quad . \tag{B.23}$$

Consider now any $\mu \in (0, \mu^*)$. Utilizing arguments similar to those used in the proof of Theorem 4.1, it follows from (B.17), (B.18), (B.20), (B.23) that, system (B.15) asymptotically tracks 0 to within the compact set

$$\tilde{C}_\mu \triangleq \{(x, e) \in \mathbb{R}^n \times \mathbb{R}^l \colon U_\mu(x, e) \le d_\mu^2\} \tag{B.24}$$

where

$$d_\mu \triangleq [\lambda_{max}(Q^{-1}P)]^{1/2}[\tilde{a}_\mu + (\tilde{a}_\mu^2 + 4\,\lambda_\mu b)^{1/2}]/2\lambda_\mu \ . \tag{B.25}$$

Letting

$$C_\mu \triangleq \{(x, y) \in \mathbb{R}^n \times \mathbb{R}^l \colon \ ||x||_P \le d_\mu \ , \ ||y||_R \le \delta_\mu\} \tag{B.26}$$

where

$$\delta_\mu \triangleq \mu^{-1/4} d_\mu + c[1 + \lambda_{max}(P^{-1})^{1/2} d_\mu]\lambda_{max}(R)^{1/2} \ , \tag{B.27}$$

it follows from (B.14), (B.12b), (B.17), (B.24) and the stability properties of system (B.15) that, system (B.1) asymptotically tracks 0 to within the compact set $C_\mu$.

Part (ii) of the theorem is proven by noting that

$$B_0 = \{x \in \mathbb{R}^n \colon \ ||x||_P \le d_0\} \ , \tag{B.28a}$$

$$B_\mu = \{x \in \mathbb{R}^n \colon \ ||x||_P \le d_\mu\} \ , \tag{B.28b}$$

and

$$\lim_{\mu \to 0} d_\mu = d_0 \ . \tag{B.29}$$

**Calculation B.1.** It follows from (B.7), (B.16b), (B.16d) that

$$\begin{aligned} x^T P\tilde{f}(t, x, e, \mu) &= x^T P\bar{f}(t, x) + x^T P[\tilde{f}(t, x, e, 0) - \bar{f}(t, x)] + x^T P[\tilde{f}(t, x, e, \mu) - \tilde{f}(t, x, e, 0)] \\ &\le -[1 - \tilde{o}_1(\mu)\lambda_1\lambda_3]||x||_Q^2 + \lambda_2\lambda_3[c + \tilde{o}_1(\mu)]||x||_Q||e||_S \\ &\quad + [a + \tilde{o}_1(\mu)\lambda_3]||x||_Q + b \end{aligned} \tag{B.30}$$

where

$$\lambda_1 \triangleq \lambda_{max}(Q^{-1})^{1/2} \ , \ \lambda_2 \triangleq \lambda_{max}(S^{-1})^{1/2} \ , \ \lambda_3 \triangleq \lambda_{max}(Q^{-1}P^2)^{1/2} \ .$$

It follows from (B.16c), (B.16e) that

$$e^T R\tilde{g}(t, x, e, \mu) = e^T R\tilde{g}(t, x, e, 0) + e^T R[\tilde{g}(t, x, e, \mu) - \tilde{g}(t, x, e, 0)]$$

$$\leq -[1 - \mu\tilde{o}_2(\mu)\lambda_2\lambda_4]||e||_S^2 + \mu\tilde{o}_2(\mu)\lambda_1\lambda_4||x||_Q||e||_S + \mu\tilde{O}_2(\mu)\lambda_4||e||_S \quad (B.31)$$

where

$$\lambda_4 \triangleq \lambda_{max}(S^{-1}R^2)^{1/2} \ .$$

Substituting (B.30) and (B.31) into (B.19), we can write

$$L_\mu(t, x, e) \leq -2 \begin{bmatrix} ||x||_Q \\ k_\mu ||e||_S \end{bmatrix}^T M_\mu \begin{bmatrix} ||x||_Q \\ k_\mu ||e||_S \end{bmatrix} + 2v_\mu^T \begin{bmatrix} ||x||_Q \\ k_\mu ||e||_S \end{bmatrix} + 2b \quad (B.32)$$

where $k_\mu \triangleq \mu^{1/4} k^{1/2}$ ,

$$v_\mu \triangleq \begin{bmatrix} a + \tilde{o}_1(\mu)\lambda_3 \\ \mu^{1/4}\tilde{o}_2(\mu)\lambda_4 k^{-1/2} \end{bmatrix}, \quad M_\mu \triangleq \begin{bmatrix} M_{11}(\mu) & \mu^{-1/4}M_{12}(\mu) \\ \mu^{-1/4}M_{21}(\mu) & \mu^{-1}M_{22}(\mu) \end{bmatrix},$$

with

$$M_{11}(\mu) \triangleq 1 - \tilde{o}_1(\mu)\lambda_1\lambda_3 \ ,$$
$$M_{12}(\mu) \triangleq M_{21}(\mu) \triangleq -\tfrac{1}{2} k^{-1/2}[\lambda_2\lambda_3(c + \tilde{o}_1(\mu)) + \mu^{1/2}\tilde{o}_2(\mu)\lambda_1\lambda_4] \ ,$$
$$M_{22}(\mu) \triangleq k^{-1}[1 - \mu\tilde{o}_2(\mu)\lambda_2\lambda_4] \ .$$

Recalling (B.21a), inequality (B.32) implies that

$$L_\mu(t, x, e) \leq -2\lambda_\mu W_\mu(x, e) + 2\tilde{a}_\mu W_\mu(x, e)^{1/2} + 2b$$

where

$$\lambda_\mu \triangleq \lambda_{min}(M_\mu) \ , \quad \tilde{a}_\mu \triangleq ||v_\mu|| \ ;$$

also,

$$\lim_{\mu\to 0} \lambda_\mu = 1, \quad \lim_{\mu\to 0} \tilde{a}_\mu = a \ .$$

## APPENDIX C: PROOF OF THEOREM 5.1

**Proof of Theorem 5.1.** Substituting (3.1) and (2.15b) into (2.15a), the feedback-controlled flexible model can be described by

$$M(\omega)\ddot{q} = \hat{\chi}(t, q, \dot{q}, \omega) - \mu^{-2}\Theta^T K^o \Theta q - \mu^{-1}\Theta^T C^o \Theta \dot{q} \tag{C.1a}$$

where

$$\hat{\chi}(t, q, \dot{q}, \omega) \triangleq \chi(t, q, \dot{q}, p(t, D(t, q, \dot{q}, \omega)), \omega) \ . \tag{C.1b}$$

Similarly, substituting (3.1) and (2.10b) into (2.10a), the feedback-controlled "rigid" model can be described by

$$\overline{M}(\omega)\ddot{\phi} = \hat{\overline{\chi}}(t, \phi, \dot{\phi}, \omega) \tag{C.2a}$$

where

$$\hat{\overline{\chi}}(t, \phi, \dot{\phi}, \omega) = \overline{\chi}(t, \phi, \dot{\phi}, p(t, \overline{D}(t, \phi, \dot{\phi}, \omega)), \omega); \tag{C.2b}$$

also, utilizing (2.12a) - (2.12c) and (C.1b),

$$\overline{M}(\omega) = U^T M(\omega) U \ , \tag{C.3a}$$

$$\hat{\overline{\chi}}(t, \phi, \dot{\phi}, \omega) = U^T \hat{\chi}(t, U^T\phi, U^T\dot{\phi}, \omega) \ . \tag{C.3b}$$

Recall the new coordinate vectors defined by (2.8) where the matrices $\Phi$ and V are chosen to satisfy (2.6) for the given matrices $\Theta$ and U. Premultiplying (C.1a) by $U^T$ and $V^T$ and utilizing (2.9), the flexible model can be described by

$$\tilde{M}_{11}(\omega)\ddot{\phi} + \tilde{M}_{12}(\omega)\ddot{\theta} = \tilde{\chi}_1(t, \phi, \theta, \dot{\phi}, \dot{\theta}, \omega) \tag{C.4a}$$

$$\tilde{M}_{21}(\omega)\ddot{\phi} + \tilde{M}_{22}(\omega)\ddot{\theta} = \tilde{\chi}_2(t, \phi, \theta, \dot{\phi}, \dot{\theta}, \omega) - \mu^{-2}K^o\theta - \mu^{-1}C^o\dot{\theta} \tag{C.4b}$$

where

$$\tilde{M}_{11} = U^T M U, \ \tilde{M}_{12} = U^T M V \ , \ \tilde{M}_{21} = V^T M U \ , \ \tilde{M}_{22} = V^T M V, \tag{C.5a}$$

$$\tilde{\chi}_1(t, \phi, \theta, \dot{\phi}, \dot{\theta}, \omega) = U^T \hat{\chi}(t, U\phi + V\theta, U\dot{\phi} + V\dot{\theta}, \omega) \ , \tag{C.5b}$$

$$\tilde{\chi}_2(t, \phi, \theta, \dot{\phi}, \dot{\theta}, \omega) = V^T \hat{\chi}(t, U\phi + V\theta, U\dot{\phi} + V\dot{\theta}, \omega) \ . \tag{C.5c}$$

It should be clear from (C.3) and (C.5) that

$$\overline{M} = \tilde{M}_{11} \ , \tag{C.6a}$$

$$\hat{\overline{\chi}}(t, \phi, \dot{\phi}, \omega) = \tilde{\chi}_1(t, \phi, 0, \dot{\phi}_1, 0, \omega) \ . \tag{C.6b}$$

Recall state vector $x \in \mathbb{R}^{2\overline{N}}$ defined by (3.2), i.e.

$$x = \begin{bmatrix} x_1 \\ x_2 \end{bmatrix} \triangleq \begin{bmatrix} \phi \\ \dot{\phi} \end{bmatrix} , \tag{C.7}$$

and introduce a new state vector $y \in \mathbb{R}^{2L}$ defined by

$$y = \begin{bmatrix} y_1 \\ y_2 \end{bmatrix} \triangleq \begin{bmatrix} \mu^{-2}\theta \\ \mu^{-1}\dot{\theta} \end{bmatrix} . \tag{C.8}$$

Defining matrices $\tilde{N}_{11}(\omega) \in \mathbb{R}^{\overline{N}\times\overline{N}}$, $\tilde{N}_{12}(\omega) \in \mathbb{R}^{\overline{N}\times L}$, $\tilde{N}_{21} \in \mathbb{R}^{L\times\overline{N}}$, and $\tilde{N}_{22} \in \mathbb{R}^{L\times L}$ by

$$\begin{bmatrix} \tilde{N}_{11} & \tilde{N}_{12} \\ \tilde{N}_{21} & \tilde{N}_{22} \end{bmatrix} = \begin{bmatrix} \tilde{M}_{11} & \tilde{M}_{12} \\ \tilde{M}_{21} & \tilde{M}_{22} \end{bmatrix}^{-1} , \tag{C.9}$$

and utilizing (C.4), a state-space representation of the flexible system is given by

$$\dot{x} = f(t, x, y, \mu, \omega) \tag{C.10a}$$

$$\mu\dot{y} = g(t, x, y, \mu, \omega) \tag{C.10b}$$

where

$$f(t, x, y, \mu, \omega) \triangleq \begin{bmatrix} x_2 \\ f_2(t, x, y, \mu, \omega) \end{bmatrix} \tag{C.11a}$$

$$g(t, x, y, \mu, \omega) \triangleq \begin{bmatrix} y_2 \\ g_2(t, x, y, \mu, \omega) \end{bmatrix} \tag{C.11b}$$

with

$$f_2(t, x, y, \mu, \omega) \triangleq \tilde{N}_{11}(\omega)\tilde{\chi}_1(t, x_1, \mu^2 y_1, x_2, \mu y_2, \omega)$$
$$+ \tilde{N}_{12}(\omega)\tilde{\chi}_2(t, x_1, \mu^2 y_1, x_2, \mu y_2, \omega) - \tilde{N}_{12}(\omega)K^o y_1 - \tilde{N}_{12}(\omega)C^o y_2 ,$$
$$g_2(t, x, y, \mu, \omega) \triangleq \tilde{N}_{21}(\omega)\tilde{\chi}_1(t, x_1, \mu^2 y_1, x_2, \mu y_2, \omega)$$
$$+ \tilde{N}_{22}(\omega)\tilde{\chi}_2(t, x_1, \mu^2 y_1, x_2, \mu y_2, \omega) - \tilde{N}_{22}(\omega)K^o y_1 - \tilde{N}_{22}(\omega)C^o y_2 .$$

Also, utilizing (C.2a), a state-space representation of the "rigid" model is given by

$$\dot{x} = \overline{f}(t, x, \omega) \tag{C.12}$$

where

$$\overline{f}(t, x, \omega) = \begin{bmatrix} x_2 \\ \overline{M}(\omega)^{-1}\hat{\chi}(t, x_1, x_2, \omega) \end{bmatrix} . \tag{C.13}$$

Recalling (C.2b), (3.4a) it can readily be seen that state-representation (C.12) is exactly the same as that of (3.7), i.e.

$$\overline{f}(t, x, \omega) = \overline{F}(t, x, p(t, x, \overline{d}(t, x, \omega)), \omega) . \tag{C.14}$$

Clearly, system description (C.10) is exactly the same as (B.1), i.e., (C.10) describes a singularly perturbed system with singular perturbation parameter $\mu$. We proceed now to show that Assumptions B.1 - B.4 are satisfied. Then we apply Theorem B.1.

Assumption B.1 is assured with

$$h(t, x) = \begin{bmatrix} h_1(t, x) \\ 0 \end{bmatrix} \tag{C.15a}$$

where

$$h_1(t, x) \triangleq (K^o)^{-1}\tilde{N}_{22}(\omega)^{-1}\tilde{N}_{21}(\omega)\tilde{\chi}_1(t, x_1, 0, x_2, 0, \omega) \tag{C.15b}$$
$$+ (K^o)^{-1}\tilde{\chi}_2(t, x_1, 0, x_2, 0, \omega) .$$

Utilizing (C.11a), (C.11c), (C.15), (C.6), (C.13) and the relationship

$$\tilde{M}_{11}^{-1} = \tilde{N}_{11} - \tilde{N}_{12}\tilde{N}_{22}^{-1}\tilde{N}_{21} ,$$

one can readily show that

$$x, h(t, x), 0, \omega) = \bar{f}(t, x, \omega)$$

where $\bar{f}$ is given by (C.13). Hence, the reduced-order system associated with (C.10) is the "rigid" model (C.12) and satisfaction of Assumption B.2 is guaranteed by Assumption A2 and (C.14).

It follows from (C.11) and (B.10) that, for any $t^* \in \mathbb{R}$ and $x^* \in \mathbb{R}^n$, the boundary layer system associated with (C.10) is given by

$$\dot{\eta}(\tau) = A[\eta(\tau) - h(t^*, x^*)] \tag{C.16}$$

where

$$A \triangleq \begin{bmatrix} 0 & I \\ -\tilde{N}_{22}(\omega)K^o & -\tilde{N}_{22}(\omega)C^o \end{bmatrix} . \tag{C.17}$$

The matrix A is strict Hurwitz, i.e., all its eigenvalues have negative real parts. This can be seen as follows. Defining matrices $\hat{R}$ and $\hat{S}$ by

$$\hat{R} = \begin{bmatrix} K^o & 0 \\ 0 & \tilde{N}_{22}(\omega)^{-1} \end{bmatrix} , \quad \hat{S} = ½ \begin{bmatrix} 0 & 0 \\ 0 & C^o + C^{oT} \end{bmatrix} , \tag{C.18}$$

it can readily be seen that

$$\hat{R}A + A^T\hat{R} + 2\hat{S} = 0 \ . \tag{C.19}$$

Since $\hat{R}$ is positive definite symmetric, $\hat{S}$ is positive semi-definite symmetric, and $(\hat{S}^{½}, A)$ is observable, it follows from (C.19) that A is strict Hurwitz.

Thus, for any positive definite symmetric $S \in \mathbb{R}^{2L\times 2L}$, there exists a unique positive definite symmetric solution $R \in \mathbb{R}^{2L\times 2L}$ to

$$RA + A^TR + 2S = 0 \ . \tag{C.20}$$

This assures satisfaction of Assumption B.3.

By straightforward calculations, one can show that Assumptions A1 and A3 imply Assumption B.4.

Applying Theorem B.1 and noticing that

$$x = \Phi \zeta \quad ; \quad y = \begin{bmatrix} \mu^{-2} & 0 \\ 0 & \mu^{-1} \end{bmatrix} \Theta \zeta \ , \tag{C.21}$$

the desired result follows.

□

## REFERENCES


[1] Ambrosino, G., G. Celentano, and F. Garofalo, "Robust Model Tracking Control for a Class of Nonlinear Plants," *IEEE Trans. Automatic Contrl.*, **AC-30,** 275, 1985.

[2] Ambrosino, G., G. Celentano, and F. Garofalo, "Tracking Control of High-Performance Robots via Stabilizing Controllers for Uncertain Systems," *J. Optimiz. Theory Appl.,* **2,** 239, 1986.

[3] Barmish, B. R., M. Corless, and G. Leitmann, "A New Class of Stabilizing Controllers for Uncertain Dynamical Systems," *SIAM J. Contrl. Optimiz.,* **21,** 246, 1983.

[4] Chen, Y. H., "Robust Control of Mechanical Manipulators," *J. Dynam. Syst. Meas. Contrl.,* submitted.

[5] Corless, M., "Modelling "Flexible Constraints" in Mechanical Systems," *Proc. 20th Midwestern Mechanics Conference,* Purdue University, West Lafayette, Indiana, 1987.

[6] Corless, M., "Control of Uncertain Mechanical Systems with Robustness in the Presence of Unmodelled Flexibilities," *Proc. of the 3rd Bellman Continuum,* Sophia - Antipolis, France, 1988.

[7] Corless, M., "Stability Robustness of Linear Feedback-Controlled Mechanical Systems in the Presence of a Class of Unmodelled Flexibilities," *Proc. 27th Conf. Decision Contrl.,* Houston, Texas, 1988.

[8] Corless, M., "Tracking Controllers for Uncertain Systems: Application to a Manutec r3 Robot," *J. Dynam. Syst. Meas. Contrl.,* to appear.

[9] Corless, M., and G. Leitmann, "Continuous State Feedback Guaranteeing Uniform Ultimate Boundedness for Uncertain Dynamic Systems," *IEEE Trans. Automatic Contrl.,* **AC-26,** 1139, 1981.

[10] Corless, M., and G. Leitmann, "Controller Design for Uncertain Systems Via Lyapunov Functions," *Proc. American Contrl. Conference,* Atlanta, Georgia, 1988.

[11] Corless, M. and G. Leitmann, "Deterministic Control of Uncertain Systems: A Lyapunov Theory Approach," *in* "Deterministic Nonlinear Control of Uncertain Systems: Variable Structure and Lyapunov Control," (A. Zinober, ed.), IEE Publishers, to appear.

[12] Corless, M., G. Leitmann and E. P. Ryan, "Tracking in the Presence of Bounded Uncertainties," *Proc. 4th IMA Int. Conf. Control Theory,* Cambridge University, England, 1984.

[13] Gutman, S., "Uncertain Dynamical Systems--Lyapunov Min-Max Approach," *IEEE Trans. Automatic Contrl.,* **AC-24,** 437, 1979.

[14] Ha, I. J., and E. G. Gilbert, "Robust Tracking in Nonlinear Systems," *IEEE Trans. Automatic Contrl.,* **AC-32,** 763, 1987.

[15] Kokotovic, P. V., and H. K. Khalil, eds., "Singular Perturbations in Systems and Control," IEEE Press, New York, 1986.

[16] Kokotovic, P. V., H. K. Khalil, and J. O'Reilly, "Singular Perturbation Methods in Control: Analysis and Design," Academic Press, 1986.

[17] Madani-Esfahani, S. M., R. A. DeCarlo, M. J. Corless, and S. H. Zak, "On Deterministic Control of Uncertain Nonlinear Systems," *Proc. American Contrl. Conf.,* Seattle, Washington, 1986.

[18] Ryan, E. P., G. Leitmann and M. Corless, "Practical Stabilizability of Uncertain Dynamical Systems, Application to Robotic Tracking," *J. Optimiz. Theory Applic.,* **47,** 235, 1985.

[19] Shoureshi, R., M. J. Corless, and M. D. Roesler, "Control of Industrial Manipulators with Bounded Uncertainties," *J. Dynam. Syst. Meas. Contrl.,*

**109,** 53, 1987

[20] Slotine, J. J., "The Robust Control of Robot Manipulators," *Int. J. Robotics Research,* **4**, 49, 1985.

[21] Spong, M. W., "Modeling and Control of Elastic Joint Robots," *J. Dynam. Syst. Meas. Control,* **109,** 310, 1987.

[22] Spong, M. W., and Vidyasagar, M., "Robust Linear Compensator Design for Nonlinear Robotic Control," *Proc. IEEE Conf. Robotics and Automat.*, 1985.

# A NEW ROBUST CONTROLLER FOR LINEAR SYSTEMS WITH ARBITRARILY STRUCTURED UNCERTAINTIES

M. S. CHEN

Mechanical Engineering Department
National Taiwan University
Taipei, Taiwan 10764
Republic of China

M. TOMIZUKA

Mechanical Engineering Department
University of California
Berkeley, CA 94720
U.S.A.

## I. INTRODUCTION

In recent years, there has been increasing attention in the literature on the use of nonlinear controllers for uncertain linear systems. Such controllers, including the Ultimate Boundedness Controller [1] and the Sliding Mode Controller [2], have been shown to be very effective in counteracting system uncertainties due to external disturbances and parameter mismatches between the system and the system model. However, these controllers impose a very strict structural condition on system uncertainties; namely, the matching condition [3] in Ultimate Boundedness Control or equivalently the invariance condition [4] in Sliding Mode Control. Under this structural condition, system uncertainties must enter the state equation through the same channel as the control input. Unfortunately, in many practical cases, we encounter systems with uncertainties which do not meet the matching condition; hence, the application of these nonlinear robust controllers is limited to only a small class of systems.

There has been some research in the literature concerning systems with "unmatched" system uncertainties. In [5], system uncertainties are decomposed into a "matched" part and an "unmatched" part. It is shown that an effective control is possible as long as the norm of the "unmatched" part does not exceed a certain threshold. In [6], the authors proposed a "generalized" matching condition, which is also a structural condition but less restrictive than the matching condition. However, despite all these efforts, no method has been found to completely relax the matching condition.

In this paper, a different approach is proposed to the design of nonlinear robust controllers so that no structural condition is imposed on system uncertainties. Before presenting this new approach, we need first to understand why the matching condition arises in Ultimate Boundedness Control and Sliding Mode Control. Recall that the objective of these control schemes is to achieve regulation of all the state variables. We will show, through the following example, that the matching condition is inherently imposed by this objective. Consider a simple system

$$\dot{x}_1 = x_2 + sin(t)$$

$$\dot{x}_2 = u$$

where $u$ is the control input and $sin(t)$ is an external disturbance. Notice that the disturbance does not satisfy the matching condition. It is obvious that in this example there exists no control $u$ which can regulate both $x_1$ and $x_2$ to zero. Regulation of both $x_1$ and $x_2$ is possible only when the disturbance $sin(t)$ enters the system through the second equation; in other words, when the disturbance satisfies the matching condition. This simple observation suggests that to relax the matching condition, we will have to give up regulation of all the system state. Therefore, we propose a less restrictive control objective: regulation of the system output and boundedness of the system state. Based on this modified objective, we develop a new nonlinear robust controller without imposing matching conditions on system uncertainties; however, we require that the system be minimum phase. When the uncertain system has a relative degree larger than one, the proposed new controller uses an output derivative estimator. The output derivative estimator is first obtained in an unrealizable form, and then its approximation in a realizable form is suggested. The effectiveness of the approximated estimator is verified by a computer simulation study [7].

In this paper, we adopt the following notations: $c_1$ and $c_2$ are two constants defined by

$$c_1 = \sup_{x \in N} g(x) \qquad \text{and} \qquad c_2 = \inf_{x \in N} g'(x)$$

where $g(\cdot)$ can be any one of the functions listed in Appendix A, $N$ is some bounded open set on the real line, and $g'(x)$ denotes the differentiation of $g(x)$ with respect to $x$. We use $N^c$ to denote the complement set of $N$, $\|\cdot\|$ is the (induced) Euclidean norm of a vector (matrix), and $y^{(i)}(t)$ denotes the $i$'th differentiation of $y(t)$ with respect to time $t$.

The remainder of this paper is arranged as follows: Section II lays out the problem formulation and constructs an important stability theorem for minimum-phase systems. Section III introduces the output derivative estimator, which is to be used in Section V. The new robust controller is first presented in Section IV for systems with relative degree one, and then in Section V for systems with relative degree larger than one. Finally Section VI gives the conclusions.

## II. PRELIMINARIES

In this paper, we are concerned with the control of a linear time-invariant SISO system

$$\dot{x} = \bar{A}x + \bar{b}u + d$$
$$y = \bar{c}x \tag{1}$$

where $x \in R^n$ is the system state, $d \in R^n$ is an external disturbance, $u$ is the control input, $y$ is the system output and

$$\bar{A} = A + \Delta A, \quad \bar{b} = b + \Delta b, \quad \bar{c} = c + \Delta c$$

with $(A, b, c)$ representing nominal system matrices and $(\Delta A, \Delta b, \Delta c)$ unknown parameter mismatches. The system state $x$ is directly accessible, and we make the following assumptions on the system:

(A1) The real system $(\bar{A}, \bar{b}, \bar{c})$ is minimum phase.

(A2) Both the nominal system and the real system are of relative degree $r (\geq 1)$; therefore,

$$K \equiv cA^{r-1}b \neq 0, \quad \text{and } when\ r > 1 \quad cA^k b = 0 \qquad \text{for } k = 0, 1, \ldots, r-2$$
$$\bar{K} \equiv \bar{c}\bar{A}^{r-1}\bar{b} \neq 0, \quad \text{and } when\ r > 1 \quad \bar{c}\bar{A}^k \bar{b} = 0, \qquad \text{for } k = 0, 1, \ldots, r-2$$

(A3) The disturbance $d$ is bounded and has bounded time derivatives up to the $(r-1)$'th order.

(A4) The upper bounds on system parameter mismatches are characterized by

$$\|G_i^T\| \le \overline{G}_i \qquad i = 1, \ldots, r$$

$$|d_i| \le \overline{d}_i\ , \qquad i = 1, \ldots, r$$

$$|\dot{d}_i| \le \overline{d}_i^{\,1}\ , \qquad i = 1, \ldots, r-1$$

where

$$G_i = \overline{c}\overline{A}^i - cA^i\ , \qquad d_i = \sum_{j=1}^{i} \overline{c}\overline{A}^{i-j} d^{(j-1)}$$

and $\overline{G}_i$, $\overline{d}_i$, $\overline{d}_i^{\,1}$ are all finite positive constants. In particular, $\overline{G}_i$ and $\overline{d}_i$ are known *a priori.*

(A5) The true high-frequency gain $\overline{K}$ is assumed positive, and bounded by

$$0 < K_m \le \frac{\overline{K}}{K} \le K_M$$

where $K$ is the nominal high-frequency gain, and constants $K_m$ and $K_M$ are known *a priori.*

To establish stability of the control schemes to be presented in the sequel, we need a theorem which relates the boundedness of the state of a minimum-phase system to that of the system output and its derivatives.

**Theorem 1:** Consider the system (1) satisfying (A1) and (A3). If the control input $u$ is devised such that the system output and its first $(r-1)$ derivatives, $y, \dot{y}, \ldots, y^{(r-1)}$, are bounded, then the system state $x$ remains bounded.

**Proof:** see Appendix B.

## III. ESTIMATION OF SYSTEM OUTPUT DERIVATIVES

In this section, we introduce the output derivative estimator developed in [7]. We will show that this output derivative estimator achieves estimation of $\dot{y}, \ldots, y^{(r-1)}$ of the uncertain system (1), where $r$ is the relative degree of the system. Before constructing the output derivative estimator, we need to solve a disturbance estimation problem. Consider a one-dimensional system

$$\dot{x} = v + w \tag{2}$$

where $x \in R^1$ is the system state, $v$ is the control input we can manipulate and $w$ is an unmeasurable disturbance. In this problem, we are interested in obtaining an on-line estimate of $w$. We restrict $w$ to be of the following form

$$w(t) = \sum_{i=1}^{q} \gamma_i(t)\, w_i(t) + d(t) = \Gamma^T(t)\, W(t) \; + \; d(t)\,, \qquad 0 \le q < \infty \tag{3}$$

where $\Gamma^T(t) = [\gamma_1(t), \ldots, \gamma_q(t)]$ and $d(t)$ are unknown functions of time, and $W^T(t) = [w_1(t), \ldots, w_q(t)]$ is a measurable signal. We further assume that $w(t)$ is "strictly regular" in the sense defined below.

**Definition 1:** The disturbance $w(t)$ given in (3) is "strictly regular" if

$$|d(t)| \le D^0\,, \qquad |\dot{d}(t)| \le D^1 \qquad \text{for } all \;\; t \tag{4a}$$

$$\|\Gamma(t)\| \le \Omega^0\,, \qquad \|\dot{\Gamma}(t)\| \le \Omega^1 \qquad \text{for } all \;\; t \tag{4b}$$

$$\|W(t)\| \le \alpha \|W(t)\| + \beta \qquad \text{for } all \;\; t \tag{4c}$$

where $D^0$, $D^1$, $\Omega^0$, $\Omega^1$, $\alpha$ and $\beta$ are all finite nonnegative constants.

**Remark 1:** The "strict regularity" condition does not exclude the case where $W(t)$ may become unbounded; however, the growth rate of $W(t)$ must be cone-bounded as required by (4c).

**Remark 2:** Notice that from (4a) and (4b) we obtain the following upper bound

$$|w(t)| \le \Omega^0 \|W(t)\| + D^0 \tag{5}$$

The following theorem shows that we can construct an estimator for $w(t)$ by defining $v$ in (2) to be

$$v = -\sigma x - \rho(W)\, g(x)\,, \qquad \rho(W) = \rho_1 \|W(t)\| + \rho_2 \tag{6a}$$

where $\sigma$ is an arbitrary positive constant, $g(\cdot)$ can be any function listed in Appendix A, and $\rho_1$ and $\rho_2$ are two positive constants satisfying

$$\rho_1 > \Omega^0 \qquad \text{and} \qquad \rho_2 > D^0 \tag{6b}$$

**Theorem 2:** Consider the system (2) subject to (3) and (4). If the disturbance estimator (6) is applied to the system, then
I. there exists a finite time $T_1$ such that

$$x(t) \in N_1 \qquad \text{for } all \;\; t \ge T_1$$

where

$$N_1 = \left\{ x : |x| < n\,, \quad n = g^{-1} [\sup_{W \in R^q} \frac{\Omega^0 \|W\| + D^0}{\rho_1 \|W\| + \rho_2}] \right\}$$

Notice that since $|g(x)| \leq 1$ for all $x$, inequalities in (6b) are essential for the existence of $n$ defined in $N_1$.

II. there exists a finite time $T_2$ such that

$$\dot{x}(t) \in N_2 \qquad \text{for } all \quad t \geq T_2$$

where

$$N_2 = \left\{ \dot{x} : |\dot{x}| < n', \quad n' = \sup_{W \in R^q} \frac{(\alpha c_1 \rho_1 + \alpha \Omega^0 + \Omega^1) \| W \| + (\beta c_1 \rho_1 + \beta \Omega^0 + D^1)}{(c_2 \rho_1 \| W \| + c_2 \rho_2)} \right\}$$

**Proof:** see Appendix C.

**Remark:** We can make $n$ defined in $N_1$ arbitrarily small if $\rho_1$ and $\rho_2$ in (6a) are chosen sufficiently large. Also notice that the ratio $c_1/c_2$ is made arbitrarily small at the same time since

$$\frac{c_1}{c_2} = \begin{cases} \dfrac{n\ (n + \delta)}{\delta} & when \quad g_i = g_1 \\ n & when \quad g_i = g_2 \\ \dfrac{n}{1 - q} & when \quad g_i = g_3 \end{cases}$$

Therefore, $n'$ defined in $N_2$ also becomes arbitrarily small if $\rho_1$ and $\rho_2$ are large enough. From the above argument, Theorem 2 can now be explained as follows. The first part of the theorem states that $x$ is regulated into an open neighborhood of the origin, and this neighborhood can be made arbitrarily small if $\rho_1$ and $\rho_2$ are chosen sufficiently large. The second part of the theorem states that the disturbance estimator (6) drives $-v$ to "track" the disturbance $w$ in the following sense: given any $n'>0$, there exist sufficiently large $\rho_1$ and $\rho_2$ in (6a) such that $|w+v|$ becomes smaller than $n'$ within a finite period of time. This latter part of the theorem is especially important since it allows us to estimate potentially unbounded uncertainties such as state-dependent uncertainties in a closed-loop control system.

With Theorem 2, we are now in a position to construct an output derivative estimator, which gives us an estimate of $y^{(i)}$, $i=1,2, \ldots, r-1$ of the uncertain system (1).

We first differentiate y in (1) successively to yield

$$y^{(i)} = cA^i x + G_i x + d_i \ , \qquad i = 1, \ldots, r-1 \tag{7}$$

where $G_i$ and $d_i$ are as defined in (A4). In the above equation, the last two terms are unknown due to the system parameter mismatches and the external disturbance. In order to obtain an estimate of $y^{(i)}$, we propose using the disturbance estimator (6) to estimate the unknown term $G_i x + d_i$. However, Theorem 2 requires that $G_i x + d_i$ be "strictly regular". We show in Appendix D that the "strict regularity" condition on $G_i x + d_i$ is met if we impose a "normality" condition on the control input $u$ in (1).

**Definition 2:** The control input $u$ in (1) is a "normal" state feedback control if

$$u(t) = k_1^T(t)\,x(t) + k_2(t) \tag{8a}$$

where $k_1(t) \in R^n$ and $k_2(t) \in R^1$ are all bounded

$$\|k_1(t)\| \le \bar{k}_1 < \infty\,, \qquad |k_2(t)| \le \bar{k}_2 < \infty \tag{8b}$$

With this "normality" condition on $u$, we propose the following output derivative estimator

$$\dot{z}_i = cA^i x + v_i\,, \qquad i = 1, \ldots, r-1 \tag{9a}$$

$$v_i = \sigma e_i + \rho_i(x) g(e_i)\,, \qquad \rho_i(x) = \rho_{1i}\|x\| + \rho_{2i} \tag{9b}$$

where $\sigma$ is an arbitrary positive constant, $g(\cdot)$ is any function listed in Appendix A, $\rho_{1i}$ and $\rho_{2i}$ are two positive constants satisfying

$$\rho_{1i} > \bar{G}_i\,, \qquad \rho_{2i} > \bar{d}_i \tag{9c}$$

and $e_i$ is the estimation error defined by

$$e_i = y^{(i-1)} - z_i \tag{9d}$$

**Theorem 3 (Unrealizable Estimator):** Consider the system (1) subject to (A2) - (A4), and the output derivative estimator (9). If the control input $u$ is a "normal" state feedback control, then given any $\varepsilon_i > 0$, there exist large enough positive constants $\rho_{1i}$ and $\rho_{2i}$ in (9c) and a finite time $T_i$ such that

$$|\dot{z}_i(t) - y^{(i)}(t)| < \varepsilon_i \qquad \text{for } all\ t \ge T_i\,, \qquad i = 1, \ldots, r-1$$

**Proof:** see Appendix D.

Theorem 3 states that as long as the control input is a state feedback control with bounded feedback gains ("normal" feedback control), we can use $\dot{z}_i$ as an estimate of $y^{(i)}$ for all $i = 1, \ldots, r-1$. Unfortunately, the estimator (9) is not realizable when $i \ge 2$ because of the way $e_i$ is defined in (9d);

therefore, we suggest the following realizable estimator to approximate the unrealizable one in (9).

**Realizable Estimator:** Define another set of estimation errors

$$\hat{e}_1 = e_1 = y - z_1$$

$$\hat{e}_i = \dot{z}_{i-1} - z_i \,, \qquad i = 2, \ldots, r-1 \tag{9d'}$$

and use $\hat{e}_i$ instead of $e_i$ in the implementation of the unrealizable estimator (9). The use of $\hat{e}_i$ to approximate $e_i$ is motivated by the following observation: since $\hat{e}_1 = e_1$, Theorem 3 holds when $i$=1; i.e., $\dot{z}_1$ can be made arbitrarily close to $\dot{y}$ within a finite period of time. In the estimation of $\ddot{y}$ (i=2), we need $\dot{y}$ as a reference to calculate $e_2$. Since $\dot{y}$ is not accessible, we use $\dot{z}_1$ as its approximate to calculate $\hat{e}_2$ in (9d'). As a result, $\hat{e}_2$ can be infinitely close to $e_2$. We then expect that Theorem 3 still holds for $i$=2. By continuing this argument up to $i=r-1$, we expect that the proposed realizable estimator approximates the unrealizable estimator effectively. No formal proof has been given for the results of the suggested realizable estimator; however, various simulation studies have verified the effectiveness of this approximation scheme.

## IV. ROBUST CONTROLLER FOR SYSTEMS WITH RELATIVE DEGREE ONE

We will first design a robust controller for systems with relative degree one. The reason for treating this case first is that when the relative degree $r$ is one, the design of a robust controller is much easier than the case when $r$ is larger than one.

When $r$=1, we use the control law:

$$u = -\frac{\sigma y + cAx + \rho(x) g(y)}{K} \tag{10a}$$

$$\rho(x) = \frac{\rho_1 |cAx| + \rho_2 \|x\| + \rho_3}{K_m} \tag{10b}$$

$$\rho_1 > h \,, \quad \rho_2 > \overline{G}_1 \quad \text{and} \quad \rho_3 > \overline{d}_1 \tag{10c}$$

where $\sigma$ is an arbitrary positive constant, $K$ is the nominal high-frequency gain defined in (A2), $h = max(|1-K_M|, |1-K_m|)$ and $\overline{G}_1$ and $\overline{d}_1$ are bounds given in (A4).

**Theorem 4:** Consider the system (1) subject to (A1) - (A5). If the system is of relative degree one and the control law (10) is applied to the system, then there exists a finite time $T$ such that

$$y(t) \in M \qquad \text{for } all \quad t \geq T$$

where

$$M = \left\{ y : |y| < n \ , \ n = g^{-1}\left(\sup_{x \in R^n} \frac{h \cdot |cAx| + \overline{G}_1 \|x\| + \overline{d}_1}{\rho_1 |cAx| + \rho_2 \|x\| + \rho_3}\right) \right\}$$

Furthermore, the system state x and the control input u remain bounded. Note that since $|g(y)| \leq 1$ for all $y$, it is essential that the inequalities in (10c) hold so that $n$ defined in $M$ exists.

**Proof:** see Appendix E.

# V. ROBUST CONTROLLER FOR SYSTEMS WITH RELATIVE DEGREE LARGER THAN ONE

We now consider systems with relative degree larger than one. To design a robust control, we first construct a variable

$$s = y^{(r-1)} + \lambda_{r-2} y^{(r-2)} + \cdots + \lambda_0 y \tag{11}$$

where $\lambda_i$'s (i=0 ,..., r-2) are chosen such that when $s$ equals zero, Eq.(11) defines an asymptotically stable differential equation of $y$. Thus, the convergence of $s$ to zero ensures the convergence of $y$ to zero.

Then, we synthesize the control input $u$ to regulate the variable $s$. However, since $s$ is not accessible, we use the output derivative estimator introduced in Section III to define an accessible variable

$$\hat{s} = \dot{z}_{r-1} + \lambda_{r-2}\dot{z}_{r-2} + \cdots + \lambda_1 \dot{z}_1 + \lambda_0 z_1 \tag{12}$$

where $\dot{z}_i$ is the estimate of $y^{(i)}$ obtained from the estimator (9). The control input $u$ is chosen as

$$u = -\frac{\sigma\hat{s} + \hat{q}(x) + \hat{\rho}(x) g(\hat{s})}{K} \tag{13a}$$

$$\hat{q}(x) = cA^r x + \lambda_{r-2}\dot{z}_{r-1} + \cdots + \lambda_0 \dot{z}_1 \tag{13b}$$

$$\hat{\rho}(x) = \frac{\rho_1 |\hat{q}(x)| + \rho_2 \|x\| + \rho_3}{K_m} \tag{13c}$$

$$\rho_1 > h \ , \quad \rho_2 > \overline{G}_r \quad \text{and} \quad \rho_3 > \overline{d}_r \tag{13d}$$

where $\sigma$, $g(\cdot)$, $h$, $K$ and $K_m$ are as defined in (10).

**Theorem 5 (Unrealizable Robust Controller):** Consider the system (1) which is subject to (A1) - (A5) and with relative degree $r$ larger than one. If the control law (12) and (13) together with the output derivative estimator (9) are applied to the system, then the system state $x$ and the control input $u$ remain bounded. Furthermore, the system output $y$ can be made arbitrarily close to zero if sufficiently large $\rho_j$ ($j$=1,2,3) in (13d) and $\rho_{1i}$, $\rho_{2i}$ ($i$=1,..,$r-1$) in (9c) are used.

**Proof:** see Appendix F.

**Realizable Robust Controller:** The controller stated in Theorem 5 is unrealizable since the output derivative estimator (9) is unrealizable. Following the approximation scheme used in Section III, in real implementation of this unrealizable controller, we replace $e_i$ in (9d) by $\hat{e}_i$ in (9d'). Although no formal stability proof is given for this approximation scheme, its effectiveness is verified by a computer simulation study [7].

## VI. CONCLUSIONS

In this paper, a new robust controller for linear systems is presented. Given the knowledge of a nominal state space model as well as quantitative bounds on system uncertainties, the new controller achieves regulation of the system output and ensures boundedness of the system state. Unlike the sliding Mode Controller or the Ultimate Boundedness Controller, the proposed robust controller does not impose any structural condition on system uncertainties. However, the system must be minimum phase. When the relative degree of the system is larger than one, the proposed controller is first obtained in an unrealizable form, and then its approximation in a realizable form is suggested. The robust controller in the approximated form is verified to be effective by a computer simulation study [7].

# APPENDIX A

Table 1 Smooth Functions for Continuous Control

| | $\lvert s \rvert > \varepsilon$ | $\lvert s \rvert \le \varepsilon$ |
|---|---|---|
| $g_1(s)$ | $\dfrac{s}{\lvert s \rvert + \delta}$ | $\dfrac{s}{\lvert s \rvert + \delta}$ |
| $g_2(s)$ | $sign(s)$ | $\dfrac{s}{\varepsilon}$ |
| $g_3(s)$ | $sign(s)$ | $\varepsilon^{q-1} s^{1-q} sign(s)$ |

In Table 1, the $sign(\cdot)$ function is defined as

$$sign(s) = \begin{cases} 1 & s \ge 0 \\ -1 & s < 0 \end{cases}$$

$\varepsilon$ and $\delta$ are small positive numbers and $q \in [0,1)$. $g_2(\cdot)$ is usually called the "saturation" function. Notice that when $q=0$, $g_3(\cdot)$ reduces to $g_2(\cdot)$. A careful examination of these functions shows that they are all continuous odd functions; i.e., $g(-s) = -g(s)$, and their values increase monotonically from -1 to +1.

# APPENDIX B

Proof of Theorem 1: without loss of generality, we assume that the system (1) is in the controllable canonical form; i.e.,

$$\bar{A} = \begin{bmatrix} 0 & 1 & 0 & . & 0 \\ 0 & 0 & 1 & . & 0 \\ \cdot & \cdot & \cdot & . & \cdot \\ 0 & 0 & 0 & . & 1 \\ -a_0 & -a_1 & -a_2 & . & -a_{n-1} \end{bmatrix} \qquad \bar{b} = \begin{bmatrix} 0 \\ 0 \\ \cdot \\ 0 \\ \bar{k} \end{bmatrix} \qquad d = \begin{bmatrix} d_1 \\ d_2 \\ \cdot \\ d_{n-1} \\ d_n \end{bmatrix}$$

$$\bar{c} = [b_0, b_1, \ldots, b_{n-r-1}, 1, 0, \ldots, 0] \tag{B.1}$$

The transfer function of the system is

$$H(s) = \bar{k} \frac{B(s)}{A(s)}$$

where

$$B(s) = s^{n-r} + b_{n-r-1} s^{n-r-1} +, \ldots, + b_1 s + b_0 \tag{B.2}$$

$$A(s) = s^n + a_{n-1}s^{n-1} +, \ldots, + a_1 s + a_0$$

We also assume that $A(s)$ and $B(s)$ are coprime.

Choose a new set of coordinate $t_i$'s, with

$$t_1 = \overline{c}x$$

$$t_2 = \overline{c}\overline{A}x$$

$$\cdots$$

$$t_r = \overline{c}\overline{A}^{r-1}x$$

$$t_{r+1} = x_1$$

$$t_{r+2} = x_2$$

$$\cdots$$

$$t_n = x_{n-r} \tag{B.3}$$

Using the expressions of $\overline{c}$ and $\overline{A}$ in (B.1), the coordinate transformation matrix $T$, for $t = Tx$, is found to be

$$T = \begin{bmatrix} X_{r\times(n-r)} & I_r \\ I_{n-r} & 0_{(n-r)\times r} \end{bmatrix}$$

where $I$ is the identity matrix, $O$ is the zero matrix and $X$ is some constant matrix. Notice that $T$ nonsingular.

To simplify expressions, define

$$\zeta = \begin{bmatrix} t_1 \\ \cdot \\ \cdot \\ t_r \end{bmatrix}, \qquad \eta = \begin{bmatrix} t_{r+1} \\ \cdot \\ \cdot \\ t_n \end{bmatrix}$$

In the new coordinates defined by (B.3), the system (1) has the following representation:

$$\dot{t}_1 = t_2 + \overline{c}d$$

$$\dot{t}_2 = t_3 + \overline{c}\overline{A}d$$

$$\cdots$$

$$\dot{t}_{r-1} = t_r + \overline{c}\overline{A}^{r-2}d$$

$$\dot{t}_r = R\zeta + S\eta + \overline{k}u + \overline{c}\overline{A}^{r-1}d$$

$$\dot{\eta} = Q\eta + Py + V \tag{B.4}$$

where $R^T \in R^r$, $S^T \in R^{n-r}$ and

$$Q = \begin{bmatrix} 0 & 1 & 0 & . & 0 \\ 0 & 0 & 1 & . & 0 \\ . & . & . & . & . \\ 0 & 0 & 0 & . & 1 \\ -b_0 & -b_1 & -b_2 & . & -b_{n-r-1} \end{bmatrix} \qquad P = \begin{bmatrix} 0 \\ 0 \\ . \\ 0 \\ 1 \end{bmatrix} \qquad V = \begin{bmatrix} d_1 \\ d_2 \\ . \\ d_{n-r-1} \\ d_{n-r} \end{bmatrix} \tag{B.5}$$

where $d_i$'s are the components of $d$. We notice that $Q$ is a companion matrix with $B(s)$ as its characteristic polynomial. Since the system (1) is minimum phase, all the eigenvalues of $Q$ are in the open left half plane. The boundedness of $\eta$ then follows from (B.4) and the boundedness of $y$ and $V$. We further notice that

$$t_1 = y$$

$$t_2 = \dot{y} - \bar{c}d$$

$$t_3 = \ddot{y} - \bar{c}\bar{A}d - \bar{c}\dot{d}$$

$$\cdots$$

$$t_r = y^{(r-1)} - \bar{c}\bar{A}^{r-2}d - \bar{c}\bar{A}^{r-3}\dot{d} - \cdots - \bar{c}d^{(r-2)}$$

From the boundedness assumption of $y, \dot{y}, \ldots, y^{(r-1)}$ and $d^{(i)}$, $i = 0,1, \ldots, r-2$, we conclude the boundedness of $\zeta$.

Since $\eta$ and $\zeta$ are all bounded and the transformation matrix $T$ is nonsingular, the system state $x$ is also bounded. This concludes the proof.

# APPENDIX C

Proof of Theorem 2: let $V_1 = \frac{1}{2} x^2$, and take the derivative of $V_1$

$$\dot{V}_1 = x\,\dot{x}$$

$$= x\,(-\sigma x - \rho(W)g(x) + \Gamma^T W + d)$$

$$\leq -\sigma x^2 - |x|\rho(W)\left( g(|x|) - \frac{\Omega^0\|W\| + D^0}{\rho_1\|W\| + \rho_2} \right)$$

$$\leq -\sigma n^2 \qquad \text{for all } x \in N_1^c$$

The last inequality shows that as long as $|x| \geq n$, $V_1$ decreases with a nonzero rate. Therefore, we conclude that after a finite period of time, say $T_1$ ($\leq V_1(0)/(\sigma n^2)$), $x$ is driven into $N_1$ and stays inside $N_1$ thereafter.

To study how $\dot{x}$ behaves, let $V_2 = \frac{1}{2}\dot{x}^2$ and differentiate $V_2$

$$\dot{V}_2 = \dot{x}\,\ddot{x}$$

$$= \dot{x}\,(-\sigma\dot{x} - \rho(W)g'(x)\dot{x} - \rho_1 g(x)\frac{d}{dt}\|W\| + \dot{\Gamma}^T W + \Gamma^T \dot{W} + \dot{d})$$

Notice that for any finite dimensional vector $W$

$$\left|\frac{d}{dt}\|W\|\right| \le \|\dot{W}\|, \qquad \text{for } all\ W \ne 0 \tag{C.1}$$

Inequality (C.1) does not hold for $W$=0; however, in the sequel we assume that (C.1) is true for all $W \in R^q$. The legitimacy of this assumption will be clarified at the end of this proof. Making use of (C.1), we obtain

$$\dot{V}_2 \le -\sigma\dot{x}^2 - |\dot{x}|^2\rho(W)g'(x) + |\dot{x}|\rho_1\|\dot{W}\|g(x) + \Omega^1|\dot{x}|\cdot\|W\|$$
$$+ \Omega^0|\dot{x}|\cdot\|\dot{W}\| + |\dot{x}|D^1$$

We know from part I of Theorem 2 that $x$ enters $N_1$ after $t>T_1$. This implies that after $t > T_1$, we have

$$\dot{V}_2 \le -\sigma\dot{x}^2 - |\dot{x}|^2\rho(W)c_2 + |\dot{x}|\rho_1\|\dot{W}\|c_1 + \Omega^1|\dot{x}|\cdot\|W\|$$
$$+ \Omega^0|\dot{x}|\cdot\|\dot{W}\| + |\dot{x}|D^1$$

Finally, it follows from the strict regularity assumption of $W$ that

$$\dot{V}_2 \le -\sigma\dot{x}^2 - |\dot{x}|\rho(W)c_2\,(|\dot{x}| - \frac{(f\alpha+\Omega^1)\|W\|+(f\beta+D^1)}{c_2\,(\rho_1\|W\|+\rho_2)})$$

$$\le -\sigma n'^2 \qquad \text{for } all\ \dot{x} \in N_2^c \tag{C.2}$$

where $f = \rho_1 c_1 + \Omega^0$. We conclude from (C.2) that $\dot{x}$ is driven into $N_2$ after a finite period of time, say $T_2\,(\le T_1 + V_2(0)/(\sigma n'^2))$, and stays inside $N_2$ thereafter. This concludes the proof.

**Remark:** To solve the differentiability problem of $\|W\|$ at $W$=0 in Equation (C.1), we replace $\rho(W)$ in (6a) by

$$\overline{\rho}(W) = \rho_1\phi(W) + \rho_2 \tag{C.3}$$

where

$$\phi(W) = \begin{cases} \|W\| & \|W\| \ge \varepsilon \\ \dfrac{1}{2\varepsilon}W^T W + \dfrac{\varepsilon}{2} & \|W\| < \varepsilon \end{cases} \tag{C.4}$$

and $\varepsilon$ is an arbitrary positive constant. Notice that $\phi(W)$ is differentiable and satisfies

$$(i) \quad \phi(W) \geq \|W\| \qquad \text{for all } W$$

$$(ii) \quad \|\frac{d}{dW}\phi(W)\| \leq 1 \qquad \text{for all } W$$

By using (C.3) in the control law (6a), we can show [7] that Theorem 2 is still true except that $N_1$ is replaced by

$$N'_1 = \left\{ x \,:\, |x| < n \,, \quad n = g^{-1}\,[\sup_{W \in R^q} \frac{\Omega^0 \|W\| + D^0}{\rho_1 \phi(W) + \rho_2}] \right\}$$

Since this result is valid for any value of $\varepsilon > 0$ in (C.4), by letting $\varepsilon$ be an infinitesimally small number, we actually make $\phi(\cdot)$ infinitely close to the $|\cdot|$ function used in (6a), and $N'_1$ infinitely close to $N_1$ in Theorem 2. For this reason, we use $\|\cdot\|$ instead of the $\phi(\cdot)$ function throughout this paper, and allow Eq.(C.1) to be used for all $Z$.

# APPENDIX D

Proof of Theorem 3: we first show that $G_i x + d_i$ in (7) is "strictly regular". Substituting (8a) into (1), we obtain

$$\dot{x} = \bar{A}x + \bar{b}u + d$$

$$= (\bar{A} + \bar{b}k_1^T(t))\,x + \bar{b}k_2(t) + d(t)$$

Taking the norm of the above equation, and noticing the boundedness of $d(t)$ and inequalities (8b), we obtain

$$\|\dot{x}\| \leq P\,\|x\| + Q$$

where

$$P \equiv \sup_{k_1(t)} \|\bar{A} + \bar{b}k_1^T(t)\| < \infty$$

$$Q \equiv \sup_{k_2(t), d(t)} \|\bar{b}k_2(t) + d(t)\| < \infty$$

Therefore, $G_i x + d_i$ is indeed "strictly regular".

From (7) and (9a), the dynamics of $e_i$ are governed by

$$\dot{e}_i = -v_i + G_i x + d_i \,, \qquad i = 1, \ldots, r-1$$

Since $G_i x + d_i$ is "strictly regular", we invoke Theorem 2 to conclude that

$v_i - G_i x - d_i (= \dot{z}_i - y^{(i)})$ can be made arbitrarily small within a finite period of time if sufficiently large $\rho_{1i}$ and $\rho_{2i}$ in (9c) are used. This concludes the proof.

## APPENDIX E.

Proof of Theorem 4: to study how the magnitude of $y$ varies with time, we first obtain an expression for $\dot{y}$

$$\begin{aligned} \dot{y} &= \bar{c}\bar{A}x + \bar{c}\bar{b}u + \bar{c}\bar{d} \\ &= cAx + \bar{K}u + G_1 x + d_1 \end{aligned}$$

Define $V = \frac{1}{2} y^2$, and take the derivative of $V$

$$\begin{aligned} \dot{V} &= y\,\dot{y} \\ &= -\frac{\bar{K}}{K}\sigma y^2 - \frac{\bar{K}}{K}\rho(x) y g(y) + y(1 - \frac{\bar{K}}{K})cAx + yG_1 x + yd_1 \end{aligned}$$

It follows from Assumptions (A4) and (A5) that

$$\begin{aligned} \dot{V} &\le -\sigma K_m y^2 - K_m \rho(x) |y| g(|y|) + |y| h |cAx| + |y| \bar{G}_1 \|x\| + |y| \bar{d}_1 \\ &\le -\sigma K_m y^2 - K_m \rho(x) |y| ( g(|y|) - \frac{h |cAx| + \bar{G}_1 \|x\| + \bar{d}_1}{K_m \rho(x)} ) \\ &\le -\sigma K_m n^2 \qquad \text{for all } y \in M^c \end{aligned} \tag{E.1}$$

where $M$ is defined in the statement of Theorem 4. Equation (E.1) shows that as long as $|y| \ge n$, the magnitude of $y$ decreases with a nonzero rate. Therefore, we conclude that $y$ enters $M$ within a finite period of time, say $T$ $(\le V(0)/(\sigma K_m n^2))$, and stays permanently inside $M$ after its first entry.

Since the system (1) is minimum phase and of relative degree one, we invoke Theorem 1 to conclude the boundedness of the state $x$. The boundedness of the control input $u$ then follows from Eq.(10a). This concludes the proof.

## APPENDIX F

Proof of Theorem 5: a careful examination of (9) and (13) shows that the control input u is indeed a "normal" state feedback control as defined in Definition 2. It then follows from Theorem 3 that

$$|\dot{z}_i - y^{(i)}| < \varepsilon_i \qquad i = 1, \ldots, r-1 \tag{F.1}$$

within a finite period of time, and $\varepsilon_i$ can be made arbitrarily small if sufficiently large $\rho_{1i}$ and $\rho_{2i}$ in (9c) are used.

Define another control input

$$u^o = -\frac{\sigma s + q(x) + \rho(x) g(s)}{K} \tag{F.2a}$$

$$q(x) = cA^r x + \lambda_{r-2} y^{(r-1)} + \cdots + \lambda_0 \dot{y} \tag{F.2b}$$

$$\rho(x) = \frac{\rho_1 |q(x)| + \rho_2 \|x\| + \rho_3}{K_m} \tag{F.2c}$$

where $s$, $g(\cdot)$, $h$, $\rho_1$, $\rho_2$ and $\rho_3$ are as defined in (11) and (12).

The two sets of variables defined in (11), (12) and (F.2) are related by the following equations:

$$\hat{s} = s + \mu_1 \tag{F.3a}$$

$$\hat{q}(x) = q(x) + \mu_2 \tag{F.3b}$$

$$\hat{\rho}(x) = \rho(x) + \mu_3 \tag{F.3c}$$

$$u^0 = u + \mu_4 + \mu_5 \rho(x) \tag{F.3d}$$

where $\mu_i$'s ($i$=1,..,5) approach zero when $\varepsilon_j$'s ($j$=1,..,$r-1$) approach zero (see Appendix G).

We now show that the control law (13) regulates the variable $s$ to almost zero. Differentiate the system output $r$ times to yield

$$y^{(r)} = \overline{cA}^r x + \overline{K} u + d_r$$
$$= cA^r x + \overline{K} u + G_r x + d_r$$

Define $V = ½ s^2$, and take the derivative of $V$

$$\dot{V} = s\,(y^{(r)} + \lambda_{r-2} y^{(r-1)} + \cdots + \lambda_0 \dot{y})$$
$$= s\,(q(x) + \overline{K} u + G_r x + d_r)$$
$$= s\,(q(x) + \overline{K} u^o + \overline{K}\mu_4 + \overline{K}\mu_5 \rho(x) + G_r x + d_r)$$
$$= s\,(-\frac{\overline{K}}{K}\sigma s - \frac{\overline{K}}{K}\rho(x) g(s) + (1 - \frac{\overline{K}}{K}) q(x) + G_r x + d_r$$
$$+ \overline{K}\mu_4 + \overline{K}\mu_5 \rho(x)\,)$$

Using inequalities in (A4) and (A5), we obtain

$$\dot{V} \le -K_m \sigma s^2 - K_m \rho(x) |s| \left( g(|s|) - \frac{h\,|q(x)| + \overline{G}_r \|x\| + \overline{d}_r + \overline{K}\mu_4 + \overline{K}\mu_5 \rho(x)}{K_m \rho(x)} \right)$$

$$\le -K_m \sigma n^2 \qquad \text{for } all \quad s \in M_1^c \tag{F.4}$$

where $M_1$ is defined by

$$M_1 = \left\{ s \,:\, |s| < n \;, \quad n = g^{-1}\left( \sup_{x \in R^n} \frac{h\,|q(x)| + \overline{G}_r \|x\| + \overline{d}_r + \overline{K}\mu_4 + \overline{K}\mu_5 \rho(x)}{K_m \rho(x)} \right) \right\}$$

From Eq.(F.4), we conclude that $s$ enters $M_1$ within a finite period of time, say $T_1$ ($\le V(0)/(\sigma K_m n^2)$), and stays permanently inside $M_1$ after its first entry; in other words,

$$|s(t)| < n \qquad \text{for } all \quad t \ge T_1 \tag{F.5}$$

We notice that $n$ defined in $M_1$ can be made arbitrarily small if $\mu_4$ and $\mu_5$ are sufficiently small and if $\rho_i$, $i$=1,2,3 in (13d) is sufficiently large. To study how the system output $y$ behaves under the constraint (F.5), we realize the Eq.(11) in the following form:

$$\dot{\zeta} = \begin{bmatrix} 0 & 1 & 0 & . & 0 \\ 0 & 0 & 1 & . & 0 \\ \cdot & \cdot & \cdot & . & \cdot \\ 0 & 0 & 0 & . & 1 \\ -\lambda_0 & -\lambda_1 & -\lambda_2 & . & -\lambda_{r-2} \end{bmatrix} \zeta + \begin{bmatrix} 0 \\ 0 \\ \cdot \\ 0 \\ 1 \end{bmatrix} s \tag{F.6}$$

where

$$\zeta_i = y^{(i-1)} \qquad i = 1, \ldots, r$$

Equation (F.6) defines an asymptotically stable system with $s$ as the system input. Thus, regulation of $\zeta$ into a small neighborhood of the origin in $R^r$ is achieved since $n$ in (F.5) can be made arbitrarily small. In particular, "almost perfect" regulation of $y$ ($=\zeta_1$) is achieved.

From the boundedness of $\zeta$ in (F.6) and Assumption (A1), we invoke Theorem 1 to conclude the boundedness of the system state. The boundedness of the control input follows from (13). This concludes the proof.

## APPENDIX G

Subtracting (11) from (12), and taking the absolute value of $\mu_1$, we obtain

$$|\mu_1| \leq \lambda_{r-2}\varepsilon_{r-2} + \cdots + \lambda_1\varepsilon_1$$

Similarly, subtracting (F.2b) from (13b), we obtain

$$|\mu_2| \leq \lambda_{r-2}\varepsilon_{r-1} + \cdots + \lambda_0\varepsilon_1$$

Subtraction of (F.2c) from (13c) yields

$$|\mu_3| \leq \mu_2 \frac{\rho_1}{K_m}$$

Finally, it follows from (F.2a) and (13b) that

$$u^o - u = \mu_1\frac{\sigma}{K} + \mu_2\frac{1}{K} + \mu_3\frac{g(\hat{s})}{K} + \rho(x)\frac{[g(s) - g(\hat{s})]}{K} \tag{G.1}$$

From the Mean-Value Theorem [8], we have

$$g(s) - g(\hat{s}) = g'(\underline{s})(s - \hat{s}) = -g'(\underline{s})\mu_1 \tag{G.2}$$

where $\underline{s}$ lies between $s$ and $\hat{s}$. Substituting (G.2) into (G.1), we obtain

$$u^o - u = \mu_4 + \mu_5\rho(x)$$

where

$$\mu_4 = \mu_1\frac{\sigma}{K} + \mu_2\frac{1}{K} + \mu_3\frac{g(\hat{s})}{K}, \quad \mu_5 = -\mu_1\frac{g'(\underline{s})}{K}$$

**Remark:** In Eq.(G.2), we assume differentiability of $g(s)$ from $s$ to $\hat{s}$ (excluding end points) in invoking the Mean-Value Theorem. For $g_1(s)$ in Table 1, the function is differentiable for all $s \in R^1$; for $g_2(s)$ and $g_3(s)$, they are differentiable only for $s \in (-\varepsilon, \varepsilon)$. Therefore, when either $|s| \geq \varepsilon$ or $|\hat{s}| \geq \varepsilon$, Eq.(G.2) can no longer be used. However, we can solve this problem easily by a slight change of Eq.(G.1).

We first observe that from the definition of $M_1$ in Appendix F, $n$ in (F.5) satisfies $n < \varepsilon$; that is, $|s| < \varepsilon$ after a finite period of time. If $\hat{s}$ also falls between $(-\varepsilon, \varepsilon)$, Equation (G.2) is legitimate; if not, choose $\bar{s}$ such that $|\bar{s}| = \varepsilon$ and $\bar{s} \cdot \hat{s} > 0$, and rewrite the last term in Eq.(G.1)

$$\rho(x)\frac{[g(s) - g(\hat{s})]}{K} = \rho(x)\frac{[g(s) - g(\bar{s}) + g(\bar{s}) - g(\hat{s})]}{K}$$

$$= \rho(x)\frac{[g(s) - g(\bar{s})]}{K} \tag{G.3}$$

where we used the fact that $g(\hat{s}) = g(\bar{s}) = 1$ *or* $-1$ depending on the sign of $\hat{s}$. We can now apply the Mean-Value Theorem to Eq.(G.3) since $g(\cdot)$ is now differentiable between $s$ and $\bar{s}$.

## REFERENCES

1. G. LEITMANN, "On The Efficacy of Nonlinear Control in Uncertain Linear Systems," *J. Dynam. Syst., Meas. and Cont., 103,* 95-102 (1981).
2. V. UTKIN, "Variable Structure Systems with Sliding Modes," *IEEE Trans., Automat. Contr., AC-22,* 212-222 (1977).
3. E. P. RYAN and M. CORLESS, "Ultimate Boundedness and Asymptotical Stability of A Class of Uncertain Dynamical Systems via Continuous and Discontinuous Feedback Control," *IMA J. Math. Control and Information, 1,* 223-242 (1984).
4. B. DRAZENOVIC, "The Invariance Conditions in Variable Structure Systems," *Automatica, 5,* 287-295 (1969).
5. B. R. BARMISH and G. LEITMANN, "On Ultimate Boundedness Control of Uncertain Systems in The Absence of Matching Conditions," *IEEE Trans., Automat. Contr., AC-27,* 153-158 (1982).
6. J. S. THORP and B. R. BARMISH, "On Guaranteed Stability of Uncertain Linear Systems via Linear control," *J. Optimiz. Theor. and Appl., 35,* 559-579 (1981).
7. M. S. CHEN, "On The Design of Nonlinear Robust Controllers for Linear Systems," Mechanical Engineering Department, University of California, Berkeley, Ph.D. Dissertation, 1989.
8. W. RUDIN, "Principles of Mathematical Analysis," McGraw-Hill, New York, 1976.

# CONTROL AND OBSERVATION OF UNCERTAIN SYSTEMS: A VARIABLE STRUCTURE SYSTEMS APPROACH

Stefen Hui
Department of Mathematical Sciences
San Diego State University
San Diego, CA 92182-0314

and

Stanislaw H. Żak
School of Electrical Engineering
Purdue University
West Lafayette, IN 47907

## 1. Introduction

Variable structure control (VSC), the control of dynamical systems with discontinuous state feedback controllers, have been developed over the last 25 years. See [1], [2], and [6] for surveys. See also [10]. This theory rests on the concept of changing the structure of the controller in response to the changing states of the system to obtain a desired response. This is accomplished by the use of a high speed switching control law which forces the trajectories of the system onto a chosen manifold, where they are maintained thereafter. The system is insensitive to certain parameter variations and disturbances while the trajectories are on the manifold. If the state vector is not accessible, then a suitable estimate must be used. The estimate should not degrade the overall performance of the system.

The design process involves three steps. The first is the design of the controller assuming the availability of the state vector. The second step is the design of an estimator, also known as an observer, of the state vector. The observer should use only the input and output of the plant ([13]). The last step is to combine the first two steps in that the control strategy uses the estimate instead of the true state vector.

Our emphasis will be on bounded controllers since physically realizable controllers must be bounded.

In Section 2 we give some background material and the description of the class of systems that we will analyze. The next three sections correspond to the steps of the design process. Section 3 will be devoted to the study of our bounded controllers. In Section 4 we will discuss the state observation problem and in Section 5 we will combine the observer and controller. In Section 6 we consider boundary layer controllers and observers. Section 7 summarizes our results.

## 2. System Description and Background Material

The class of systems under consideration in this chapter is modeled by

$$\dot{x} = Ax + B(u + \xi)$$
$$y = Cx ,$$

where $x \in \mathbb{R}^n$, $u \in \mathbb{R}^m$, $y \in \mathbb{R}^p$, $p \geq m$, and the matrices A, B, C have appropriate dimensions. We assume that (A,B) is completely controllable, and that (A,C) is completely observable. The vector $\xi$ represents the uncertainty of the system and is a function of time and state. It is assumed that $||\xi|| \leq \eta$, where $\eta$ is a positive constant.

In this chapter, we use the following notation. If $x \in \mathbb{R}^n$, then $||x||$ denotes the Euclidean norm, that is, $||x|| = (x_1^2 + ... + x_n^2)^{1/2}$. If A is a matrix, then $||A||$ is the spectral norm defined by

$$||A|| = \sup \{||Ax|| \mid ||x|| \leq 1\} .$$

For any square matrix A, we let $\lambda_{min}(A)$ be the minimum eigenvalue of A and $\lambda_{max}(A)$ be the maximum eigenvalue of A.

With these notations,

$$||A||^2 = \lambda_{max}(A^T A) .$$

Rayleigh principle states that if P is a real symmetric positive definite matrix, then

$$\lambda_{min}(P) \, ||x||^2 \leq x^T P x \leq \lambda_{max}(P) \, ||x||^2 \ .$$

## 3. Variable Structure Control

An important concept in variable structure control is that of an attractive manifold on which certain desired dynamical behaviors are guaranteed. Trajectories of the system should be steered towards the manifold and subsequently constrained to remain on it.

**Definition 3.1.** ([2])

A domain $\Delta$ in the manifold $\{x \mid \sigma(x) = 0\}$ is a sliding mode domain if for each $\epsilon > 0$ there exists a $\delta > 0$ such that any trajectory starting in the n-dimensional $\delta$-neighborhood of $\Delta$ may leave the n-dimensional $\epsilon$-neighborhood of $\Delta$ only through the n-dimensional $\epsilon$-neighborhood of the boundary of $\Delta$.

We next describe the manifold which is used in this chapter. Suppose

$$S = \begin{bmatrix} s_1 \\ \vdots \\ s_m \end{bmatrix} \in \mathbb{R}^{m \times n} \ ,$$

where

$$s_i \in \mathbb{R}^{1 \times n} \ .$$

We assume that S is of full rank. For $x \in \mathbb{R}^n$, let

$$\sigma(x) = \begin{bmatrix} \sigma_1(x) \\ \vdots \\ \sigma_m(x) \end{bmatrix} = \begin{bmatrix} s_1 x \\ \vdots \\ s_m x \end{bmatrix} = Sx \ .$$

We assume that SB is nonsingular. The manifold we use is $\{x \mid \sigma(x) = 0\}$. When a trajectory $x(t)$ is on the manifold we have

$$Sx = 0$$

and

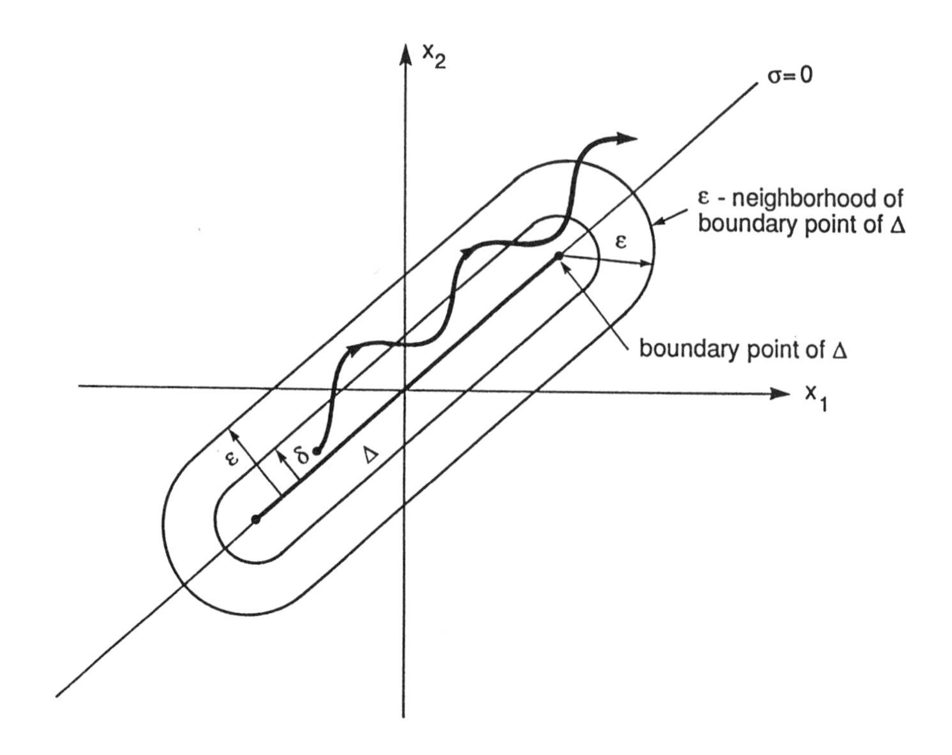


Fig. 1. Two-dimensional illustration of sliding mode domain.

$$S\dot{x} = S(Ax + Bu + B\xi) = 0 .$$

Since SB is nonsingular, we can solve for u:

$$u = -(SB)^{-1}SAx - \xi .$$

This u is called the equivalent control and is denoted by $u_{eq}$ ([1], [17). The use of this controller forces the trajectory to lie on a level surface of $\sigma(x)$. When the equivalent control is used, the system is called the equivalent system and the state is governed by

$$\dot{x} = [I_n - B(SB)^{-1}S]Ax .$$

It follows that while the system is in the sliding mode, the governing equations are:

$$Sx = 0 ,$$
$$\dot{x} = [I_n - B(SB)^{-1}S]\, Ax .$$

Observe that the dimension of the system in sliding is reduced to n−m.

There are two basic steps in the design of a variable structure controller:

(1) The design of the switching surface so that the behavior of the system has certain prescribed properties on the surface. For example, the switching surface will be designed so the system is asymptotically stable on the surface.

(2) The design of the control strategy to steer the system to the switching surface and to maintain it there.

We base the design of the switching surface on a method developed by El-Ghezawi et al. [5].

Consider the equivalent system

$$\dot{x} = [I_n - B(SB)^{-1}S]Ax .$$

It is easy to see that $B(SB)^{-1}S$ is a projector and has rank m. Hence $I_n - B(SB)^{-1}S$ is also a projector with rank n−m ([8]). Therefore the matrix $A_{eq} = [I_n - B(SB)^{-1}S]A$ in the equivalent system can have at most n−m nonzero eigenvalues. Our goal is to choose S so that the nonzero eigenvalues of $A_{eq}$ are prescribed negative real numbers and the corresponding eigenvectors $\{w_1,...,w_{n-m}\}$ are to be chosen to lie on the switching surface. Let $W = [w_1 ... w_{n-m}]$; note that $W \in \mathbb{R}^{n\times(n-m)}$. It is well known that complete controllability of the pair (A,B) is equivalent to the existence of a controller of the form $u = -Kx$ so that the eigenvalues of $A - BK$ can be arbitrarily assigned [7]. Our equivalent system has the form

$$\dot{x} = Ax - B[(SB)^{-1}SA]\, x.$$

If we let $K = (SB)^{-1}SA$, we need $A - BK$ to have n−m prescribed negative eigenvalues $\{\lambda_1,...,\lambda_{n-m}\}$ and n−m corresponding eigenvectors $\{w_1,...,w_{n-m}\}$. This is equivalent to

$$(A - BK)W = WJ$$

where $J = \mathrm{diag}[\lambda_1, \ldots, \lambda_{n-m}]$.

Denote by R(T) the range of the operator T. Since we require SB to be nonsingular and $SW = 0$, we must have

$$R(B) \cap R(W) = \{0\} .$$

It then follows that ([5]) we should choose the generalized inverses $B^g$, $W^g$ of B, W so that

$$B^g W = 0 \tag{3.1a}$$

and

$$W^g B = 0 . \tag{3.1b}$$

We choose $\{w_1, \ldots, w_{n-m}\}$ so that (3.1b) holds. Observe that $R(B) \cap R(W) = \{0\}$ follows from (3.1b). We can now construct S. Let $W^{\perp} \in \mathbb{R}^{m \times n}$ be any full rank annihalator of W, that is $W^{\perp} W = 0$. Since a necessary condition for $Sx = 0$ to be a switching surface is $SW = 0$, we see that $\Gamma W^{\perp}$ is a candidate for any nonsingular $\Gamma \in \mathbb{R}^{m \times m}$. Note that since $R(W) \cap R(B) = \{0\}$, $W^{\perp} B$ is invertible. We let $S = (W^{\perp} B)^{-1} W^{\perp}$. It is easy to see that $SB = I_m$ and hence $(W^{\perp} B)^{-1} W^{\perp}$ is a generalized inverse of B. If we let $B^g = S$ in (3.3a), the condition is satisfied. We then choose the switching surface to be

$$\Omega = \{x \in \mathbb{R}^n \mid Sx = 0\} .$$

With the above information on the switching surface, we discuss a transformation ([12]) which decomposes the system into "fast" and "slow" subsystems. For an alternative approach see [9]. Let $M \in \mathbb{R}^{n \times n}$ be defined by

$$M = \begin{bmatrix} W^g \\ S \end{bmatrix} .$$

Note that M is invertible with $M^{-1} = [W \;\; B]$. Introduce the new coordinates

$$\hat{x} = Mx .$$

Let $z = W^g x$ and $y = Sx$. Then $\hat{x} = \begin{bmatrix} z \\ y \end{bmatrix}$. In the new coordinates, the system becomes

$$\dot{\hat{x}} = MAM^{-1}\hat{x} + MB(u + \xi) .$$

We write

$$MAM^{-1} = \begin{bmatrix} A_{11} & A_{12} \\ A_{21} & A_{22} \end{bmatrix}$$

where $A_{11} \in \mathbb{R}^{(n-m)\times(n-m)}$, $A_{22} \in \mathbb{R}^{m\times m}$. Note that

$$MB = \begin{bmatrix} 0 \\ I_m \end{bmatrix} .$$

Hence

$$\begin{aligned} \dot{z} &= A_{11} z + A_{12} y \\ \dot{y} &= A_{21} z + A_{22} y + u + \xi . \end{aligned}$$

Observe that $y = \sigma$ and that $A_{21} z + A_{22} y = SAM^{-1}\hat{x} = SAx = - u_{eq}$ for the nominal system, $\dot{x} = A + Bu$. The slow subsystem is governed by the first equation and the fast subsystem is governed by the second equation which we can write as:

$$\begin{cases} \dot{z} = A_{11} z + A_{12} \sigma \\ \dot{\sigma} = - u_{eq} + u + \xi \quad . \end{cases} \tag{3.2}$$

From the condition

$$(A - BK)W = WJ ,$$

we obtain

$$W^g AW = J$$

since $W^g B = 0$ and $W^g W = I$. We know that $A_{11} = W^g AW$, and therefore $A_{11} = J$. A simple condition which guarantees that the trajectories are tending towards the surface $\Omega$ is $\frac{d}{dt}\|\sigma(x)\|^2 = 2\sigma^T\dot{\sigma} < 0$. With our switching surface, we would have a (compact) sliding mode domain $\Delta$ if $\sigma^T\dot{\sigma} < 0$ in a neighborhood of $\Delta$. With this condition in mind, we now proceed to design the controller.

In general, a variable structure controller varies its structure depending on the position of the state relative to the switching surface and has the form:

$$u_i = \begin{cases} u_i^+(x) & \text{if} \quad \sigma_i(x) > 0 \\ u_i^-(x) & \text{if} \quad \sigma_i(x) < 0 . \end{cases}$$

Consider the controller

$$u = -\mu \frac{\sigma}{\|\sigma\|}, \quad \mu > 0 .$$

Clearly u is bounded by $\mu$. We view $\mu$ as a design parameter. The closed-loop system has the form

$$\dot{z} = A_{11}z + A_{12}\sigma$$

$$\dot{\sigma} = A_{21}z + A_{22}\sigma - \mu\frac{\sigma}{\|\sigma\|} + \xi .$$

We use the following notation:

$$a_{ij} = \|A_{ij}\| \quad \text{for} \quad i = 1,2, \quad j = 1,2,$$

and

$$\lambda = \min\ \{|\lambda_1|, \ldots, |\lambda_{n-m}|\} .$$

**Lemma 3.1.** Suppose $\|z\| \neq 0$, $\|\sigma\| \neq 0$. Then

$$\frac{d}{dt}\|\sigma\| \leq a_{21}\|z\| + a_{22}\|\sigma\| - \mu + \eta$$

and

$$\frac{d}{dt}\|z\| \leq -\lambda\|z\| + a_{12}\|\sigma\| .$$

**Proof.** By straightforward estimation,

$$\frac{d}{dt}\,||\sigma|| = \frac{\sigma^T\dot{\sigma}}{||\sigma||} = \frac{\sigma^T}{||\sigma||}\,A_{21}z + \sigma^T A_{22}\,\frac{\sigma}{||\sigma||} - \sigma^T\,\mu\,\frac{\sigma}{||\sigma||^2} + \frac{\sigma^T}{||\sigma||}\,\xi$$

$$\leq a_{21}||z|| + a_{22}||\sigma|| - \mu + \eta\,.$$

Similarly,

$$\frac{d}{dt}\,||z|| = \frac{z^T\dot{z}}{||z||} = z^T A_{11}\,\frac{z}{||z||} + \frac{z^T}{||z||}\,A_{12}\sigma$$

$$\leq -\lambda||z|| + a_{12}||\sigma||\,.$$

Here we used the fact that

$$A_{11} = \mathrm{diag}\,\{\lambda_1\,,\,\dots\,,\,\lambda_{n-m}\},\qquad \lambda_i < 0.$$

□

We next give a region of asymptotic stability with sliding. From now on we absorb $\eta$ into our design parameter $\mu$. Observe that in the region $\{(z,\sigma)\ |\ a_{21}||z|| + a_{22}||\sigma|| < \mu\}$ we have $\frac{d}{dt}\,||\sigma|| \leq 0$. Hence a trajectory reaching the switching surface $\Omega$ from this region cannot leave the switching surface. One would expect that the intersection of this region with the switching surface is a sliding mode domain and indeed this is the case. Let

$$\Sigma_1 = \{(z,\sigma)\ |\ ||\sigma|| < \frac{\lambda\mu}{a_{12}a_{21} + \lambda a_{22}}\,,\ z\in\mathbb{R}^{n-m}\,,\ \sigma\in\mathbb{R}^m\}\,,$$

$$\Sigma_2 = \{(z,\sigma)\ |\ a_{21}||z|| + a_{22}||\sigma|| < \mu\,,\ z\in\mathbb{R}^{n-m}\,,\ \sigma\in\mathbb{R}^m\}\,,$$

and

$$\Sigma = \Sigma_1 \cap \Sigma_2\,.$$

**Theorem 3.2.** ([31]) A trajectory that starts in $\Sigma$ stays in $\Sigma$ and reaches the switching surface in finite time, which implies that $\Sigma$ is a region of asymptotic stability with sliding.

**Proof.** Let

$$N_1 = \{(z,\sigma) \mid a_{12}\|\sigma\| \geq \lambda\|z\|\}$$

and

$$N_2 = \{(z,\sigma) \mid a_{12}\|\sigma\| < \lambda\|z\|\}.$$

(see Fig. 2) From the remarks before the theorem, we only need to consider

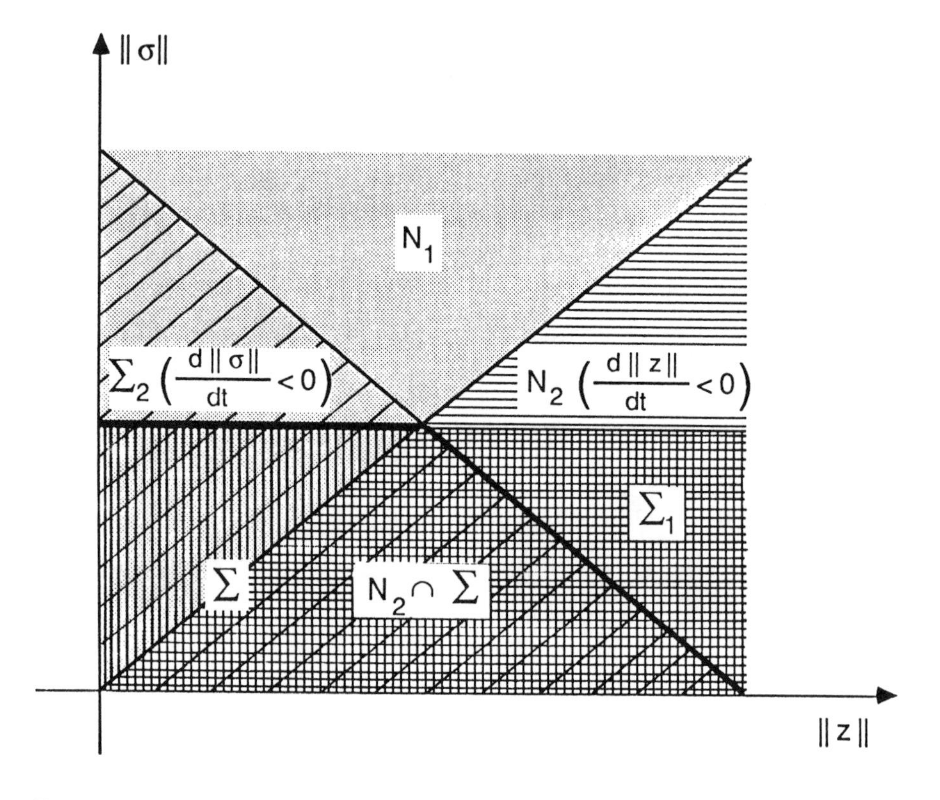


Fig. 2. Illustration of the regions used in the proof of Theorem 3.2.

the portion of the trajectory off the switching surface. By Lemma 3.1 we have $\frac{d\|\sigma\|}{dt} < 0$ in $\Sigma$. Hence if $(z(t_0), \sigma(t_0)) \in \Sigma$, we have $(z(t), \sigma(t)) \in \Sigma_1$ for $t \geq t_0$. For $(z(t), \sigma(t)) \in N_1 \cap \Sigma \subset N_1 \cap \Sigma_1$, we have

$$a_{21}||z|| + a_{22}||\sigma||$$

$$\leq \frac{a_{21}a_{12}}{\lambda}||\sigma|| + a_{22}||\sigma||$$

$$\leq \frac{a_{21}a_{12}}{\lambda}||\sigma(t_0)|| + a_{22}||\sigma(t_0)|| = (\frac{a_{12}a_{21} + \lambda a_{22}}{\lambda})||\sigma(t_0)||$$

$$< \mu - \epsilon \quad \text{for some } \epsilon > 0 .$$

Thus we can conclude that a trajectory $(z(t), \sigma(t))$ can leave $\Sigma$ only through $N_2 \cap \Sigma$. However we have $\frac{d||\sigma||}{dt} < 0$ in $\Sigma$ and $\frac{d||z||}{dt} < 0$ in $N_2$, and hence $a_{21}||z|| + a_{22}||\sigma||$ is a decreasing function in $N_2 \cap \Sigma$. Therefore a trajectory cannot leave $N_2 \cap \Sigma$. Hence a trajectory which starts in $\Sigma$ stays in $\Sigma$.

Suppose $(z(t), \sigma(t)) \in N_1 \cap \Sigma$ for $t_1 \leq t \leq t_2$. Then we have

$$\frac{d||\sigma||}{dt} \leq a_{21}||z|| + a_{22}||\sigma|| - \mu$$

$$\leq \left(\frac{a_{12}a_{21}}{\lambda} + a_{22}\right)||\sigma|| - \mu .$$

Let $k = \frac{a_{12}a_{21}}{\lambda} + a_{22}$. Since $||\sigma||$ is decreasing in $\Sigma$, we have

$$\frac{d||\sigma||}{dt} \leq k\,||\sigma(t_1)|| - \mu .$$

Hence

$$||\sigma(t_1)|| - ||\sigma(t_2)|| \geq (\mu - k||\sigma(t_1)||)(t_2 - t_1)$$

$$\geq (\mu - k\,||\sigma(t_0)||)(t_2 - t_1) .$$

We conclude that a trajectory cannot spend an infinite amount of time in $N_1 \cap \Sigma$ with $||\sigma(t)|| > 0$.

We claim that if $(z(t_1), \sigma(t_1)) \in N_2 \cap \Sigma$, $(z(t_2), \sigma(t_2)) \in N_2 \cap \Sigma$, and $t_1 < t_2$, then $||z(t_2)|| < ||z(t_1)||$. By Lemma 3.1, the claim is clear if $(z(t), \sigma(t)) \in N_2 \cap \Sigma$ for $t_1 \leq t \leq t_2$. Otherwise, suppose $(z(t), \sigma(t)) \in N_2 \cap \Sigma$ for $t_1 < t \leq T_1$, $T_2 \leq t < t_2$, and $(z(T_1), \sigma(T_1))$,

$(z(T_2), \sigma(T_2))$ are on the boundary of $N_2 \cap \Sigma$. Therefore

$$\frac{\|z(T_1)\|}{\|\sigma(T_1)\|} = \frac{\|z(T_2)\|}{\|\sigma(T_2)\|} .$$

Since $\|\sigma(t)\|$ decreases in $\Sigma$, we have $\|z(T_2)\| < \|z(T_1)\|$ and we can conclude that $\|z(t_2)\| < \|z(t_1)\|$ as in the case where the whole segment is in $N_2 \cap \Sigma$.

Hence if $(z(t_0), \sigma(t_0)) \in N_2 \cap \Sigma$, we have for $t \geq t_0$,

$$\frac{d\|\sigma\|}{dt} \leq a_{21} \|z(t_0)\| + a_{22}\|\sigma(t_0)\| - \mu < 0 .$$

Therefore, if $(z(t), \sigma(t)) \in N_2 \cap \Sigma$ for $t_1 \leq t \leq t_2$ then

$$\|\sigma(t_1)\| - \|\sigma(t_2)\| \geq (\mu - a_{21}\|z(t_0)\| - a_{22}\|\sigma(t_0)\|)(t_2 - t_1).$$

We can conclude that a trajectory cannot spend an infinite amount of time in $N_2 \cap \Sigma$ with $\sigma(t) \neq 0$.

Thus we must reach the switching surface $\sigma = 0$ in finite time if we start in $\Sigma$.

□

From the above, we can give explicit estimates of the time it takes to reach the switching surface starting in $\Sigma$.

**Corollary 3.3.** Let

$$\beta = \min\left\{\mu - (\frac{a_{12} a_{21}}{\lambda} + a_{22})\|\sigma(t_0)\|,\ \mu - a_{21}\|z(t_0)\| - a_{22}\|\sigma(t_0)\|\right\}.$$

Starting at $(z(t_0), \sigma(t_0)) \in \Sigma$, we must reach $\sigma = 0$ in

$$t \leq t_0 + \frac{\|\sigma(t_0)\|}{\beta} .$$

**Example 1.** Consider the following system:

$$\dot{x} = \begin{bmatrix} 3 & 1 & 1 \\ -6 & 1 & 0 \\ 0 & 0 & 3 \end{bmatrix} x + \begin{bmatrix} 0 & 0 \\ 1 & 1 \\ 0 & 1 \end{bmatrix} (u + \xi)$$

with $||u|| \leq 10 + \eta$. Suppose the desired eigenvalue of the reduced order system is $-3$ with corresponding eigenvector $w_1 = \begin{bmatrix} 1 \\ -6 \\ 0 \end{bmatrix}$. One can check that

$$W^g = w_1^g = [1\ 0\ 0], \quad S = B^g = \begin{bmatrix} 6 & 1 & -1 \\ 0 & 0 & 1 \end{bmatrix}$$

satisfy (3.1). The transformation matrix is

$$M = \begin{bmatrix} 1 & 0 & 0 \\ 6 & 1 & -1 \\ 0 & 0 & 1 \end{bmatrix}.$$

The system in the new coordinates has the form:

$$\dot{z} = -3z + \sigma$$

$$\dot{\sigma} = \begin{bmatrix} -30 \\ 0 \end{bmatrix} z + \begin{bmatrix} 7 & 10 \\ 0 & 1 \end{bmatrix} \sigma - (10 + \eta) \frac{\sigma}{||\sigma||} + \xi\,.$$

In this case, $\mu$=10, $\lambda$=3, $a_{12}$=1, $a_{21}$=30, $a_{22}$=12.23. Hence

$$\Sigma = \{(z,y) \mid ||y|| < \frac{30}{66.69},\ 30||z|| + 12.23||y|| < 10\}\,.$$

Using a Lyapunov function argument, we can give another region of asymptotic stability.

**Theorem 3.4.** ([31]) A region of asymptotic stability of the system is

$$\mathscr{R} = \{(z,\sigma) \mid \alpha||z|| + \beta||\sigma|| < \mu\}$$

where $\beta = \frac{1}{2}\left[(a_{22}-\lambda) + \sqrt{(a_{22}+\lambda)^2+4a_{12}a_{21}}\right]$ and $\alpha = \beta a_{21}/(\beta+\lambda)$.

**Proof.** Let V be the positive definite function defined by

$$V(z,\sigma) = \alpha||z|| + \beta||\sigma|| .$$

The Lyapunov derivative is

$$\dot{V}(z,\sigma) = \alpha\frac{z^T\dot{z}}{||z||} + \beta\frac{\sigma^T\dot{\sigma}}{||\sigma||} .$$

By Lemma 3.1, we have

$$\dot{V}(z,\sigma) \leq \alpha(-\lambda||z||+a_{12}||\sigma||) + \beta(a_{21}||z||+a_{22}||\sigma||-\mu)$$

$$= (-\lambda\alpha+\beta a_{21})||z|| + (\alpha a_{12}+a_{22}\beta)||\sigma|| - \mu\beta .$$

Using the values of $\alpha$ and $\beta$, we have

$$\dot{V}(z,\sigma) \leq \beta[\alpha||z|| + \beta||\sigma|| - \mu] .$$

The right hand side is less than 0 in $\mathscr{R}$. This finishes the proof if $z(t)\neq 0$ and $\sigma(t)\neq 0$.

Otherwise observe that $\beta \geq a_{22}$ and $\alpha \leq a_{21}$. By Lemma 3.1 we have on $\mathscr{R}\bigcap\{z=0\}$

$$\frac{d||\sigma||}{dt} \leq a_{22}\ ||\sigma|| - \mu \leq \beta\ ||\sigma|| - \mu < 0$$

and on $\mathscr{R}\bigcap\{\sigma=0\}$

$$\frac{d||z||}{dt} \leq -\lambda\ ||z|| < 0 .$$

Hence $V(z,\sigma)$ decreases on the critical surfaces also and we are done.

□

As a consequence of Theorems 3.2 and 3.4 we give a new region of asymptotic stability with sliding.

**Theorem 3.5.** ([31]) Let

$$\mathscr{R}_1 = \{(z,\sigma)\ |\alpha||z|| + \beta||\sigma|| < \mu,\ ||\sigma|| \geq \frac{\lambda}{a_{12}}||z||\}$$

$$\bigcup\ \{(z,\sigma)\ |a_{21}||z|| + a_{22}||\sigma|| < \mu,\ ||\sigma|| \leq \frac{\lambda}{a_{12}}||z||\} .$$

Then $\mathscr{R}_1$ is a region of asymptotic stability with sliding.

**Proof.** We use the same notation as in Theorems 3.2 and 3.4. Observe that

$$\mathscr{R}_1 = \Sigma \bigcup (\mathscr{R} \bigcap N_1)\,.$$

For a trajectory that starts in $\mathscr{R} \bigcap N_1$ to reach the switching surface $\{\sigma{=}0\}$, it must pass through $\Sigma$, which is a region of asymptotic stability with sliding. See Fig. 3.

□

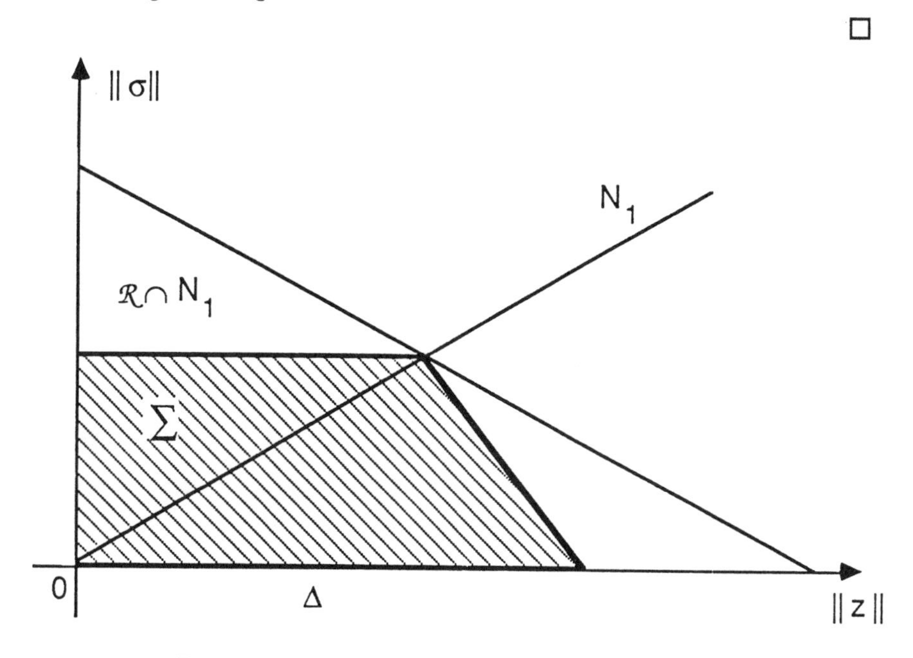


Fig. 3. Illustration of the proof of Theorem 3.5.

**Example 1** (continued). We have found

$$\Sigma = \{(z,\sigma)\,|30||z|| + 12.23||\sigma|| < 10,\ ||\sigma|| < 0.45\}\,.$$

We have $\alpha{=}25.2$ and $\beta{=}15.8$. Therefore

$$\mathscr{R} = \{(z,\sigma)\,|25.2||z|| + 15.8||\sigma|| < 10\}$$

and

$$\mathscr{R}_1 = \Sigma \bigcup (\mathscr{R} \bigcap N_1)\,.$$

Recall that $N_1 = \{(z,\sigma)\,|\,||\sigma|| > 1.34||z||\}$.

## 4. State Observation of Uncertains Systems

Consider the system modeled by

$$\dot{x} = Ax + B(u + \xi)$$
$$y = Cx .$$

Our goal in this Section is to construct an observer, that is, an estimator of the state. Let $\bar{x}$ be the estimate of x obtained from the observer. We denote the error by e(t), that is,

$$e(t) = \bar{x}(t) - x(t) .$$

The observability of (A,C) implies the existence of a matrix K so that $A - KC$ has prescribed eigenvalues in the open left half plane, and this implies that for any $Q = Q^T > 0$, we can solve

$$(A - KC)^T P + P(A - KC) = -Q$$

with $P = P^T > 0$. We choose Q, if possible, so that for some $F \in \mathbb{R}^{m \times p}$ we have

$$FC = B^T P .$$

We need this technical condition for the construction of the observer and it has a nice system theoretic interpretation [22], [23]. If such Q, F exist for some F then the function

$$G(s) = FC[sI - A + KC]^{-1} B$$

is positive real. Recall that a function G(s) is positive real if G(s) has elements analytic in the open right half plane and $G(s) + G^*(s) \geq 0$ in the open right half plane. Here the superscript asterisk indicates the conjugate transpose.

We define the function

$$E(e,\rho) = \begin{cases} \dfrac{BFCe}{\|FCe\|}\rho & \text{for } FCe \neq 0 \\ 0 & \text{for } FCe = 0 . \end{cases}$$

We now introduce the observer ([18])

$$\dot{\overline{x}} = (A - KC)\overline{x} - E(e,\rho) + Ky + Bu .$$

**Theorem 4.1** If $\rho \geq \eta$, then the function $\overline{x}$ is an asymptotic estimate of the state x, that is,

$$\lim_{t\to\infty} e(t) = \lim_{t\to\infty} (\overline{x}(t) - x(t)) = 0 .$$

**Proof.** It is clear that the error satisfies, for $FCe \neq 0$,

$$\dot{e} = (A - KC)e - \frac{BFCe}{||FCe||}\rho - B\xi .$$

Consider the positive definite function

$$V(e) = e^T Pe .$$

The Lyapunov derivative is

$$\dot{V}(e) = e^T[(A - KC)^T P + P(A - KC)]e$$

$$- 2\, e^T \frac{PBFCe}{||FCe||}\rho - 2\, e^T PB\xi .$$

Using the fact that $FC = B^T P$, we have

$$\begin{aligned} \dot{V}(e) &= -\, e^T Qe - 2\, ||FCe||\rho - 2e^T C^T F^T \xi \\ &\leq -\, e^T Qe - 2\, ||FCe||\rho + 2||FCe||\, \eta \\ &< -\, e^T Qe < 0 \quad \text{for} \quad \rho \geq \eta . \end{aligned}$$

For $FCe = 0$, we have

$$\dot{V}(e) = -\, e^T Qe < 0 .$$

Therefore

$$\lim_{t\to\infty} e(t) = 0 .$$

□

We now show that the error actually tends to zero exponentially. Let $P = H^T H$, where H is nonsingular. Then from the proof, we have

$$\frac{1}{2}\frac{d}{dt}\|He\|^2 \le -e^T Q e \le -\lambda_{\min}(Q)\|e\|^2$$

$$\le -\lambda_{\min}(Q)\frac{\|He\|^2}{\|H\|^2} = -\frac{\lambda_{\min}(Q)}{\lambda_{\max}(P)}\|He\|^2 .$$

This shows that the error decays exponentially in the coordinates He and hence in all coordinates. Interesting studies on the ratio $\lambda_{\min}(Q)/\lambda_{\max}(P)$ can be found in [21]. It is shown that $\lambda_{\min}(Q)/\lambda_{\max}(P)$ achieves a maximum when Q is the identity matrix.

We next investigate the sliding properties of the error trajectories with respect to the naturally defined switching surface $\{e \mid FCe = 0\}$, which is the same as $\{e \mid B^T P e = 0\}$. Let $\phi(e) = B^T P e = FCe$. Note that we can write $\dot{e}$ as

$$\dot{e} = (A - KC)e - \frac{B\phi}{\|\phi\|}\rho - B\,\xi .$$

We thus have

$$\frac{1}{2}\frac{d}{dt}\|\phi\|^2 = \phi^T\dot{\phi} = \phi^T B^T P\dot{e}$$

$$= \phi^T B^T P(A - KC)e - \frac{\phi^T B^T P B\phi}{\|\phi\|}\rho - \phi^T B^T P B\xi$$

$$\le \|\phi\|\,\|B^T P(A - KC)e\| - \lambda_{\min}(B^T P B)\|\phi\|\rho + \lambda_{\max}(B^T P B)\|\phi\|\eta .$$

So $\phi^T\dot{\phi} < 0$ if

$$\frac{\lambda_{\max}(B^T P B)}{\lambda_{\min}(B^T P B)}\eta + \frac{\|B^T P(A - KC)e\|}{\lambda_{\min}(B^T P B)} < \rho .$$

In special cases where $PBB^T P(A - KC) + (A - KC)^T PBB^T P$ is negative semidefinite, then

$$\phi^T B^T P(A - KC)e$$

$$= \frac{1}{2}e^T\{PBB^T P(A - KC) + (A - KC)^T PBB^T P\}e \le 0 .$$

Therefore in this case we have $\phi^T\dot{\phi} < 0$ if

$$\frac{\lambda_{\max}(B^TPB)}{\lambda_{\min}(B^TPB)}\,\eta < \rho\,.$$

Let

$$\gamma_\rho = \frac{1}{\|B^TP(A-KC)\|}\,\{\lambda_{\min}(B^TPB)\rho - \lambda_{\max}(B^TPB)\eta\}\,.$$

We choose the design parameter $\rho$ so that $\gamma_\rho > 0$. It is then clear that $\phi^T\dot{\phi} < 0$ in the ball

$$B_\rho = \{e \mid \|e\| < \gamma_\rho\}\,.$$

From the proof of Theorem 4.1, we know that the sets $\{e \mid e^TPe < R\}$ are regions of asymptotic stability. Let $R_\rho = \sup\,\{R \mid \{e \mid e^TPe < R\} \subset B_\rho\}$. Therefore we have

**Theorem 4.2.** The domain

$$D = \{e \mid e^TPe < R_\rho\}$$

is a region of asymptotic stability with sliding and

$$\Delta = \{e \mid \phi(e) = 0\} \cap \{e \mid e^TPe < R_\rho\}$$

is a sliding domain for the error trajectories.

We have also the following theorem concerning the hitting time on the switching surface by an error trajectory.

**Theorem 4.3.** A trajectory starting in $D = \{e \mid e^TPe < R_\rho\}$ will reach the switching surface $\{e \mid \phi(e) = 0\}$ in finite time.

**Proof.** If $e(t_1) \in D \cap \{e \mid \phi(e) = 0\}$, then $e(t) \in D \cap \{e \mid \phi(e) = 0\}$ for $t \geq t_1$. Hence we only need to consider the case when $\phi(e) \neq 0$. Suppose $e(t_0) \in D \backslash \{e \mid \phi(e) = 0\}$. From the proof of Theorem 4.2 we have

$$\frac{d\|\phi\|}{dt} = \frac{\phi^T\dot{\phi}}{\|\phi\|} \leq \|B^TP(A-KC)\|\;\{\|e\| - \gamma_\rho\}\,.$$

However $e(t_0) \in D$ implies that $e^T(t)P\,e(t) < R_\rho - \epsilon_1$ for some $\epsilon_1 > 0$ and $t \geq t_0$. By the definition of $R_\rho$, we have $\|e(t)\| < \gamma_\rho - \epsilon_2$ for some

$\epsilon_2 > 0$ and $t \geq t_0$. Therefore

$$\frac{d\|\phi\|}{dt} \leq -\epsilon_2 \|B^T P(A-KC)\| \triangleq -\epsilon \,.$$

It is then easy to see that for $t_2 > t_1$, we have

$$\|\phi(e(t_1))\| - \|\phi(e(t_2))\| \geq \epsilon(t_2 - t_1) \,.$$

Therefore $\phi(e(t))$ must be zero in finite time.

□

## 5. Observer-Controller Synthesis

In the previous two sections we discussed controllers and observers separately. In this Section, the observer will be used to provide an estimate of the state as the input to the controller, and in this case, one would not expect the state to stay on the switching surface. Instead, the state will "oscillate" about the surface. Our goal is to show that the estimate $\bar{x}$ from the observer will stay on the surface $\{\bar{x} \mid \sigma(\bar{x}) = 0\}$. Since we know that $\bar{x}$ tends to x, we have x close to the switching surface when $\bar{x}$ is sliding.

Our system and observer have the form:

$$\dot{x} = Ax + Bu + B\xi$$
$$y = Cx$$

$$\dot{\bar{x}} = (A-KC)\bar{x} - E(e,\rho) + Ky + Bu \,.$$

To make $\bar{x}$ slide, we need $S\bar{x} = 0$ and $S\dot{\bar{x}} = 0$, and more explicitly,

$$0 = S\dot{\bar{x}} = S(A-KC)\bar{x} - SE(e,\rho) + SKC(\bar{x} - e) + SBu \,.$$

Solving for u, we have

$$u_{eq} = -(SB)^{-1}SA\bar{x} + (SB)^{-1}SE(e,\rho) + (SB)^{-1}SKCe.$$

Substituting $u_{eq}$ into the observer equation yields

$$S\bar{x} = 0 \,,$$

$$\dot{\bar{x}} = [I - B(SB)^{-1}S]\,(A\bar{x} - E(e,\rho) - KCe)\ ,$$

and

$$\dot{e} = (A - KC)e - E(e,\rho) - B\xi\ ,$$

which describes the combined system in sliding. To find a sliding mode domain for $\bar{x}$, we compute

$$\begin{aligned}\frac{1}{2}\frac{d}{dt}\,||S\bar{x}||^2 &= \bar{x}^T S^T S\dot{\bar{x}} \\ &= -\,\mu||S\bar{x}|| + \bar{x}^T S^T S(A\bar{x} - KCe - E(e,\rho)) \\ &\le ||S\bar{x}||\,(||SA\bar{x}|| - \mu + ||SKCe|| + \rho)\ .\end{aligned}$$

Here we used the fact that $SB = I_m$.

In real physical systems, one expects to know the bound on the error. So we assume that $||SKC\,e(t)|| \le \beta$ for $t \ge t_0$. Hence we have

$$\frac{1}{2}\frac{d}{dt}\,||S\bar{x}||^2 \le ||S\bar{x}||\,\{||SA\bar{x}|| - \mu + \rho + \beta\}\ .$$

Therefore if the controller gain $\mu$ is greater than $\rho + \beta$, then a sliding mode domain is given by

$$\{\bar{x} \mid S\bar{x} = 0\} \cap \{\bar{x} \mid ||SA\bar{x}|| < \mu - \rho - \beta\}\ .$$

## 6. Boundary Layer Controllers and Observers

Previously we had assumed that the controller and the observer can switch with infinitely high frequency. In practice, very high frequencies may excite unmodeled high frequency modes and degrade the performance of the controller. To circumvent this problem, boundary layer controllers and observers can be ultilized ([16], [18], [24], [29]). The boundary layer version of our controller takes the form

$$u(\sigma(x)) = \begin{cases} -\mu \dfrac{\sigma(x)}{\|\sigma(x)\|} & \text{for} \quad \|\sigma(x)\| \geq \epsilon_c \\ \\ -\mu \dfrac{\sigma(x)}{\epsilon_c} & \text{for} \quad \|\sigma(x)\| < \epsilon_c . \end{cases}$$

The boundary layer observer has the same structure as before except we now use

$$E(e,\rho) = \begin{cases} \dfrac{BFCe}{\|FCe\|}\rho & \text{for} \quad \|FCe\| \geq \epsilon_0 \\ \\ \dfrac{BFCe}{\epsilon_0}\rho & \text{for} \quad \|FCe\| < \epsilon_0 . \end{cases}$$

We need the following definitions ([**16**]).

**Definition 6.1.** The ball $B_\delta = \{x \mid \|x\| \leq \delta\}$ is uniformly stable if for all $\eta > \delta$, there is $r(\eta) > 0$ so that if $\|x(t_0)\| \leq r(\eta)$, then $\|x(t)\| \leq \eta$ for $t \geq t_0$.

**Definition 6.2.** A system is uniformly bounded if for any given $r > 0$, there exist $d(r)$ so that for each solution x which satisfies $\|x(t_0)\| \leq r$, we have $\|x(t)\| \leq d(r) < \infty$ for $t \geq t_0$.

**Definition 6.3.** A system is uniformly ultimately bounded with respect to the ball B if for every $r > 0$, there is $T(r) \geq t_0$ such that for each solution $x(t)$ with $\|x(t_0)\| \leq r$, we have $x(t) \in B$ for $t \geq T(r)$.

With the boundary layer controller, the closed-loop system may only be uniformly ultimately bounded with respect to some ball depending on $\epsilon_c$ and similarly the error trajectories of the boundary layer observer may only be uniformly ultimately bounded with respect to some ball depending on $\epsilon_0$. The precise formulation, with the same notation as in Section 3, is

**Theorem 6.1.** A trajectory starting in

$$\Sigma_3 = \mathscr{R}_1 \backslash \{(\sigma, z) \, | a_{21} || z || > \mu - a_{22} \epsilon_c\}$$

will reach the region

$$\mathscr{R}_2 = \{(\sigma, z) \, | || \sigma || < \epsilon_c, \; \lambda || z || < a_{12} \epsilon_c\}$$

in finite time and stay in $\mathscr{R}_2$ thereafter. We call such regions attractive regions.

**Proof.** Let

$$L = \{(\sigma, z) \, | || \sigma || < \epsilon_c\}$$

be the boundary layer. From Lemma 3.1, we know that $\frac{d||\sigma||}{dt} < 0$ in $\Sigma_3 \backslash L$, and in $N_2 \cap \Sigma_3$ we have $\frac{d||z||}{dt} < 0$ (see Fig. 4). The conclusion now follows from Theorem 3.5.

□

Observe that the boundary layer in $\Sigma_3$ is also an attractive region and that if the starting points are uniformly bounded away from the boundary of $\Sigma_3$, the trajectories will reach $\mathscr{R}_2$ uniformly. Note also that, as expected, as $\epsilon_c \rightarrow 0$ we have $\Sigma_3 \rightarrow \mathscr{R}_1$.

For the error of the boundary layer observer, we show that the trajectories can be forced into a ball whose radius depends on $\epsilon_0$ and tends to zero with $\epsilon_0$. Let

$$V(e) = e^T P e$$

be as in Section 4. From that Section, we have

$$\dot{V}(e) = -e^T Q e - 2e^T P E(e, \rho) - 2e^T P B \varsigma \, .$$

From the definition of the newly defined $E(e, \rho)$, we have

$$\dot{V}(e) \leq \begin{cases} -e^T Q e & \text{for} \quad ||FCe|| \geq \epsilon_0 \\ -e^T Q e + 2\epsilon_0 \eta & \text{for} \quad ||FCe|| < \epsilon_0 \end{cases}$$

For the second inequality, we used the fact that

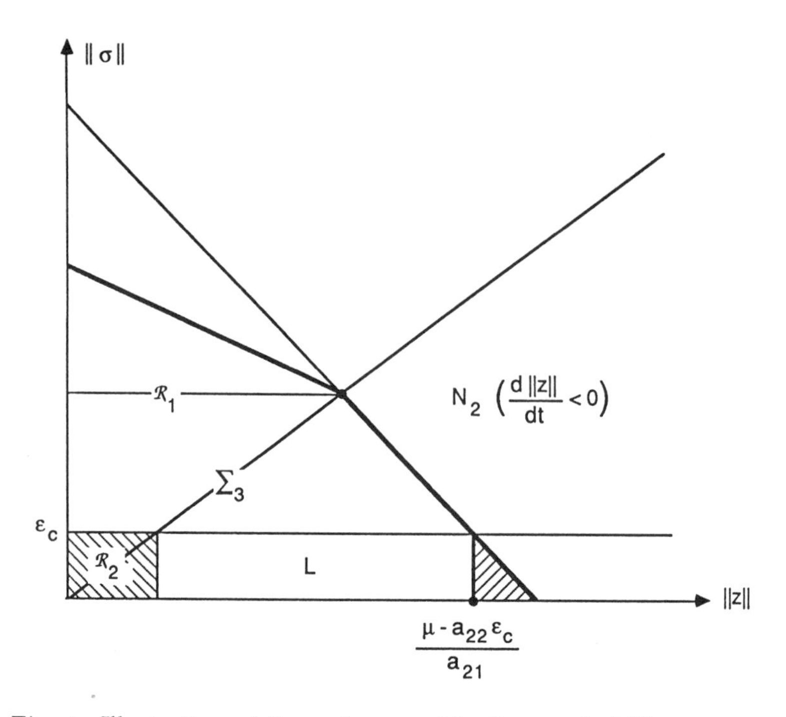


Fig. 4. Illustration of the regions used in the proof of Theorem 6.1.

$$e^T PBFCe = ||FCe||^2 \geq 0 .$$

Since

$$-e^T Qe + 2\epsilon_0 \eta \leq -\lambda_{min}(Q)||e||^2 + 2\epsilon_0 \eta ,$$

we can conclude that $\dot{V}(e) < 0$ for

$$||e|| > \sqrt{\frac{2\epsilon_0 \eta}{\lambda_{min}(Q)}} \triangleq R .$$

Observe that as $\epsilon_0 \longrightarrow 0$ we have $R \longrightarrow 0$. Let **U** be the smallest elliptical region of the form $\{e \,|e^T Pe < r\}$ which contains $B_R$. Since $\dot{V}(e) < 0$ outside of $B_R$, any trajectory that starts in **U** stays in **U**. We can further conclude that **U** is an attractive region (see Fig. 5). We can summarize the above discussion in

**Theorem 6.2.** The error of the boundary layer observer is uniformly ultimately bounded.

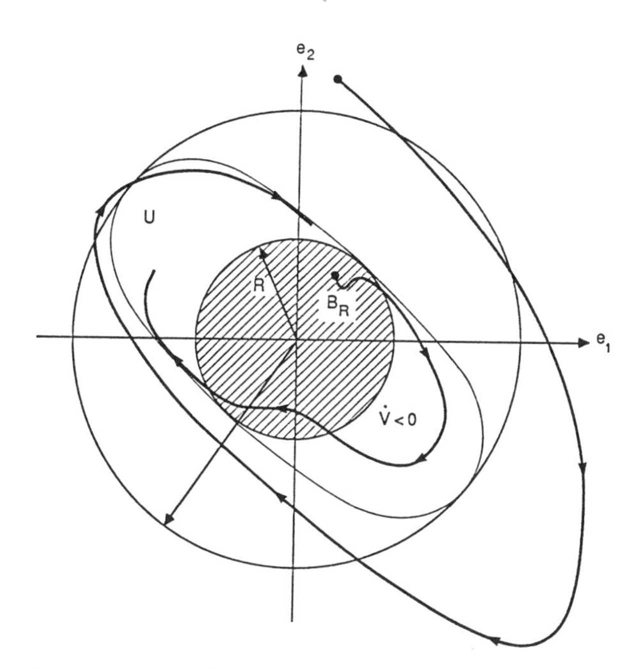


Fig. 5. Two-dimensional illustration of the regions used in the discussion of the boundary layer observer.

**Example 2.**

Consider a dynamical system modeled by the equations

$$\dot{x} = \begin{bmatrix} -x_2 \\ x_1 - 2\sin x_1 \end{bmatrix} + \begin{bmatrix} 0 \\ 1 \end{bmatrix} u ,$$

$$y = \begin{bmatrix} -1 & 1 \end{bmatrix} x .$$

We represent the equations describing the system as

$$\dot{x} = \begin{bmatrix} 0 & -1 \\ 1 & 0 \end{bmatrix} x + \begin{bmatrix} 0 \\ 1 \end{bmatrix} (u - 2\sin x_1) ,$$

$$y = \begin{bmatrix} -1 & 1 \end{bmatrix} x .$$

The system can then be viewed as an uncertain dynamical system with the uncertainty

$$\xi = -\ 2\sin x_1\ .$$

Note that $\|\xi\| \le 2 = \eta$. We design a variable structure observer for this system. We require the eigenvalues of the matrix A-KC to be $-1$ and $-2$. If we take $K=[-1\ \ 2]^T$ then

$$A - KC = \begin{bmatrix} -1 & 0 \\ 3 & -2 \end{bmatrix}.$$

Let

$$Q = \begin{bmatrix} 10 & -6 \\ -6 & 4 \end{bmatrix}.$$

Solving the Lyapunov matrix equation $(A-KC)^T P + P(A-KC) = -Q$ yields

$$P = \begin{bmatrix} 2 & -1 \\ -1 & 1 \end{bmatrix}.$$

Note that $PB=C^T$ and we can take $F=1$. Hence the observer has the form

$$\dot{\bar{x}} = \begin{bmatrix} -1 & 0 \\ 3 & -1 \end{bmatrix}\bar{x} - \begin{bmatrix} 0 \\ 1 \end{bmatrix}\frac{(-e_1+e_2)}{\|-e_1+e_2\|}\,\rho + \begin{bmatrix} -1 \\ 2 \end{bmatrix}y + \begin{bmatrix} 0 \\ 1 \end{bmatrix}u\ .$$

The equation governing the error is

$$\dot{e} = \begin{bmatrix} -1 & 0 \\ 3 & -2 \end{bmatrix}e - \begin{bmatrix} 0 \\ 1 \end{bmatrix}\frac{(-e_1+e_2)}{\|-e_1+e_2\|}\,\rho - \begin{bmatrix} 0 \\ 1 \end{bmatrix}\xi\ .$$

For a naturally defined switching surface $\{e\,|-e_1+e_2=0\}$ to be globally attractive it is enough to choose $\rho > \eta + \|B^T PAe\|$. We now compute $\epsilon_0$ so that the ball $\{e\,|\,\|e\| \le 1\}$ is attractive for the error of the boundary layer observer. Let **U** be the largest elliptical region of the form $\{e\,|\,e^T\begin{bmatrix} 2 & -1 \\ -1 & 1 \end{bmatrix}e \le \tau\}$ contained in the ball of radius one. Then it is well known that the largest ball contained in **U** has radius $R = \sqrt{\lambda_{min}(P)/\lambda_{max}(P)}$. Hence

$$\epsilon_0 = \frac{\lambda_{\min}(Q)R^2}{2\eta} = 0.011 \ .$$

The function $E(e,\rho)$ for the boundary layer observer is

$$E(e,\rho) = \begin{cases} \begin{bmatrix} 0 \\ 1 \end{bmatrix} \dfrac{-e_1+e_2}{||e_1+e_2||}\rho & \text{for} \quad ||-e_1+e_2|| \geq 0.011 \\ \begin{bmatrix} 0 \\ 1 \end{bmatrix} \dfrac{-e_1+e_2}{0.011}\rho & \text{for} \quad ||-e_1+e_2|| < 0.011 \end{cases}$$

We can let $\rho=2$ in the observer.

## 7. Conclusions

In this chapter we proposed an approach to the control and observation of a class of uncertain systems. This approach can also be applied to a class of nonlinear systems with matched nonlinearities. For other approaches to the control aspect of uncertain or nonlinear systems see [3], [4], [11], [15], [20], and for observation see [14], [19], [25]-[28], [30].

## References

1. V.I. Utkin, "Sliding Modes and Their Application in Variable Structure Systems," Moscow, Mir Publishers, 1978.

2. V. I. Utkin, "Variable structure systems with sliding modes," IEEE Trans. Automat. Contr., Vol. AC-22, No. 2, pp. 212-222, 1977.

3. K.-K.D. Young, P. V. Kokotović, and V. I. Utkin, "A singular perturbation analysis of high gain feedback systems," IEEE Trans. Automat. Contr., Vol. AC-22, No. 6, pp. 931-938, 1977.

4. V. I. Utkin, "Application of equivalent control method to the systems with large feedback gain," IEEE Trans. Automat. Contr., Vol. AC-23, No. 3, pp. 484-486, 1978.

5. O. M. E. El-Ghezawi, A. S. I. Zinober and S. A. Billings, "Analysis and design of variable structure systems using a geometric approach," Int. J. Control, Vol. 38, No. 3, pp. 657-671, 1983.

6. R. A. DeCarlo, S. H. Żak, and G. P. Matthews, "Variable structure control of nonlinear multivariable systems: A tutorial," Proceedings of the IEEE, Vol. 76, No. 3, pp. 212-232, 1988.

7. W. M. Wonham, "Linear Multivariable Control; A Geometric Approach," New York, Springer Verlag, 1979.

8. A. Ben-Israel, and T. N. E. Greville, "Generalized Inverses," New York, Wiley, 1974.

9. P. V. Kokotović, "A Riccati equation for block-diagonalization of ill-conditioned systems," IEEE Trans. Automat. Contr., Vol. AC-20, No. 6, pp. 812-814, 1975.

10. A. F. Filippov, "Differential equation with discontinuous right-hand sides," American Mathematical Society Translations, Series 2, Vol. 42, pp. 199-231, 1964.

11. H. H. Rosenbrock, "A method of investigating stability," IFAC Proceedings, pp. 590-594, Basel, Switzerland, 1963.

12. S. M. Madani-Esfahani and S. H. Żak, "Variable structure control of dynamical systems with bounded controllers," Proc. 1987 American Control Conf., Minneapolis, MN, pp. 90-95, June 10-12, 1987.

13. D. G. Luenberger, "An introduction to observers," IEEE Trans. Automat. Contr., Vol. AC-16, No. 6, pp. 596-602, Dec. 1971.

14. K-K. D. Young, "Analysis and synthesis of high gain and variable structure feedback systems," Tech. Report DC-7, Coordinated Science Lab., Univ. of Illinois-Urbana, Illinois, Nov. 1977.

15. L. G. Chouinard, J. P. Dauer, and G. Leitmann, "Properties of matrices used in uncertain linear control systems," SIAM J. Control and Optimization, Vol. 23, No. 3, pp. 381-389, May 1985.

16. M. J. Corless and G. Leitmann, "Continuous state feedback guaranteeing uniform ultimate boundedness for uncertain dynamical systems," IEEE Trans. Automat. Contr., Vol. AC-26, No. 5, pp. 1139-1144, Oct. 1981.

17. B. Draženović, "The invariance conditions in variable structure systems," Automatica, Vol. 5, No. 3, pp. 287-295, May 1969.

18. B. L. Walcott and S. H. Żak, "State observation of nonlinear uncertain dynamical systems," IEEE Trans. Automat. Contr., Vol. AC-32, No. 2, pp. 166-170, Feb. 1987.

19. A. J. Krener and W. Respondek, "Nonlinear observers with linearizable error dynamics," SIAM J. Control and Optimization, Vol. 23, No. 2, pp. 197-216, March 1985.

20. A. G. Bondarev, S. A. Bondarev, N. E. Kostyleva, and V. I. Utkin, "Sliding modes in systems with asymptotic state observers," Automation and Remote Control, Vol. 46, No. 6, Pt. 1, pp. 679-684, June 1985.

21. R. V. Patel and M. Toda, "Quantitative measures of robustness for multivariable systems," Proc. Joint Automatic Control Conf. (JACC), pp. TP8-A, San Francisco, CA, 1980.

22. A. R. Galimidi and B. R. Barmish, "The constrained Lyapunov problem and its application to robust output feedback stabilization," IEEE Trans. Automat. Contr., Vol. AC-31, No. 5, pp. 410-419, May 1986.

23. A. Steinberg and M. Corless, "Output feedback stabilization of uncertain dynamical systems," IEEE Trans. Automat. Contr., Vol. AC-30, No. 10, pp. 1025-1027, Oct. 1985.

24. S. Gutman and Z. Palmor, "Properties of min-max controllers in uncertain dynamical systems," SIAM J. Control and Optimization, Vol. 20, No. 6, pp. 850-861, Nov. 1982.

25. D. Bestle and M. Zeitz, "Canonical form observer design for nonlinear time-variable systems," Int. J. Control, Vol. 38, No. 2, pp. 419-431.

26. M. Zeitz, "The extended Luenberger observer for nonlinear systems," Systems & Control Letters, Vol. 9, No. 2, pp. 149-156, Aug. 1987.

27. F. E. Thau, "Observing the state of non-linear dynamic systems," Int. J. Control, Vol. 17, No. 3, pp. 471-479, 1973.

28. S. R. Kou, D. L. Elliott, and T. J. Tarn, "Exponential observers for nonlinear dynamic systems," Information and Control, Vol. 29, No. 3, pp. 204-216, Nov. 1975.

29. J.-J. E. Slotine, J. K. Hedrick, and E. A. Misawa, "On sliding observers for nonlinear systems," Proc. American Control Conf., Seattle, WA, pp. 1794-1800, June 18-20, 1986.

30. B. L. Walcott, M. J. Corless, and S. H. Żak, "Comparative study of non-linear state-observation techniques," Int. J. Control, Vol. 45, No. 6, pp. 2109-2132, 1987.

31. S. M. Madani-Esfahani, S. Hui, and S. H. Żak, "On the estimation of sliding domains and stability regions of variable structure control systems with bounded controllers," submitted for publication.

CONTROL AND DYNAMIC SYSTEMS, VOL. 34

# DISCRETIZATION CHAOS: FEEDBACK CONTROL AND TRANSITION TO CHAOS†

WALTER J. GRANTHAM and AMIT M. ATHALYE

Department of Mechanical and Materials Engineering
Washington State University
Pullman, Washington 99164–2920

## I. INTRODUCTION

A chaotic system is a nonlinear deterministic dynamical system whose behavior is erratic and irregular and so sensitive to small changes in initial conditions that it is impossible to predict precisely the motion of the system. In effect, a chaotic system is a deterministic system exhibiting essentially random motion. This paradox is but one of many fascinating, and in some cases revolutionary, properties associated with chaotic dynamical systems.

Examples of chaos include the path of a lightening bolt, the time between drops from a dripping faucet, the motion of a jetting balloon after release, and many others. At a more basic level chaos has been observed both experimentally and in mathematical models for many different types of otherwise normal dynamical systems. In fact, the existence of chaotic behavior can be found in a wide spectrum of scientific fields, such as economics, engineering, physics, chemistry, and life sciences [1, 2]. Convection of a fluid heated from below [3], stirred chemical reactor systems [4], forced oscillators [5, 6], biological population dynamics [7–9], and Newton's method for finding roots of a polynomial [2] are but a few examples of dynamical systems that can exhibit chaotic behavior.

† *This research was supported by NASA–Ames Research Center under Interchange No. NCA 2–219 and by NSF and Australian Dept. of Science and Technology travel grants.*

Chaos is a fairly recent and very rapidly expanding area of dynamical systems research. In 1892 the French mathematician Poincaré [10] showed that a Hamiltonian system of differential equations for a three–body problem in celestial mechanics had solutions that were extremely sensitive to small changes in initial conditions. In 1963 Lorenz [3] also observed this phenomenon, in connection with his attempts to model weather using a simplified reduction of the Navier–Stokes equations. In the past ten years a considerable body of knowledge about chaotic dynamical systems has been developed, including several excellent texts dealing with chaos [11–15].

Current knowledge about chaotic systems is by no means complete and many important areas remain to be studied. One of these is the problem of developing feedback controllers for chaotic systems. Research on the control of chaotic systems is in its infancy, with results confined mostly to the occurrence of chaos in discrete–time control systems. Ushio and Hirai [16] investigated sufficient conditions for chaos in a class of linear sampled–data control systems with nonlinear controllers. They showed that chaotic motion occurs for a sufficiently large sampling time interval. Chaotic behavior in feedback systems was also reported by Cook [17,18]. Rubio, *et. al.* [19] investigated chaotic motion in an adaptive control system by studying its power spectra.

In this article we present an introduction to some aspects of chaotic dynamical systems, from the viewpoint of control systems engineering, and we explore some topics related to the design of feedback controllers for chaotic systems. To focus and isolate our discussions, we will be concerned with two example situations in which chaotic motion is caused solely by converting a continuous–time dynamical system into a simpler discrete-time system. That is, we will consider chaos that occurs in a discrete–time analog for a continuous–time system for cases where the continuous–time system does not exhibit any chaos at all.

As illustrative examples we will consider two different discretization schemes applied to nonchaotic Lotka–Volterra population models. One discretization scheme is a common method used by population biologists to convert a continuous-time model to an exponential discrete–time model. The other discretization scheme is, in some sense, representative of numerical integration procedures applied to systems of nonlinear differential equations.

We shall see that both discretization procedures can produce chaos where none exists in the corresponding continuous–time system. For the numerical integration procedure, we will also introduce a stabilizing feedback controller and show that asymptotic stability in the continuous–time system is not sufficient to eliminate chaos in the discrete–time system. Preliminary results for the two discretization schemes were presented previously in [20]. In this article we provide a more detailed discussion of the topics of chaos and feedback control and we present additional results, including a study of the transition to chaos as the overall gain of the feedback controller is varied. In addition, we also discuss some practical aspects of the computation of Lyapunov exponents and other measures of chaos applied to feedback control systems, where the right–hand sides of the differential equations do not satisfy the smoothness assumptions usually invoked in dynamical systems theory. For control systems applications, involving such devices as on–off controllers or saturating controllers, these smoothness assumptions often are not satisfied. Failure to account for this basic difference in scope between control systems theory and standard dynamical systems theory can lead to erroneous results concerning whether or not a system is chaotic.

The article is structured as follows. In Section II we discuss continuous-time and discrete–time system models and in Section III we present two different approaches for obtaining discrete–time models from a continuous-time system. In Section IV we discuss several topics related to chaotic dynamical systems. In Section V we develop an "exponential discretization" model for a Lotka-Volterra two–species competition system The discrete-time model is the same one considered by May [7], who showed that it is chaotic for certain parameter values. We prove that the continuous–time system is not chaotic for any model parameter values and we present a state-space simulation of the discrete–time competition model to verify that the discrete–time system can exhibit chaotic behavior. In Section VI we consider a continuous-time Lotka–Volterra prey–predator system and a corresponding discrete–time "pseudo-Euler" numerical integration model used by Peitgen and Richter [2]. Again, we prove that the continuous-time system is nonchaotic for all parameter values and we compare simulation results for the nonchaotic continuous–time model and the chaotic discrete–time model.

In Section VII we develop a saturating state feedback controller for the continuous–time prey–predator system. We prove that this controller yields an asymptotically stable equilibrium point for the continuous–time prey-predator system and that all initial population states get transferred asymptotically to this equilibrium point by the feedback controller. In Section VIII we present a sequence of simulation results for the controller applied to the pseudo–Euler discrete-time integration system for the prey–predator model. These results show that chaos can occur in the discrete–time model even though the continuous–time model is always asymptotically stable. In Section IX we present a detailed study of a transition to chaos for the discrete–time version of the prey–predator system with feedback control. Section X summarizes the results presented in the article.

## II. SYSTEM MODELS

In the modeling, analysis, and control of dynamical systems, various types of system models may be employed, ranging downward in complexity from systems of partial differential equations, to ordinary differential equations, to difference equations, and finally to algebraic equations for equilibrium conditions. Other models that are often employed, but will not be discussed here, include stochastic systems and systems with time delays.

Frequently a system at a higher level of complexity will be approximated by a model at a lower level of complexity. For example, Lorenz's system of three chaotic differential equations [3] is a simplification of the Navier–Stokes partial differential equations for Rayleigh-Bénard convection (for a short derivation, see [13]). Similarly, in biological population systems and other resource management systems, ordinary differential equation models are often approximated by algebraic discrete-time difference equation models. This type of approximation also occurs when numerical integration algorithms such as Runge–Kutta are used to simulate systems of differential equations on a digital computer.

When a simplified approximation model is used, such as ordinary differential equations instead of partial differential equations or difference equations instead of differential equations, and chaos is observed in the simplified model, the question naturally arises as to whether or not chaos would occur in the original system or in a more complex model. The answer is that it may or it may not. Experiments on actual physical systems [21] verify that chaos can occur in Rayleigh–Bénard convection, as well as in other physical systems [2]. On the other hand, for the ordinary differential equation models and their simplified difference equation analogs considered in this article, chaos occurs only in the simplified model.

We will be concerned with a comparison of two types of models for dynamical systems with control inputs. We will suppose that the original system is governed by differential equations and we will consider two different approaches to developing simplified models based on algebraic difference equations.

### A. Continuous–Time Models

A *continuous–time dynamical system* model is a system of $n$ ordinary differential equations of the form

$$\dot{x}(t) = g[x(t)] \ , \tag{1}$$

where $x(t) \in \mathbb{R}^n$ is the state of the system at time $t$, and $(\dot{\ })$ denotes the time derivative $d()/dt$. Except where noted otherwise, we will assume that $g(x)$ is continuous and has continuous first-order partial derivatives for all $x$.

From the theory of ordinary differential equations [22] there exists a unique solution $x(t) = \xi(t,x_0)$ to (1) for each initial state $x_0 = \xi(0,x_0)$. In addition, the solution is a continuous and continuously differentiable function of the initial state $x_0$. For the dynamical systems that we will consider, the solutions $x(t)$ exist for all $t \in (-\infty, +\infty)$.

A *continuous–time control system* model is a system of $n$ differential equations, with control inputs, of the form

$$\dot{x} = f(x,u) \ , \tag{2}$$

where $\boldsymbol{u} \in \Omega \subseteq \mathbb{R}^m$ is the control input vector (to be specified) and $\Omega$ is a given control constraint set. We assume that $\boldsymbol{f}(\boldsymbol{x},\boldsymbol{u})$ is continuous and has continuous first-order partial derivatives for all $\boldsymbol{x} \in \mathbb{R}^n$ and all $\boldsymbol{u} \in \Omega$.

To generate solutions to (2) we first must specify a control input function $\boldsymbol{u}$, either as a closed–loop feedback control $\boldsymbol{u} = \boldsymbol{u}(\boldsymbol{x})$, or as an open-loop control $\boldsymbol{u} = \boldsymbol{u}(t)$. We will be concerned primarily with the case of closed-loop controls. However, for completeness and for subsequent comparison of procedures for computing Lyapunov exponents we will also examine the simpler case of open–loop controls.

If a control input function is chosen as an open–loop control $\boldsymbol{u}(t)$ then (2) becomes a nonautonomous system of ordinary differential equations. We will restrict open–loop controls to be at worst piecewise continuous functions of time. This is sufficient to ensure that solutions $\boldsymbol{x}(t)$ to (2) for $\boldsymbol{u} = \boldsymbol{u}(t)$ exist and that the solutions are continuous and continuously differentiable functions of the initial state. The principal difference between the properties of these solutions and those of autonomous dynamical systems of the form (1) is that for piecewise continuous open-loop controls $\boldsymbol{u}(t)$ the corresponding solutions $\boldsymbol{x}(t)$ to (2) only satisfy (2) "almost everywhere," that is, except at isolated *switching times* where $\boldsymbol{u}(t)$ has a jump discontinuity. Usually at such times the derivative $\dot{\boldsymbol{x}}(t)$ is also discontinuous, so that (2) is not well-defined, and the solution trajectory $\boldsymbol{x}(t)$ plotted in state space may have a *corner*, corresponding to the discontinuous change in the velocity vector $\dot{\boldsymbol{x}}(t)$. This phenomenon poses no real difficulty and solutions are simply pieced together across switching times by using the state at the switching time as the new initial condition for re–starting the solution after the switching time.

We will be concerned primarily with the case of closed–loop feedback controls of the form $\boldsymbol{u} = \boldsymbol{u}(\boldsymbol{x})$, subject to the control constraints $\boldsymbol{u}[\boldsymbol{x}(t)] \in \Omega$. Once such a control has been specified the control system (2) becomes an autonomous continuous–time dynamical system of the form (1), given by

$$\dot{\boldsymbol{x}} = \boldsymbol{g}(\boldsymbol{x}) \triangleq \boldsymbol{f}[\boldsymbol{x},\boldsymbol{u}(\boldsymbol{x})] \ . \tag{3}$$

Note that open–loop controls $\boldsymbol{u}(t)$ can be accommodated in the framework of (3) by adopting the convention that if time $t$ appears explicitly in the differential equations (3) then it is treated as one of the elements of the state vector $\boldsymbol{x}$. For example, $x_n = t$, with the differential equation $\dot{x}_n = 1$.

## B. Discontinuous Feedback Controls

The fundamental difference between a feedback control system (3) and a standard dynamical system (1) is that in general we do not require that $\boldsymbol{u}(\boldsymbol{x})$ be continuous or differentiable. Thus the function $\boldsymbol{g}(\boldsymbol{x})$ in (3) may not be continuous. This possibility will require a redefinition of the notion of a solution for discontinuous differential equations.

In a general control system setting we restrict the class of admissible feedback control functions $\boldsymbol{u}(\boldsymbol{x})$ to those that satisfy the control constraints $\boldsymbol{u}[\boldsymbol{x}(t)] \in \Omega$ with $\boldsymbol{u}(\boldsymbol{x})$ being piecewise continuous and piecewise differentiable. More precisely, we require that there exists a *partition* of state space (a collection of disjoint open sets $\mathcal{R}_i \subseteq \mathbb{R}^n$ with $\cup\{\overline{\mathcal{R}_i}\} = \mathbb{R}^n$, where $(\overline{\ })$ denotes the closure of a set) such that $\boldsymbol{u}(\boldsymbol{x})$ is continuous and continuously differentiable on each subset $\mathcal{R}_i$ of the partition. This allows $\boldsymbol{u}(\boldsymbol{x})$ or its partial derivatives to be discontinuous across *switching surfaces* defined by the boundaries of the partition subsets.

Within each region $\mathcal{R}_i$ the control $\boldsymbol{u}(\boldsymbol{x})$ and the right-hand sides of the differential equations (3) are continuous and differentiable. Thus inside each region $\mathcal{R}_i$ standard solutions exist, as discussed for (1). However, as a trajectory reaches the boundary of $\mathcal{R}_i$ the differential equations become discontinuous and the concept of a "solution" needs to be extended to handle this case.

To define solutions for discontinuous differential equations, we begin by defining the *forward–time derivative* (or simply the "velocity" vector) as

$$\dot{\boldsymbol{x}}(t) \triangleq \lim_{\Delta t \downarrow 0} \frac{\boldsymbol{x}(t + \Delta t) - \boldsymbol{x}(t)}{\Delta t} \tag{4}$$

and letting $\dot{\boldsymbol{x}}(t-)$ and $\dot{\boldsymbol{x}}(t+)$ denote the forward–time velocity vector just before and just after time $t$, respectively. Similarly we let $\boldsymbol{x}(t-)$ and $\boldsymbol{x}(t+)$ denote the "position" vector $\boldsymbol{x}(t)$ before and after time $t$, respectively.

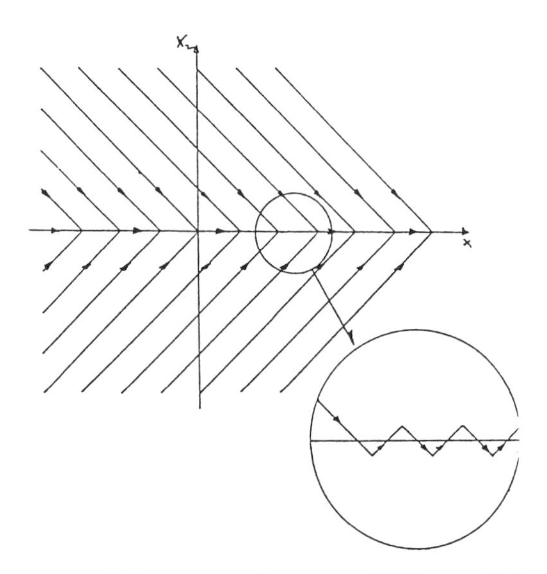

Figure 1. Solutions in the sense of Filippov (Chatter).

A function $x(t)$ is a *solution* to (3) in the sense of Filippov [23] if

$$\dot{x}(t) = \alpha\dot{x}(t-) + (1-\alpha)\dot{x}(t+) \tag{5}$$

for some $\alpha \in [0,1]$, where $\dot{x}(t-) = g[x(t-)]$ and $\dot{x}(t+) = g[x(t+)]$.

Note that if $g(x)$ is continuous at $x = x(t)$ then it follows from (4) that $\dot{x}(t-) = \dot{x}(t+) = g[x(t)]$ and (5) is equivalent to (3). More generally, (5) states that the (forward time) velocity vector must be a convex combination of the velocity vectors just before and after time $t$. The geometry of (5) is illustrated in Figure 1. This figure also illustrates the fact that, unlike solutions to (1) which are unique forward and backward in time, the Filippov solutions to (3) may not be unique, in this case backward in time.

When a solution $x(t)$ to (3) reaches a switching surface, one of two things happens: either $x(t)$ penetrates the switching surface or it moves along the switching surface, as illustrated in Figure 1. To be more specific, assume that a trajectory $x(t)$ penetrates slightly to the other side of the switching surface. Let $\dot{x}(t-) = g[x(t-)]$ and $\dot{x}(t+) = g[x(t+)]$ denote the forward-time velocity vector just before and just after penetrating the switching surface, respectively. If $\dot{x}(t-)$ and $\dot{x}(t+)$ both point to the same side of the switching surface, then $x(t)$ will in fact penetrate the surface and the control $u[x(t)]$ will simply experience a jump discontinuity, as in the case of a piecewise continuous open-loop control $u(t)$. On the other hand, if $\dot{x}(t-)$ points through the switching surface and $\dot{x}(t+)$ points back to the opposite (original) side of the surface, as illustrated in Figure 1, the control will *chatter* back and forth between the values associated with $\dot{x}(t-)$ and $\dot{x}(t+)$ on opposite sides of the switching surface. This repeatedly sends the state $x(t)$ back toward the switching surface, as illustrated in the expanded view in Figure 1, and the solution $x(t)$ will *slide* along the switching surface. The effective velocity vector $\dot{x}(t)$, which is not equal to either of the values $\dot{x}(t-)$ or $\dot{x}(t+)$, is determined by (5), with $\alpha$ chosen so that $\dot{x}(t)$ is tangent to the surface.

Note that along the "sliding mode" trajectory illustrated in Figure 1, the velocity vector $\dot{x}(t)$ given by (5) may not correspond to any of the velocity vectors $\dot{x} = f(x,u)$ that could normally be achieved at $x = x(t)$ by choosing some control vector $u \in \Omega$. This phenomenon should convince the reader that discontinuous differential equations can present difficulties that are not encountered with continuous differential equations.

In the examples that we will consider, the chaotic motion that we encounter will *not* be due to discontinuities. The reason is that the differential equations (3) for the example systems will not have discontinuous right-hand sides. However, some auxiliary differential equations of interest *will* have discontinuous right–hand sides.

Since we have assumed in (3) that $f(x,u)$ and its partial derivatives with respect to $x$ and $u$ are continuous, the function $g(x) = f[x,u(x)]$ will be continuous (differentiable) if $u(x)$ is continuous (differentiable). The feedback control $u(x)$ that we will use for the application considered in this article is a saturating linear feedback control. As a consequence, the control $u(x)$ will be continuous, but only piecewise continuously differentiable. Therefore $g(x)$ will be continuous but its partial derivatives will only be piecewise continuous. Hence the complexities associated with discontinuous differential equations will not occur in the basic feedback control system (3). However, some difficulties will occur in the corresponding linearized state perturbation equations used to compute Lyapunov exponents, since the linearized differential equations will only be piecewise continuous.

## C. Discrete–Time Models

A *discrete–time dynamic system* model is a system of $n$ algebraic equations of the form

$$\boldsymbol{x}(k+1) = \boldsymbol{G}[\boldsymbol{x}(k)] , \tag{6}$$

where $\boldsymbol{x}(k) \in \mathbb{R}^m$ denotes the state at time $t_k = k\Delta t$, $\Delta t$ is a specified time step (not necessarily small), and $k = 0, 1, 2, \ldots$ is the discrete–time index. Except where noted otherwise we will assume that $\boldsymbol{G}(\boldsymbol{x})$ and its first-order partial derivatives are continuous for all $\boldsymbol{x}$.

A *discrete–time control system* model is a system of $n$ algebraic difference equations of the form

$$\boldsymbol{x}(k+1) = \boldsymbol{F}[\boldsymbol{x}(k),\boldsymbol{u}(k)] , \tag{7}$$

where $\boldsymbol{u}(k) \in \Omega \subseteq \mathbb{R}^m$ denotes the control vector at time $t_k = k\Delta t$ and the function $\boldsymbol{F}(\boldsymbol{x},\boldsymbol{u})$ is assumed to be continuous and continuously differentiable for all $\boldsymbol{x} \in \mathbb{R}^n$ and all $\boldsymbol{u} \in \Omega$.

When a feedback control $\boldsymbol{u}(k) = \boldsymbol{u}[\boldsymbol{x}(k)]$ is specified, (7) becomes a discrete–time dynamical system of the form (6), given by

$$\boldsymbol{x}(k+1) = \boldsymbol{G}[\boldsymbol{x}(k)] \triangleq \boldsymbol{F}\{\boldsymbol{x}(k),\boldsymbol{u}[\boldsymbol{x}(k)]\} . \tag{8}$$

An open–loop control $\boldsymbol{u}(k) = \boldsymbol{u}(t_k)$ can be accommodated by having $t_k$ be an element of the state vector. For example: $x_n = t_k$ with the difference equation $x_n(k+1) = x_n(k) + \Delta t$ or, in terms of the index $k$, $x_n = k$ with the difference equation $x_n(k+1) = x_n(k) + 1$.

Note that the function $\boldsymbol{G}(\boldsymbol{x}) = \boldsymbol{F}[\boldsymbol{x},\boldsymbol{u}(\boldsymbol{x})]$ is continuous (differentiable) if $\boldsymbol{u}(\boldsymbol{x})$ is continuous (differentiable). For the application considered in this article, the feedback control $\boldsymbol{u}(\boldsymbol{x})$ is continuous but it is only piecewise continuously differentiable, as is the resulting function $\boldsymbol{G}(\boldsymbol{x})$. This will create a difficulty later because a corresponding set of linearized discrete–time equations needed for computing Lyapunov exponents will be discontinuous.

## III. TWO DISCRETIZATION METHODS

In this section we present two different approaches for converting a continuous–time system of differential equations into a corresponding system of discrete–time difference equations.

The applications that we will consider later deal with classical Lotka-Volterra population dynamics models. For several reasons, a population biologist may decide to model a dynamical population system using discrete-time difference equations instead of continuous-time differential equations. One reason is that a solution $x(k)$ to a system of nonlinear algebraic equations of the form (6) is much easier to generate (at least forward in time) than is a solution $x(t)$ to a system of nonlinear differential equations of the form (1). Another reason cited by population biologists is a belief that many biological systems are in fact discrete-event systems. For example, a species may reproduce only once per year. Another practical reason for using a discrete-time model is that a field biologist may not be able to observe a species on a continuous basis. For example, it may be difficult for the biologist to be in the field on a daily or even monthly basis, or a species may migrate annually and only be available for observation once a year.

For a population biologist a discrete–time model yields a prediction of the population at the next cycle based on the population at the previous cycle. Although the length of a discretization cycle may vary from system to system, the cycle length is often very large compared to an approximate continuous–time period, such as yearly versus daily.

In this article we will consider the effects of two different discretization approaches applied to classical Lotka–Volterra continuous-time biological population models. It has been shown previously [2, 7] that both discretization approaches can produce discrete–time trajectories that are drastically different than those of the corresponding continuous–time system. In particular, both approaches may produce chaotic motion and related phenomena, even though these phenomena do not occur in the continuous–time model. We will show that the chaotic results for the discrete–time system can continue to occur even if a stabilizing feedback controller is applied to the continuous–time model.

### A. Exponential Discretization

To illustrate one common approach for developing a discrete–time model from a continuous–time model, consider the scalar system

$$\dot{x} = ax \ . \tag{9}$$

The solution to this differential equation is given by

$$x(t) = x(0)e^{at} \ ,$$

where $x(0)$ is the initial value of $x$. Thus the continuous–time system (9) is exactly equivalent to the discrete–time system

$$x(k+1) = x(k)e^{a\Delta t} \ , \tag{10}$$

where $\Delta t$ is the discrete time step. Note that both of the models (9) and (10) have the property that $x$ decays (or grows) exponentially.

In a more general setting, consider the scalar system

$$\dot{x} = xh(x) \ . \tag{11}$$

Following the same procedure used to integrate (9), we can re–write (11) and then integrate the result as

$$\int \frac{dx}{x} = \int h[x(t)]dt \ . \tag{12}$$

Now, if we approximate $h[x(t)]$ as a constant from time $t_k = k\Delta t$ to time $t_{k+1} = (k+1)\Delta t$, integration yields the discrete–time model

$$x(k+1) = x(k)exp\{ \ h[x(k)]\Delta t \ \} \ . \tag{13}$$

This type of exponential discrete–time model has been employed previously for biological systems [7,8,24,25] and it has two properties that appeal to population biologists: 1) the population varies exponentially with time, which agrees with many observed population systems, and 2) the population always remains positive, which agrees with the original continuous–time model. In addition, note that the equilibrium points (constant state solutions) for the continuous–time system (11), corresponding to $x(t) \equiv 0$ or $h[x(t)] \equiv 0$, remain as fixed points for the discrete–time model (13).

The main disadvantage of the discrete–time model given by (13) is that the nonequilibrium solutions for (13) are not the same as those for the continuous–time model (11). In general the function $h(x)$ is not constant over a discrete–time cycle. Reproduction may occur only once during a cycle, but deaths occur throughout the cycle. This difference can be important for large $\Delta t$ (e.g., one year versus one day) and, as we shall see later, the discrete–time model can produce fundamentally different predictions of population size than those associated with the continuous–time model.

## B. Pseudo–Euler Integration

Even if a continuous–time model is employed, the process of generating a solution $\mathbf{x}(t)$ to a system of nonlinear differential equations generally requires the use of some numerical integration scheme. This is equivalent to converting a continuous–time model to a corresponding discrete–time model.

The algorithm that we will use was introduced in [2] to study chaotic motion in the same discrete–time Lotka–Volterra system that we will study. The algorithm, which we call the "pseudo–Euler" method, is not a standard numerical integration scheme. To indicate why it is not, and to provide a rationale for the name, we first present the classical Euler and second–order Runge-Kutta integration procedures.

The simplest numerical integration scheme is Euler's method. Applied to (1), or to (3) with $\mathbf{g}(\mathbf{x}) = \mathbf{f}[\mathbf{x},\mathbf{u}(\mathbf{x})]$, Euler's method yields the corresponding discrete–time system

$$\mathbf{x}(k+1) = G[\mathbf{x}(k)] \triangleq \mathbf{x}(k) + \mathbf{g}[\mathbf{x}(k)]\Delta t \,. \tag{14}$$

For example, applying Euler's Method to the continuous–time dynamical system (9) yields the discrete–time model

$$x(k+1) = x(k)\{\ 1 + h[x(k)]\Delta t\ \}\ . \tag{15}$$

Note that, unlike the exponential discrete–time model (13), the result in (15) does *not* insure that the population will remain positive. In fact if $h[x(k)]$ is sufficiently negative the predicted population will become negative!

Euler's method is seldom used in practice, because it is not very accurate. Geometrically, it does not sense any curvature along a trajectory $\boldsymbol{x}(t)$ and numerically it has a per–step error of $O(\Delta t^2)$. The most commonly used numerical integration algorithms are fourth–order Runge–Kutta procedures [26], which have a per–step error of $O(\Delta t^5)$. These algorithms are fairly easy to program, but they are more complicated than is necessary for our demonstration purposes.

The numerical integration algorithm that we will investigate is related to second–order Runge–Kutta procedures. The general second–order Runge-Kutta algorithm, applied to a feedback control system (3), is defined by the discrete–time system

$$\boldsymbol{x}(k+1) = G[\boldsymbol{x}(k)] \triangleq \boldsymbol{x}(k) + \bar{\boldsymbol{g}}[\boldsymbol{x}(k)]\Delta t\ , \tag{16}$$

where $\bar{\boldsymbol{g}}(\boldsymbol{x})$ is a weighted average velocity vector, defined by

$$\bar{\boldsymbol{g}}(\boldsymbol{x}) \triangleq \alpha\boldsymbol{g}(\boldsymbol{x}) + \beta\boldsymbol{g}[\boldsymbol{x} + \boldsymbol{g}(\boldsymbol{x})\Delta\tau] \tag{17}$$

with $\boldsymbol{g}(\boldsymbol{x}) \triangleq \boldsymbol{f}[\boldsymbol{x},\boldsymbol{u}(\boldsymbol{x})]$, and the parameters $\alpha$, $\beta$, and $\Delta\tau$, in addition to $\Delta t$, are to be chosen subject to certain constraints.

For a given time step $\Delta t$, the parameters $\alpha$, $\beta$, and $\Delta\tau$ are determined by requiring that the discrete–time approximate solution $\boldsymbol{x}(k+1) \approx \boldsymbol{x}(t_{k+1})$ at time $t_{k+1} = t_k + \Delta t$ must agree with a Taylor series expansion of the actual solution $\boldsymbol{x}(t_k + \Delta t)$ through terms of second–order in $\Delta t$, assuming that both solutions agree at time $t_k$. Expanding (17) to first order in $\Delta\tau$, substituting the result into (16), and then expanding $x(t_k + \Delta t)$ to second order in $\Delta t$ and equating coefficients of corresponding terms yields the conditions

$$\begin{aligned} &\alpha + \beta = 1 \\ &\beta \Delta\tau = \Delta t/2 \ . \end{aligned} \tag{18}$$

Note that the second–order Taylor series conditions (18) assume that $g(x)$ is continuous and continuously differentiable at $x = x(t_k)$ and that $x(t)$ is twice continuously differentiable at $t = t_k$. As we have seen, these conditions may not be satisfied for differential equations with discontinuous right–hand sides. At times and states where these conditions are satisfied, second–order Runge–Kutta procedures have a per–step error of $O(\Delta t^3)$.

If we choose $\alpha = \beta$ then (18) yields $\alpha = \beta = 1/2$ and $\Delta\tau = \Delta t$ and we have the modified Euler method, given by

$$\begin{aligned} x(k+1) &= G[x(k)] \\ &\triangleq x(k) + \frac{\Delta t}{2}\Big[g[x(k)] + g\{x(k)+g[x(k)]\Delta t\}\Big] \end{aligned} \tag{19}$$

With this background we present the *pseudo–Euler integration* method that we will use for the remainder of this study. The algorithm is defined by

$$\begin{aligned} x(k+1) &= G[x(k)] \\ &\triangleq x(k) + \frac{\Delta t}{2}\Big[g[x(k)] + g\{x(k)+g[x(k)]\Delta\tau\}\Big], \end{aligned} \tag{20}$$

where $\Delta t$ and $\Delta\tau$ are two independent parameters and $g(x) \triangleq f[x,u(x)]$.

This algorithm has the same structure as second–order Runge–Kutta methods, but it only satisfies the Taylor series matching conditions (18) if $\Delta\tau = \Delta t$, in which case the procedure becomes the modified Euler method. While no attempt is made to satisfy the Taylor series conditions (18), we also note that the pseudo–Euler algorithm makes no assumptions about the continuity or differentiability of either $g(x)$ or $x(t)$. If $g(x)$ and $x(t)$ do satisfy the same smoothness assumptions as in second–order Runge-Kutta methods, then pseudo–Euler integration has per–step error of $O(\Delta t \cdot [\Delta t - \Delta\tau - \Delta\tau^2])$.

This error magnitude is the same as for second–order Runge–Kutta methods if $\Delta\tau$ and $\Delta t$ are of the same order of magnitude (and have the same sign). It should also be noted that if $\Delta\tau = 0$ then the pseudo-Euler method reduces to the standard first–order Euler method, with a per-step error of $O(\Delta t^2)$.

## IV. CHAOTIC SYSTEMS

To discuss various topics related to chaotic dynamical systems consider a feedback control system of the form

$$\dot{x} = g(x) \triangleq f[x,u(x)] \tag{21}$$

We will usually assume that $g(x)$ is continuous and differentiable for all $x$. Deviations from these assumptions will be addressed as special cases.

As we have discussed previously, a unique solution $x(t) = \xi(t,x_0)$ exists for each initial state $\xi(0,x_0) = x_0$. Furthermore the solution $x(t)$ is a continuous and continuously differentiable function of the initial state $x_0$. Mathematically this means that the state $x(t)$ is absolutely predictable, forward or backward in time and even over long time intervals, and that small (differential) changes in the initial state produce small (differential) changes in the state at other times.

However, in 1892 Poincaré [10] showed that, despite the mathematical predictability of solutions to (21), there exist dynamical systems whose solutions are so sensitive to changes in initial conditions that the initial state would have to be known exactly in order to predict the state accurately over long time intervals. Since any physical measurement of the state contains noise, prediction of the future state for such systems is essentially impossible. The term *chaos* is used to denote this type of unpredictable motion.

In terms of dynamical systems theory, the existence of chaotic motion is one of the most remarkable and intriguing discoveries of the past century. The philosophical implications of chaotic motion challenge the basic scientific methodology. Since the time of Sir Isaac Newton the scientific approach to the analysis of any physical phenomenon has been to formulate a model for the process, validate the model experimentally, and then use the model to predict the behavior of the physical system. In a chaotic system the motion is essentially unpredictable, so the basic scientific premise of being able to predict the future state of the physical system based on a model of the system does not hold. Fortunately, recent research [27, 28] has shown that there often exists an underlying structure in chaotic motion and that generic mechanisms exist by which chaos tends to occur [13].

It should also be noted that chaotic motion in a continuous–time system (21) does not mean that a numerical simulation of the system will be unrepeatable. It may be inaccurate, but within the finite–word–length limitations inherent in digital computers, a given numerical simulation is completely repeatable for the same numeric initial conditions. However, since numerical integration procedures only yield approximate solutions to differential equations, errors due to roundoff and truncation will cause different simulation algorithms to yield different solutions for a chaotic system. This is particularly true for systems of discontinuous differential equations, where different simulation and root finding procedures may yield, for example, different numerical estimates for switching times and states.

## A. Qualitative Features of Chaos

In dynamical systems theory a basic defining property of a chaotic system is that solutions to (21) have a *sensitive dependence on initial conditions*: small differences in initial conditions grow exponentially with time. Lyapunov exponents, which we will discuss shortly, provide quantitative measures of this sensitivity.

Another property of chaotic systems is related to certain special types of solutions to (21), particularly those that are attractors, to which nearby trajectories converge. More precisely [12], an *attractor* is a closed invariant set $\mathcal{A}$ such that, for almost all initial conditions (that is, except possibly on a set of Lebesgue measure zero) in some neighborhood of $\mathcal{A}$, the corresponding solutions $x(t)$ to (21) converge to $\mathcal{A}$ as $t \to \infty$. The set of all initial states $x(0)$ for which $x(t) \to \mathcal{A}$ is called the *basin of attraction* for $\mathcal{A}$.

As an example, consider a constant–state solution $x(t) \equiv \hat{x}$, called an equilibrium point. If an equilibrium point is asymptotically stable, then neighboring trajectories $x(t)$ approach it asymptotically as $t \to \infty$ and the equilibrium point is an attractor. If an equilibrium solution is globally asymptotically stable, then its basin of attraction is all of state space. If a system contains more than one attracting equilibrium point then each will have its own corresponding basin of attraction.

A periodic solution $x(t) = x(t + T)$, of period $T$, is one for which the solution repeats itself. A periodic solution that has no periodic neighbors is called a limit cycle. If all neighboring solutions asymptotically approach the limit cycle then the limit cycle is an attractor, with an associated basin of attraction.

A nearly periodic [29] solution $x(t)$ is one which approaches periodicity as $t \to \infty$, such as a trajectory that asymptotically approaches a limit cycle. A quasiperiodic [30] (or almost periodic [29]) solution is one that is not periodic but comes arbitrarily close to being periodic after a sufficiently long time. A classic example, related to motion on a torus, is a function such as

$$x(t) = A_1 \sin \omega_1 t + A_2 \sin \omega_2 t$$

with incommensurate frequencies, meaning that there do not exist integers $n_1$ and $n_2$ such that $n_1\omega_1 + n_2\omega_2 = 0$. That is, $\omega_1/\omega_2$ is an irrational number.

Normally, the only types of bounded attractors associated with (21) are equilibrium points, limit cycles, and (to a lesser extent) quasiperiodic orbits [30]. Chaotic systems however also possess *strange attractors* [27, 31, 32]. These are bounded attracting trajectories that are not equilibrium points and are neither periodic nor quasiperiodic orbits. They are called strange attractors partly because they have the unusual property that they are confined to a bounded region but they never repeat themselves. This in itself is not enough to warrant the label "strange". For example, in a three-dimensional system a quasiperiodic solution may exist which wraps itself around the surface of a torus, covering the torus but never actually repeating itself.

The truly strange property of strange attractors is that they are *fractals* [33]. That is, they are geometric objects that have a fractional dimension.

In the usual notion of dimension, a point has dimension zero, a curve has dimension one, a plane has dimension two, and so forth. Thus an equilibrium point for (21) is zero–dimensional and any nonequilibrium solution $x(t)$ is theoretically a one–dimensional curve in an $n$–dimensional state space, since it only takes one parameter value $t$ to specify points along a solution.

In a chaotic system a solution corresponding to a strange attractor wanders around erratically in such a way that it covers more than a one-dimensional region. As an example of a system becoming chaotic, suppose that a system of three differential equations generates a quasiperiodic solution that covers a torus. The trajectory is thus a two–dimensional object. As a parameter in the differential equations is changed the solution may produce a chaotic strange attractor which smears out and covers a region whose "dimension" is a fraction, greater than two. The existence of such objects requires an extension of the notion of dimension and many such extensions have been proposed [34]. One of these extensions, related to Lyapunov exponents, will be employed later for our control system example.

In a chaotic system with a strange attractor, not only is the attractor a fractal but also its basin of attraction typically has a *fractal basin boundary* [6, 35]. In addition, chaotic attractors tend to have a *self–similarity* [13] independent of viewing scale. For example, if a Poincaré map, or cross section, of a strange attractor exhibits a banded structure of parallel lines, then the same type of structure will also tend to appear at smaller scales.

## B. Lyapunov Exponents

In a chaotic dynamical system, sensitivity with respect to initial conditions means that the distance between neighboring trajectories grows exponentially with time. Lyapunov exponents measure these growth rates in nonlinear systems, in the same way that eigenvalues do for linear systems.

To present the basic idea of a Lyapunov exponent, let $\delta(t)$ denote the distance between two trajectories for a (continuous-time or discrete–time) dynamical system. If $\delta(0)$ is arbitrarily small and if

$$\delta(t) \to \delta(0)e^{\sigma t} \qquad \text{as } t \to \infty , \tag{22}$$

then $\sigma$ is called a Lyapunov exponent. The distance between trajectories grows, shrinks, or remains constant depending on whether $\sigma$ is positive, negative, or zero, respectively.

For a nonlinear dynamical system in an $n$–dimensional state space there are $n$ Lyapunov exponents, corresponding to $n$ orthogonal directions, just as there are $n$ eigenvalues for an $n$–dimensional linear system. To define these Lyapunov exponents somewhat more precisely than in (22), consider an $n$–dimensional ellipsoid centered at a point $x(t)$ along a trajectory for the dynamical system. Let $\delta_i(t)$ denote the length of the $i$–th semi–axes of the ellipsoid and choose the initial ellipsoid as an $n$–sphere of radius $\delta_0$, with $\delta_i(0) = \delta_0$, $i = 1, ..., n$.

Assuming that the following limits exist [36], the *Lyapunov exponents* $\sigma_i$, $i = 1, ..., n$, are defined by

$$\sigma_i \triangleq \lim_{t\to\infty} \left[ \lim_{\delta_i(0)\to 0} \left\{ \frac{1}{t} \ell n \left[ \frac{\delta_i(t)}{\delta_i(0)} \right] \right\} \right], \tag{23}$$

with the indices ordered so that $\sigma_1 \geq \sigma_2 \geq ... \geq \sigma_n$. More rigorous definitions can be found in [12, 37], but (23) is sufficient for our purposes.

A necessary condition for a chaotic strange attractor is that the largest Lyapunov exponent $\sigma_1$ must be positive. This corresponds to a *stretching* of neighboring trajectories. However, in order for the strange attractor to remain bounded there must also exist a *folding* of trajectories, caused by nonlinear terms in the equations of motion. Because of this stretching and folding, a chaotic strange attractor can only occur in a nonlinear system and, for continuous–time systems, only in a state space of dimension three or greater. Also, at least one Lyapunov exponent must be zero [38] for a continuous-time system if the strange attractor does not contain an equilibrium point. In discrete–time systems the three–dimensional restriction and the existence of a zero Lyapunov exponent do not apply. Chaos can occur in even a one-dimensional discrete–time system with a positive Lyapunov exponent.

An alternate interpretation of Lyapunov exponents, in terms of information loss [36], can be developed simply be changing the base by which logarithms are computed. If "$e$" in (22) and "$\ell n()$" in (23) are replaced by "2" and "$\ell og_2()$", respectively, then the resulting base–2 Lyapunov exponents calculated via (23) give the *rate of information loss* (bits/sec) along the corresponding trajectory. To illustrate later a connection between Lyapunov exponents and eigenvalues, we will focus on the natural logarithm representation in (22)–(23). However, our simulation results, using the algorithm in [36], will be presented in base–2 (bits/sec).

To compute Lyapunov exponents, in a way that automatically handles the limiting process for $\delta_i(0) \to 0$ in (23), a reference trajectory is computed for the nonlinear dynamical system and the ellipsoid semi–axis lengths $\delta_i(t)$ are computed using a set of corresponding linearized equations of motion for small perturbations from the reference trajectory.

**Continuous–Time Systems.** To compute the Lyapunov exponents for a continuous-time system of the form (21), let $\mathbf{x}(t) \triangleq \xi[t,\mathbf{x}(0)]$ be a solution starting from $\mathbf{x}(0)$. Let $\bar{\mathbf{x}}(t) \triangleq \xi[t,\bar{\mathbf{x}}(0)]$ be the solution *generated by the same control function* $\mathbf{u}(\cdot)$, but starting from a point $\bar{\mathbf{x}}(0) = \mathbf{x}(0) + \epsilon\boldsymbol{\eta}(0)$ arbitrarily close to $\mathbf{x}(0)$, where $\epsilon$ is small and $\boldsymbol{\eta}(0)$ is a unit vector.

Assuming that $\mathbf{g}(\mathbf{x})$ is continuous and continuously differentiable along $\mathbf{x}(t)$, and applying Taylor's theorem to $\mathbf{g}(\mathbf{x} + \epsilon\boldsymbol{\eta})$, we have

$$\bar{\mathbf{x}}(t) = \mathbf{x}(t) + \epsilon\boldsymbol{\eta}(t) + O(\epsilon)\,, \tag{24}$$

where $O(\epsilon)/\epsilon \to \mathbf{0}$ as $\epsilon \to 0$ and the vector $\boldsymbol{\eta}(t) \in \mathbb{R}^n$ satisfies the system of $n$ linearized *continuous–time state perturbation equations*

$$\dot{\boldsymbol{\eta}} = D(t)\boldsymbol{\eta}\,, \tag{25}$$

$$D(t) \triangleq \frac{\partial \mathbf{g}[\mathbf{x}(t)]}{\partial \mathbf{x}} = \left[\frac{\partial \mathbf{f}(\mathbf{x},\mathbf{u})}{\partial \mathbf{x}} + \frac{\partial \mathbf{f}(\mathbf{x},\mathbf{u})}{\partial \mathbf{u}}\frac{\partial \mathbf{u}(\mathbf{x})}{\partial \mathbf{x}}\right], \tag{26}$$

where the terms in brackets in (26) are evaluated at $\mathbf{x} = \mathbf{x}(t)$ and $\mathbf{u} = \mathbf{u}[\mathbf{x}(t)]$.

For future reference, in connection with systems of discontinuous or nonsmooth differential equations, we note that our previous discussion on the nature of solutions for such systems (in Section II, A) implies that the linearized perturbation result in (24) also holds for discontinuous or nonsmooth systems, if the $n \times n$ matrix $\boldsymbol{D}(t)$ defined by (26) is no worse than piecewise continuous along $\boldsymbol{x}(t)$. In the example control system that we will consider, this condition will be satisfied, except for one isolated point (at the target equilibrium point). We will discuss this special case when it arises. Until then we will focus on the usual case where $\boldsymbol{g}(\boldsymbol{x})$ is continuous and continuously differentiable, which implies that $\boldsymbol{D}(t)$ is continuous.

Before we present procedures for computing Lyapunov exponents, one other aspect of the state perturbation equations needs to be discussed. In nonlinear control systems analysis we often consider the case of open–loop controls $\boldsymbol{u} = \boldsymbol{u}(t)$, with a corresponding set of open–loop state perturbation equations. For example, in developing the adjoint differential equations in Pontryagin's necessary conditions for an optimal open–loop control, we consider a control system

$$\dot{\boldsymbol{x}} = \boldsymbol{f}[\boldsymbol{x},\boldsymbol{u}(t)] \tag{27}$$

and state perturbation solutions that are due to changes in $\boldsymbol{x}(0)$, with the control function $\boldsymbol{u}(t)$, $t_0 \leq t < \infty$, and the initial time $t_0 = 0$ held fixed. In such a case, the state perturbations correspond to $\partial \boldsymbol{u}/\partial \boldsymbol{x} \equiv \boldsymbol{0}$ in (26):

$$\dot{\boldsymbol{\eta}} = \boldsymbol{D}(t)\boldsymbol{\eta} \tag{28}$$

$$\boldsymbol{D}(t) = \frac{\partial \boldsymbol{f}[\boldsymbol{x}(t),\boldsymbol{u}(t)]}{\partial \boldsymbol{x}}. \tag{29}$$

It should be emphasized that *open–loop state perturbations* (28)–(29) are *not valid for computing Lyapunov exponents.* The open–loop results will be incorrect (and may be grossly misleading), in part because the right–hand terms in (26) are missing. Also, the ability to vary the initial time is not included in the open–loop formulation. The correct, closed–loop formulation incorporates time as one of the state variables and the corresponding closed-loop state perturbations include a perturbation in the initial time. In fact, the zero Lyapunov exponent that must exist for a continuous-time strange attractor often corresponds to the time component of the state vector, as in the case of a periodic forcing function, where there is no stretching or shrinking in the time dimension. In the control system example that we will consider later, we will demonstrate the misleading results that can occur from using the incorrect, open-loop state perturbation formulation.

To calculate Lyapunov exponents, using the method in [36], we choose $n$ orthogonal unit vectors $\boldsymbol{\eta}_i(0)$, $i = 1, \ldots, n$, with $\epsilon\boldsymbol{\eta}_i(0)$ representing the $n$ semi–axes of an arbitrarily small initial (spherical) ellipsoid. Then the closed-loop state perturbation equations (25)–(26) are used to generate corresponding perturbation solutions $\boldsymbol{\eta}_i(t)$. To avoid numerical overflow, the integrations in (25) are restarted at convenient times $t = k\Delta T$ (*e.g.*, $\Delta T = \Delta t$), after the results have been reorthogonalized, $\boldsymbol{\eta}_i(t) \rightarrow \boldsymbol{\psi}_i(t)$, and renormalized, $\boldsymbol{\psi}_i(t)/\|\boldsymbol{\psi}_i(t)\| \rightarrow \boldsymbol{\eta}_i(t)$, using a Gramm-Schmidt process:

$$
\begin{aligned}
\boldsymbol{\psi}_1 &= \boldsymbol{\eta}_1 \\
\rho_1 &= \|\boldsymbol{\psi}_1\| \\
\boldsymbol{\eta}_1 &= \boldsymbol{\psi}_1/\|\boldsymbol{\psi}_1\| \\
&\vdots \\
\boldsymbol{\psi}_m &= \boldsymbol{\eta}_m - \sum_{j=1}^{m-1} <\boldsymbol{\eta}_m, \boldsymbol{\eta}_j> \boldsymbol{\eta}_j\,, \qquad m = 2, \ldots, n \\
\rho_m &= \|\boldsymbol{\psi}_m\| \\
\boldsymbol{\eta}_m &= \boldsymbol{\psi}_m/\|\boldsymbol{\psi}_m\|\,,
\end{aligned}
\tag{30}
$$

where all quantities are evaluated at time $t = k\Delta T$, $k = 1, 2, \ldots$, with the notation $<\boldsymbol{a},\boldsymbol{b}> \triangleq \boldsymbol{a}\cdot\boldsymbol{b} \triangleq \boldsymbol{a}^{\mathrm{T}}\boldsymbol{b}$ representing the dot product or scalar product of two (column) vectors $\boldsymbol{a}$ and $\boldsymbol{b}$, and $()^{\mathrm{T}}$ denoting the transpose.

At time $t_k = k\Delta T$, the lengths $\rho_i(t_k) = \|\boldsymbol{\psi}_i(t_k)\|$ of the orthogonalized vectors, prior to rescaling, are the lengths of the transformed ellipsoid semi–axes after a time interval $\Delta T$, starting from a unit sphere at time $t_{k-1}$.

From (23) and (30) the Lyapunov exponents can be calculated as

$$\sigma_i = \lim_{K\to\infty} \left\{ \frac{1}{K\Delta T} \sum_{k=1}^{K} \ell n\left[\rho_i(k\Delta T)\right] \right\}. \tag{31}$$

In deriving the result in (31) we have used

$$\frac{\delta_i(k\Delta T)}{\delta_i(0)} = \frac{\delta_i(k\Delta T)}{\delta_i[(k-1)\Delta T]} \cdot \frac{\delta_i[(k-1)\Delta T]}{\delta_i[(k-2)\Delta T]} \cdots \frac{\delta_i(\Delta T)}{\delta_i(0)} \tag{32}$$

and the fact that the differential equations (25) are homogeneous in $\boldsymbol{\eta}_i$. Thus each length ratio on the right–hand side of (32) is unchanged if we multiply both lengths by a common scale factor. Using unit vectors at the beginning of each time interval, (31) follows from (23), (30), and (32) with

$$\rho_i(j\Delta T) = \frac{\delta_i(j\Delta T)}{\delta_i[(j-1)\Delta T]}, \qquad j = 1,2, \ldots \tag{33}$$

Note that along a trajectory $\boldsymbol{x}(t)$ the solutions to (25) are given by

$$\boldsymbol{\eta}(t) = \boldsymbol{\Phi}(t,\tau)\boldsymbol{\eta}(\tau), \tag{34}$$

where the *continuous–time state transition matrix* $\boldsymbol{\Phi}(t,\tau)$, defined by (34), satisfies the $n\times n$ system of differential equations

$$\frac{\partial\boldsymbol{\Phi}(t,\tau)}{\partial t} = \boldsymbol{D}(t)\boldsymbol{\Phi}(t,\tau) \qquad \boldsymbol{\Phi}(\tau,\tau) = \boldsymbol{I}. \tag{35}$$

For

$$[\boldsymbol{\eta}_1(0) \ldots \boldsymbol{\eta}_n(0)] = \boldsymbol{I} = \boldsymbol{\Phi}(0,0)\ , \tag{36}$$

the vectors $\boldsymbol{\eta}_i(t)$ generated by (25) would simply be the columns of $\boldsymbol{\Phi}(t,0)$, if the reorthonormalization in (30) were not performed.

For an alternate representation [13] of the Lyapunov exponents let $\gamma_i(t)$ be the eigenvalues of $\boldsymbol{\Phi}(t,0)$, with corresponding unit eigenvectors $\zeta_i(t)$. Instead of the initial conditions (36), choose $\boldsymbol{\eta}_i(0) = \zeta_i(t)$ for a given time $t$. Then from (34)

$$\boldsymbol{\eta}_i(t) = \boldsymbol{\Phi}(t,0)\boldsymbol{\eta}_i(0) = \gamma_i(t)\boldsymbol{\eta}_i(0)\ .$$

From the definition (23) of Lyapunov exponents, expressed as (22), we have

$$\|\boldsymbol{\eta}_i(t)\| \to e^{\sigma_i t}\|\boldsymbol{\eta}_i(0)\| \qquad \text{as } t \to \infty\ .$$

Thus

$$|\gamma_i(t)| \to e^{\sigma_i t} \qquad \text{as } t \to \infty$$

and the Lyapunov exponents can be expressed as

$$\sigma_i = \lim_{t\to\infty} \left[\ell n\,|\gamma_i(t)|^{1/t}\right]\ , \tag{37}$$

where $\gamma_i(t)$, $i = 1, \ldots, n$, are the eigenvalues of $\boldsymbol{\Phi}(t,0)$.

As an example, consider a diagonal constant–coefficient linear system

$$\dot{\boldsymbol{x}} = \boldsymbol{A}\boldsymbol{x}\ , \qquad \boldsymbol{A} = \begin{bmatrix} \lambda_1 & & \mathbf{0} \\ & \ddots & \\ \mathbf{0} & & \lambda_n \end{bmatrix}.$$

The state transition matrix

$$\boldsymbol{\Phi}(t,0) = \begin{bmatrix} e^{\lambda_1 t} & & \mathbf{0} \\ & \ddots & \\ \mathbf{0} & & e^{\lambda_n t} \end{bmatrix}$$

has eigenvalues $\gamma_i = e^{\lambda_i t}$. Thus the Lyapunov exponents $\sigma_i = \lambda_i$ are simply the eigenvalues for the linear system matrix $\boldsymbol{A}$, ordered so that $\sigma_1 \geq \dots \geq \sigma_n$.

To compute only the largest Lyapunov exponent, instead of the entire spectrum of Lyapunov exponents computed via (30) and (31), a simplified procedure [13] can be used, involving only one perturbation vector $\boldsymbol{\eta}_1(t)$. Choose an arbitrary initial vector $\boldsymbol{\eta}_1(0)$, which can be written in terms of the eigenvectors of $\boldsymbol{\Phi}(t,0)$ as

$$\boldsymbol{\eta}_1(0) = a_1 \zeta_1(t) + \dots + a_n \zeta_n(t) \; .$$

Then

$$\boldsymbol{\eta}_1(t) = \boldsymbol{\Phi}(t,0)\boldsymbol{\eta}_1(0) = a_1 \gamma_1(t)\zeta_1(t) + \dots + a_n \gamma_n(t)\zeta_n(t)$$

and

$$\|\boldsymbol{\eta}_1(t)\| \to e^{\sigma_1 t}\|a_1 \zeta_1(t)\| \qquad \text{as } t \to \infty \; ,$$

since $\|a_i \gamma_i(t)\zeta_i(t)\| \to e^{\sigma_i t}\|a_i \zeta_i(t)\|$ and the $e^{\sigma_1 t}$ term dominates the others.

Thus we see that the first three of equations (30), along with (31), can be used to compute the largest Lyapunov exponent. As in the case of the full spectrum of Lyapunov exponents, the length of the vector $\boldsymbol{\eta}_1(t)$ is renormalized at times $t = k\Delta T$ in order to avoid numerical overflow problems.

**Discrete–Time Systems.** The procedure for computing the Lyapunov exponents for a discrete–time system of the form (8) is essentially the same as (25)–(26) and (30)–(31) for continuous–time systems. The main difference is that, along a discrete-time strange attractor trajectory $\boldsymbol{x}(k)$ with neighboring solutions

$$\bar{\boldsymbol{x}}(k) = \boldsymbol{x}(k) + \epsilon\boldsymbol{\eta}(k) + O(\epsilon) , \tag{38}$$

where $O(\epsilon)/\epsilon \to \mathbf{0}$ as $\epsilon \to 0$, the *discrete–time state perturbation equations* are

$$\boldsymbol{\eta}(k+1) = \boldsymbol{J}(k)\boldsymbol{\eta}(k) , \tag{39}$$

where $\boldsymbol{J}(k)$ is the Jacobian matrix

$$\boldsymbol{J}(k) \triangleq \frac{\partial \boldsymbol{G}[\boldsymbol{x}(k)]}{\partial \boldsymbol{x}} = \left[ \frac{\partial \boldsymbol{F}(\boldsymbol{x},\boldsymbol{u})}{\partial \boldsymbol{x}} + \frac{\partial \boldsymbol{F}(\boldsymbol{x},\boldsymbol{u})}{\partial \boldsymbol{u}} \frac{\partial \boldsymbol{u}(\boldsymbol{x})}{\partial \boldsymbol{x}} \right] , \tag{40}$$

with the terms in brackets evaluated at $\boldsymbol{x} = \boldsymbol{x}(k)$ and $\boldsymbol{u} = \boldsymbol{u}[\boldsymbol{x}(k)]$.

In addition, the state transition equations (33) become

$$\boldsymbol{\eta}(k) = \boldsymbol{\Phi}(k,0)\boldsymbol{\eta}(0) , \tag{41}$$

where the *discrete–time state transition matrix* is the sequential product of the Jacobian matrices:

$$\boldsymbol{\Phi}(k,0) = \prod_{j=0}^{k-1} \frac{\partial \boldsymbol{G}[\boldsymbol{x}(j)]}{\partial \boldsymbol{x}} = \boldsymbol{J}(k-1)\boldsymbol{J}(k-2)\cdots\boldsymbol{J}(0) . \tag{42}$$

As an example, consider the diagonalized linear system

$$\boldsymbol{x}(k+1) = \boldsymbol{A}\boldsymbol{x}(k) \qquad \boldsymbol{A} = \begin{bmatrix} \lambda_1 & & \mathbf{0} \\ & \ddots & \\ \mathbf{0} & & \lambda_n \end{bmatrix} .$$

We choose orthogonal initial vectors $[\boldsymbol{\eta}_1(0) \ldots \boldsymbol{\eta}_n(0)] = \boldsymbol{I}$ and calculate

$$\boldsymbol{\eta}_i(k) = \boldsymbol{\Phi}(k,0)\boldsymbol{\eta}_i(0) = \boldsymbol{A}^k\boldsymbol{\eta}_i(0) = \lambda_i^k\boldsymbol{\eta}_i(0).$$

Then from (23) or (37) with $t = k$ (*i.e.*, $\Delta t = 1$) we have

$$\sigma_i = \ell n |\lambda_i| ,$$

which contrasts with the results $\sigma_i = \lambda_i$ obtained for a continuous–time system with the same $\boldsymbol{A}$ matrix.

Concerning discontinuous discrete–time systems, we note that the neighboring solution results in (38)–(40) assume only that $\boldsymbol{G}(\boldsymbol{x})$ is continuous and continuously differentiable at each discrete point $\boldsymbol{x} = \boldsymbol{x}(k)$ along the reference trajectory. In the example control system that we will consider later, discontinuous changes in $\partial \boldsymbol{G}(\boldsymbol{x})/\partial \boldsymbol{x}$ can occur between $\boldsymbol{x} = \boldsymbol{x}(k)$ and $\boldsymbol{x} = \boldsymbol{x}(k+1)$. However, we will assume that no point $\boldsymbol{x}(k)$ on the reference trajectory falls exactly on a switching surface. An exception will occur in our example, at the target equilibrium point, which we will address later as a special case.

## C. Fractal Dimension

As we have indicated previously, strange attractors have a fractional dimension. Of the many possible definitions of fractional dimension [34], the simplest is the Kolomogorov capacity dimension, which is a simplification of the Hausdorff dimension. The *capacity dimension* $d_C$ of a set is defined as

$$d_C \triangleq \lim_{\epsilon \to 0} \frac{\ell n\ N(\epsilon)}{\ell n(1/\epsilon)} , \tag{43}$$

where $N(\epsilon)$ is the minimum number of $n$–dimensional cubes of side $\epsilon$ needed to cover the set. For standard Euclidean objects, such as a point, a curve, or an area, the capacity (or box–counting) dimension is integer–valued. For a point, $N(\epsilon) = 1$ and $d_C = 0$. For a curve of length $L$, $N(\epsilon) = L/\epsilon$ in the limit as $\epsilon \to 0$. Thus $d_C = 1$ for a curve. For an area $A$, $N(\epsilon) = A/\epsilon^2$ as $\epsilon \to 0$, yielding $d_C = 2$. In general however, the capacity dimension may be fractional. For example, the Cantor middle–thirds set, obtained by successively deleting the middle third of a line segment and each resulting segment, has capacity dimension $d_C = \ell n(2)/\ell n(3) = 0.6309...$ .

The *Lyapunov dimension* [34] of a trajectory is defined as

$$d_L \triangleq k + \frac{\sum_{i=1}^{k} \sigma_i}{|\sigma_{k+1}|}, \tag{44}$$

where $\sigma_i$, $i = 1,...,n$, are the Lyapunov exponents, with $\sigma_1 \geq \sigma_2 \geq \dots \sigma_n$, and $k$ is the largest integer for which $\sigma_1 + \dots + \sigma_k \geq 0$. If $\sigma_i \geq 0$ for all $i = 1,...,n$ then $d_L = n$. If $\sigma_i < 0$ for all $i = 1,...,n$ then $d_L = 0$.

The Hausdorff and capacity dimensions are usually (but not always) equal [34] and the common value is called the *fractal dimension* $d_F$. Kaplan and Yorke [39] introduced the Lyapunov dimension, defined by (44), as a lower bound on the fractal dimension, $d_F \geq d_L$, along with a conjecture that $d_F = d_L$ for most attractors in *dissipative systems*, defined by $\Sigma_{i=1}^{n}\{\sigma_i\} < 0$. This conjecture has not been proven in general, but several results [34] tend to support its validity.

## V. A LOTKA–VOLTERRA COMPETITION SYSTEM

In this section we present an example of chaotic motion caused by exponential discretization of a nonchaotic continuous–time system.

### A. Continuous–Time System

Consider a two–species Lotka–Volterra competition model given by

$$\begin{aligned} \dot{x}_1 &= r x_1(1 - x_1 - 0.5x_2) \\ \dot{x}_2 &= 2r x_2(1 - 0.5x_1 - x_2) \,, \end{aligned} \tag{45}$$

where the populations $x_i$ and the parameter $r$ are all positive.

The equilibrium point at $\hat{\boldsymbol{x}} = (2/3, 2/3)$ is asymptotically stable and its basin of attraction is the positive quadrant $\mathbb{R}^2_+ \triangleq \{(x_1, x_2) \mid x_1 > 0,\, x_2 > 0\}$. To establish this, consider the function

$$V(\boldsymbol{x}) = \hat{x}_1\left[\frac{x_1 - \hat{x}_1}{\hat{x}_1} - \ell n\left[\frac{x_1}{\hat{x}_1}\right]\right] + \hat{x}_2\left[\frac{x_2 - \hat{x}_2}{\hat{x}_2} - \ell n\left[\frac{x_2}{\hat{x}_2}\right]\right]. \tag{46}$$

We have $V(\hat{\boldsymbol{x}}) = 0$, $V(\boldsymbol{x}) > 0$ for all $\boldsymbol{x} \in \mathbb{R}^2_+ - \{\hat{\boldsymbol{x}}\}$, and at any point $\boldsymbol{x} = \boldsymbol{x}(t)$ along a trajectory we also have

$$\begin{aligned}\frac{dV[\boldsymbol{x}(t)]}{dt} &= \frac{\partial V(\boldsymbol{x})}{\partial x_1}\dot{x}_1 + \frac{\partial V(x)}{\partial x_2}\dot{x}_2 \\ &= -(\boldsymbol{x} - \hat{\boldsymbol{x}})^{\mathrm{T}} P(\boldsymbol{x} - \hat{\boldsymbol{x}}) \\ &< 0 \qquad \text{for all } \boldsymbol{x} \neq \hat{\boldsymbol{x}}\,,\end{aligned}$$

since

$$P = r\begin{bmatrix} 1 & 3/4 \\ 3/4 & 2 \end{bmatrix}$$

is positive definite. Thus $V(\boldsymbol{x})$ is a Lyapunov function for (45) on $\mathbb{R}^2_+$, which implies that $\hat{\boldsymbol{x}}$ is asymptotically stable and its basin of attraction is all of $\mathbb{R}^2_+$.

## B. Exponential Discrete–Time Competition System

Using the procedure outlined in Section III, A we develop an exponential discrete–time model for the competition system (45) by dividing each equation in (45) by the corresponding $x_i$ and treating the new right-hand sides as constants while integrating over one time cycle. For $\Delta t = 1$ we get

$$
\begin{aligned}
x_1(k+1) &= x_1(k)exp\{r[1 - x_1(k) - 0.5x_2(k)]\} \\
x_2(k+1) &= x_2(k)exp\{2r[1 - 0.5x_1(k) - x_2(k)]\} \ .
\end{aligned}
\tag{47}
$$

This discrete–time model is the same as one considered by May [7]. Note that the fixed points for the discrete–time model are the same as the equilibrium points for the continuous–time model. In [7], time plots of $x_1(k)$ show that the attractor for (47) becomes chaotic as $r$ is increased in steps from $r = 1.1$ to $r = 4.0$.

Figures 2 and 3 show plots of $x_i(k)$ for the $r = 2.5$ case considered in [7], starting from $x_1(0) = x_2(0) = 0.5$. Figure 4 shows the corresponding state-space plot of $\boldsymbol{x}(k)$, $k = 0,1,\ldots,10^6$. The discrete–time trajectory shown in Figure 4 wanders around chaotically and never repeats itself. Clearly this result does not agree with the results for the continuous–time system (45), which predict a trajectory that asymptotically converges to the equilibrium point at $\hat{x}_1 = \hat{x}_2 = 2/3$.

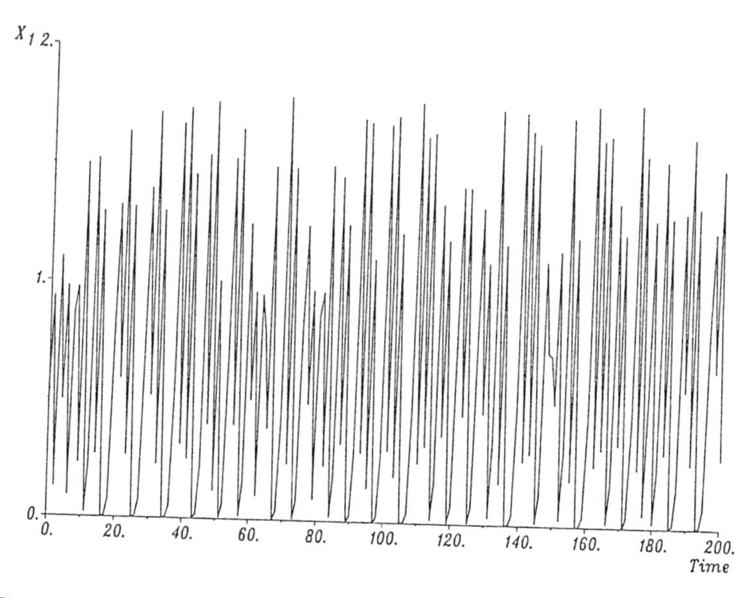


Figure 2. Exponential discrete–time competition model (Population 1).

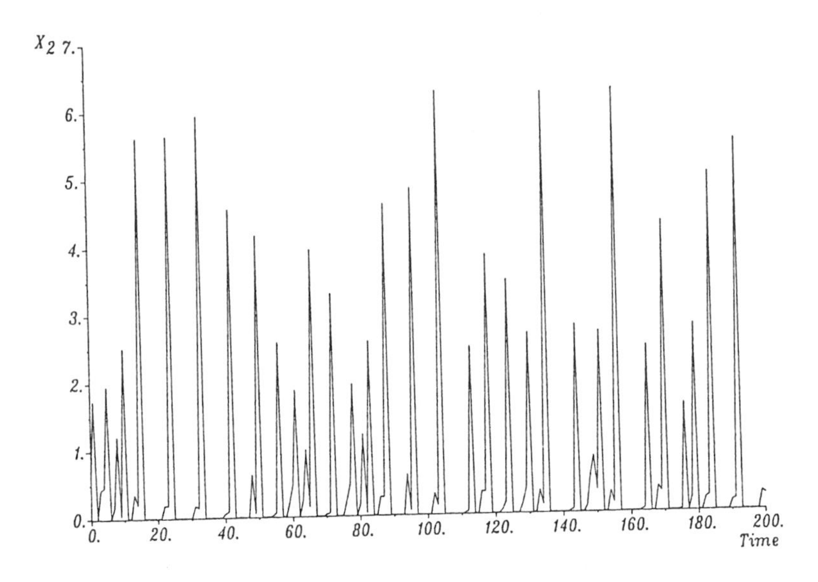


Figure 3. Exponential discrete–time competition model (Population 2).

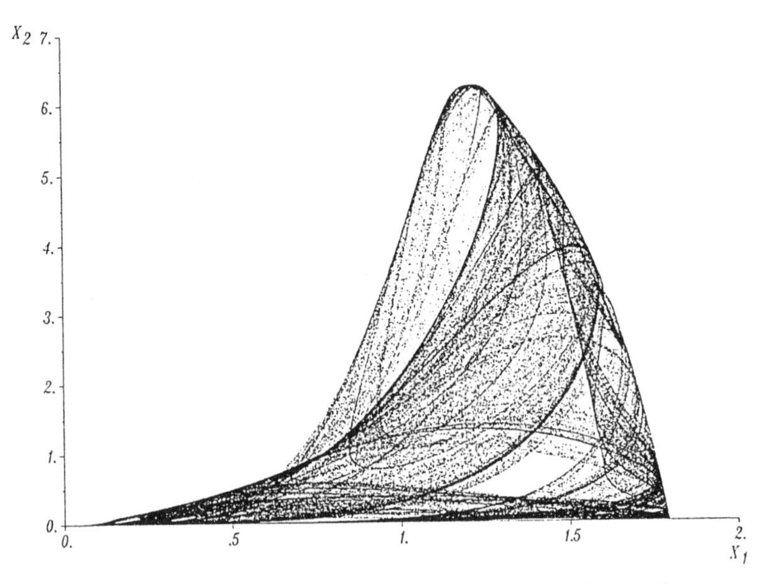


Figure 4. Exponential discrete–time competition trajectory.

## VI. A LOTKA–VOLTERRA PREY–PREDATOR SYSTEM

For the remainder of this article we investigate chaos caused solely by a numerical integration procedure. In the application that we consider, a hint of chaos occurs even for a fourth–order Runge–Kutta integration algorithm, unless the step size is very small. However, we will study a simpler case akin to second–order Runge–Kutta integration, where chaos can occur even for fairly small time steps.

### A. Continuous–Time Prey–Predator System

For illustrative purposes consider a Lotka–Volterra prey–predator system given by

$$\begin{aligned} \dot{x}_1 &= x_1(\alpha - \beta x_2) \\ \dot{x}_2 &= x_2(-\gamma + \delta x_1) \ , \end{aligned} \tag{48}$$

where $x_1$ and $x_2$ are the prey and predator population levels, respectively, and the parameters $\alpha$, $\beta$, $\gamma$, and $\delta$ are all positive. In vector notation the differential equations (48) are in the form

$$\dot{\boldsymbol{x}} = \boldsymbol{g}(\boldsymbol{x}) \ . \tag{49}$$

The equilibrium point at

$$\begin{aligned} \hat{x}_1 &= \gamma / \delta \\ \hat{x}_2 &= \alpha / \beta \end{aligned} \tag{50}$$

is stable but not asymptotically stable. This can be verified locally by noting that the linearized equations of motion near this equilibrium point,

$$\dot{\eta} = \frac{\partial g(\hat{x})}{\partial x}\,\eta = \begin{bmatrix} 0 & -\beta\gamma/\delta \\ \delta\alpha/\beta & 0 \end{bmatrix} \begin{bmatrix} \eta_1 \\ \eta_2 \end{bmatrix}, \tag{51}$$

have purely imaginary eigenvalues $\lambda = \pm\, i\sqrt{\alpha\gamma}$. More generally, the nonlinear differential equations (48) possess a first integral that can be obtained analytically. The resulting trajectories in the positive quadrant $\mathbb{R}^2_+$ of $(x_1, x_2)$ state space are all periodic orbits given by $V(x) = $ constant, where

$$V(x) = \delta\left[\frac{x_1 - \hat{x}_1}{\hat{x}_1} - \ell n\left[\frac{x_1}{\hat{x}_1}\right]\right] + \beta\left[\frac{x_2 - \hat{x}_2}{\hat{x}_2} - \ell n\left[\frac{x_2}{\hat{x}_2}\right]\right]. \tag{52}$$

Note that, unlike the continuous–time Lotka–Volterra competition model (45), the prey–predator model (48) is *not structurally stable*: small changes in the differential equations can yield qualitatively different motions [34]. For example, if one of the diagonal terms in (51) is replaced by $0 \to -\epsilon$ then the equilibrium point changes from a neutrally stable center to an asymptotically stable focus.

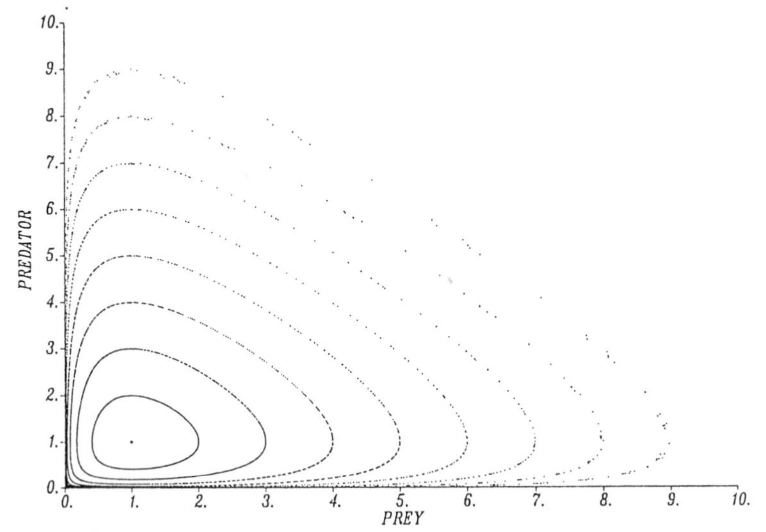


Figure 5. Continuous–time prey–predator trajectories.

Figure 5 shows a state–space plot of numerical solutions $\boldsymbol{x}(t_k) \approx \boldsymbol{x}(k\Delta t)$ to (48) from various initial conditions for the case where $\alpha = \beta = \gamma = \delta = 1$. The solutions were generated using the standard fourth–order Runge–Kutta algorithm with a time step of $\Delta t = 0.1$. Figure 5 shows the individual solution points $\boldsymbol{x}(k\Delta t)$, $k = 0,1,\ldots,100$, generated by the numerical procedure.

We purposely have not "connected the dots" in Figure 5, in order to show the cumulative errors that can occur even though the per–step error is of the order $10^{-5}$. In particular, note that at the top of the outer trajectory in Figure 5 the solution points do not lay on a smooth curve, as they would for an exact integration of (48). Figure 6 shows a single numerical trajectory $\boldsymbol{x}(t_k)$, generated by fourth–order Runge–Kutta with $\Delta t = 0.1$, $x_1(0) = 1$, $x_2(0) = 9$, and $t_k = k\Delta t$ with $k = 0,1,\ldots,10^5$. Instead of a periodic solution, which would be obtained by exact integration, fourth–order Runge–Kutta integration yields a numerical trajectory that asymptotically approaches a limit cycle. This behavior disappears if we take a small enough step size, say $\Delta t = 0.01$. However, from evidence that we develop later for a related numerical integration procedure, it may be that this behavior is a precursor to chaotic motion for fourth–order Runge–Kutta at other step sizes.

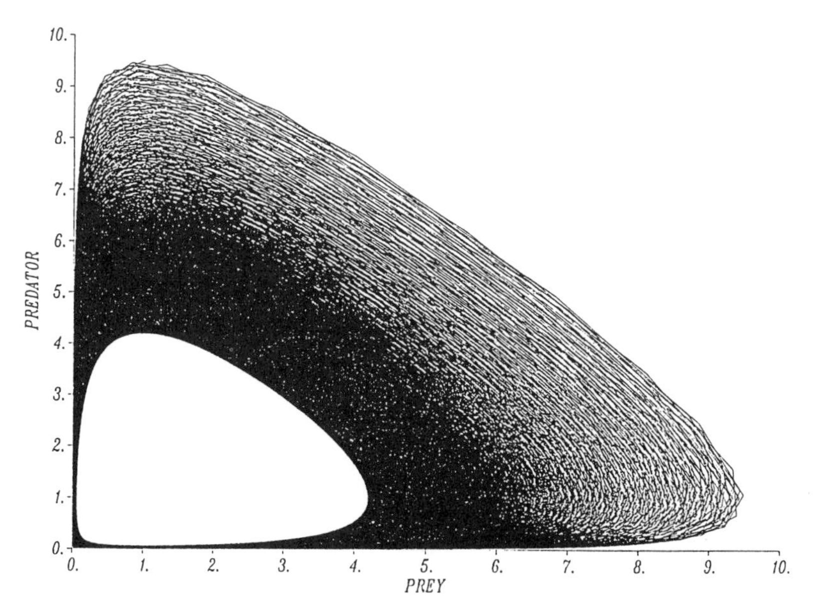


Figure 6. Long–term behavior of a fourth–order Runge–Kutta trajectory.

### B. Discrete–Time Pseudo–Euler Integration

Our purpose here is to investigate chaotic motion caused by numerical integration. For simplicity, instead of using fourth–order Runge-Kutta we will consider the "pseudo–Euler" numerical integration algorithm, akin to second–order Runge–Kutta, previously developed in Section III, B:

$$x(k+1) = G[x(k)]\ , \tag{53}$$

$$G(x) \triangleq x + \frac{\Delta t}{2}\{\ g[x] + g[x + g(x)\Delta\tau]\ \} \tag{54}$$

where $\Delta\tau$ and $\Delta t$ are parameters.

Peitgen and Richter [2] introduced the algorithm (53)–(54) and used it to study chaos in the resulting discrete–time system corresponding to the prey–predator system (48). They showed that chaos occurs in roughly an expanding parabolic band of points in $(\Delta t,\Delta\tau)$ parameter space. For $\Delta t$ values on the order of $0.01 \leq \Delta t \leq 0.8$ the chaotic region lies near the second–order Runge–Kutta condition $\Delta\tau = \Delta t$, but at values of $\Delta\tau$ below the $\Delta\tau = \Delta t$ second–order Runge–Kutta line. For $0.8 \leq \Delta t \leq 1$ chaos occurs for $\Delta\tau$ values above, below, and on the $\Delta\tau = \Delta t$ line.

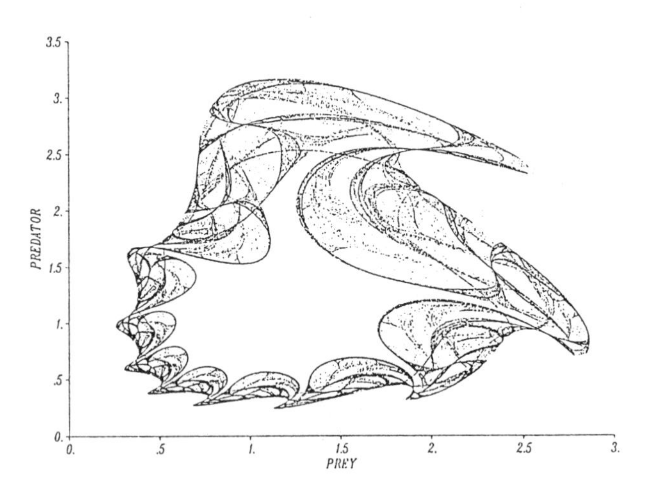


Figure 7. Chaotic motion in the pseudo–Euler prey–predator system.

Figure 7 shows the trajectory $x(k)$ for (48) generated by pseudo–Euler integration with $\Delta t = 0.7$ and $\Delta\tau = 0.63$ for the case where $\alpha = \beta = \gamma = \delta = 1$, with $x_1(0) = x_2(0) = 0.5$. This trajectory converges to a chaotic attractor that bears little resemblance to the periodic orbits in Figure 5 associated with the continuous-time system (48). Figures 8 and 9 show corresponding time plots of the prey population $x_1(k)$ and the predator population $x_2(k)$.

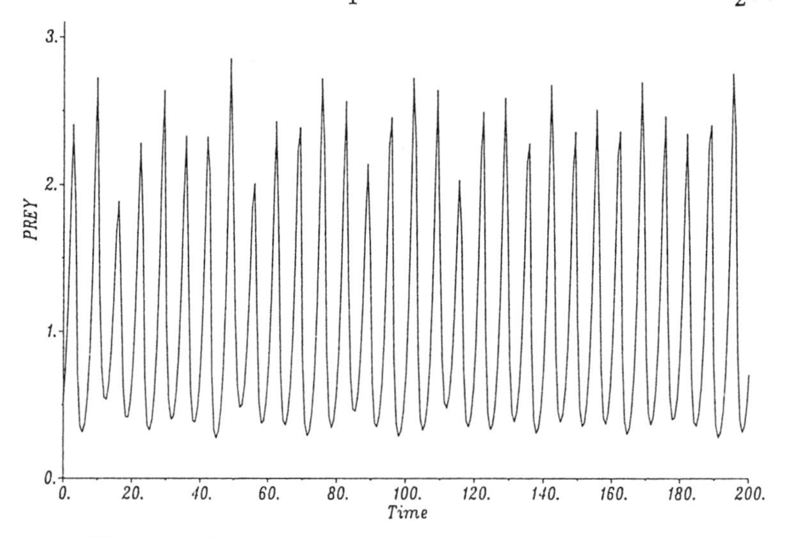


Figure 8. Prey population (pseudo–Euler simulation).

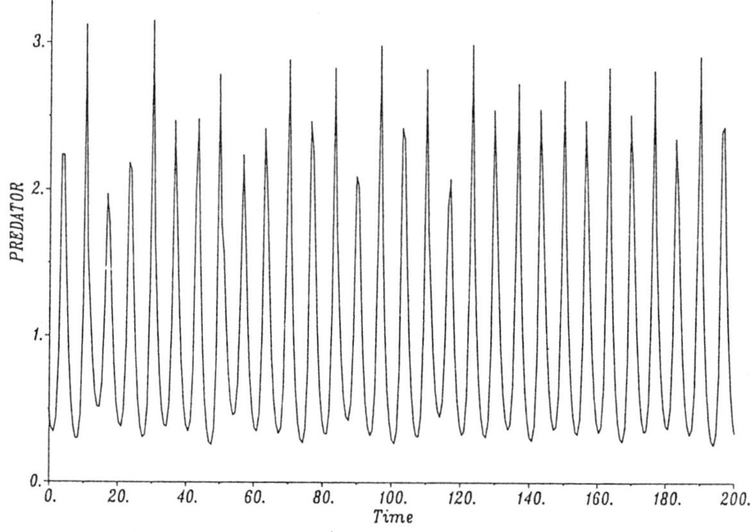


Figure 9. Predator population (pseudo–Euler simulation).

## VII. PREY–PREDATOR FEEDBACK CONTROL

To investigate the effects of a feedback control strategy applied to the continuous–time Lotka–Volterra prey–predator system (48), we add effort harvesting terms to each of equations (48), yielding a system of the form

$$\dot{\boldsymbol{x}} = \boldsymbol{f}(\boldsymbol{x},\boldsymbol{u}) \tag{55}$$

given by

$$\begin{aligned} \dot{x}_1 &= x_1[\alpha - \beta x_2 - u_1] &&= f_1(\boldsymbol{x},\boldsymbol{u}) \\ \dot{x}_2 &= x_2[-\gamma + \delta x_1 - u_2] &&= f_2(\boldsymbol{x},\boldsymbol{u})\ . \end{aligned} \tag{56}$$

The control vector $\boldsymbol{u}$ is to be chosen subject to maximum effort and no stocking constraints of the form

$$\begin{aligned} u_{1\min} &\le u_1 \le u_{1\max} \\ u_{2\min} &\le u_2 \le u_{2\max}\ , \end{aligned} \tag{57}$$

where $u_{1\min} = u_{2\min} = 0$ for no stocking. Later, for technical reasons associated with computing Lyapunov exponents, we will also consider slightly modified lower bounds $u_{1\min} = u_{2\min} = -\epsilon$, where $\epsilon \ge 0$ is very small. Until then, we consider $u_{1\min} = u_{2\min} = 0$.

### A. Linearized Control Design

We will design a feedback controller to stabilize the system (56) about the equilibrium point given by (50). In a neighborhood of this equilibrium point the linearized equations of motion are given by

$$\dot{\boldsymbol{\eta}} = \boldsymbol{A}\boldsymbol{\eta} + \boldsymbol{B}\boldsymbol{u}\ , \tag{58}$$

where $\boldsymbol{\eta} = \boldsymbol{x} - \hat{\boldsymbol{x}}$ is the state perturbation vector, $\hat{\boldsymbol{x}}$ is the equilibrium point (50) corresponding to $\boldsymbol{u} = \mathbf{0}$, and

$$A = \frac{\partial f(\hat{x},0)}{\partial x} = \begin{bmatrix} \alpha - \beta\hat{x}_2 & -\beta\hat{x}_1 \\ \delta\hat{x}_2 & -\gamma + \delta\hat{x}_1 \end{bmatrix} = \begin{bmatrix} 0 & -\beta\gamma/\delta \\ \delta\alpha/\beta & 0 \end{bmatrix} \tag{59}$$

$$B = \frac{\partial f(\hat{x},0)}{\partial u} = \begin{bmatrix} -\hat{x}_1 & 0 \\ 0 & -\hat{x}_2 \end{bmatrix} = \begin{bmatrix} -\gamma/\delta & 0 \\ 0 & -\alpha/\beta \end{bmatrix}. \tag{60}$$

Ignoring the control constraints (57) for the moment, we note that the pair $(A,B)$ is completely controllable for $\hat{\boldsymbol{x}} \neq \mathbf{0}$. Thus a linear feedback controller of the form

$$\boldsymbol{u} = K\boldsymbol{\eta}, \qquad K = \begin{bmatrix} k_{11} & k_{12} \\ k_{21} & k_{22} \end{bmatrix} \tag{61}$$

can be used to stabilize the $\hat{\boldsymbol{\eta}} = \mathbf{0}$, $\boldsymbol{u} = \mathbf{0}$ equilibrium point for the linear system (58), yielding a new linear closed–loop feedback system

$$\dot{\boldsymbol{\eta}} = \tilde{A}\boldsymbol{\eta}, \tag{62}$$

where

$$\tilde{A} = A + BK = \begin{bmatrix} -k_{11}\gamma/\delta & -(\beta + k_{12})\gamma/\delta \\ (\delta - k_{21})\alpha/\beta & -k_{22}\alpha/\beta \end{bmatrix}. \tag{63}$$

Furthermore, any desired set of eigenvalues for the closed–loop linear system (62), which are the roots of the characteristic equation

$$0 = \lambda^2 + [k_{11}\tfrac{\gamma}{\delta} + k_{22}\tfrac{\alpha}{\beta}]\lambda + \tfrac{\gamma\alpha}{\delta\beta}[k_{11}k_{22} + (\beta + k_{12})(\delta - k_{21})], \tag{64}$$

can be achieved by a suitable choice of the elements in the $K$ matrix.

From Routh's criteria, the origin will be an asymptotically stable equilibrium point for the linear system (62) if and only if all of the coefficients in (64) are positive. For simplicity, we consider $\alpha = \beta = \gamma = \delta = 1$ and choose

$$k_{11} = k_{22} = \mu > 0, \qquad k_{12} = k_{21} = 0\,, \tag{65}$$

yielding the linear feedback controller

$$u_1 = \mu\eta_1 \qquad u_2 = \mu\eta_2$$

and the eigenvalues

$$\lambda = -\mu \pm i$$

for the closed–loop linear system (62).

## B. Nonlinear Control Design

For the nonlinear system (56), with the control constraints (57), we employ the same controller as for the linearized system, but we allow the control values to saturate when they reach their upper or lower bounds. The *saturating linear feedback controller* for (56) is given explicitly by

$$u_i(x_i) = \begin{cases} u_{i\max} & \text{if } \mu(x_i - \hat{x}_i) > u_{i\max} \\ \mu(x_i - \hat{x}_i) & \text{if } u_{i\min} \le \mu(x_i - \hat{x}_i) \le u_{i\max} \\ u_{i\min} & \text{if } \mu(x_i - \hat{x}_i) < u_{i\min} \end{cases}, \tag{66}$$

where $i = 1,2$ and $\mu$ is the *gain* or *strength* of the controller.

Note that the control vector $\boldsymbol{u}(\boldsymbol{x})$ defined by (66) is continuous for all $\boldsymbol{x}$ and is continuously differentiable for almost all $\boldsymbol{x}$. That is, $\boldsymbol{u}(\boldsymbol{x})$ is only piecewise continuously differentiable. In particular, $\boldsymbol{u}(\boldsymbol{x})$ is not differentiable on the lines $\mu(x_i - \hat{x}_i) = u_{i\max}$ or $\mu(x_i - \hat{x}_i) = u_{i\min}$. On one side of such a line we have $\partial u_i/\partial x_i = \mu$ and on the other side we have $\partial u_i/\partial x_i = 0$. As a consequence, the controller is not differentiable at the equilibrium point for the case $u_{1\min} = u_{2\min} = 0$.

For the feedback controller given by (66) the prey–predator system (56) becomes a system of ordinary differential equations of the form

$$\dot{x} = g(x) \tag{67}$$

given by

$$\begin{aligned} \dot{x}_1 &= x_1[\alpha - \beta x_2 - u_1(x_1)] \\ \dot{x}_2 &= x_2[-\gamma + \delta x_1 - u_2(x_2)] \ . \end{aligned} \tag{68}$$

To verify that the feedback controller (66) yields asymptotic stability for the equilibrium point $\hat{x}$ defined by (50), consider the function $V(x)$ given in (52). We note that $V(\hat{x}) = 0$. In addition, for all $x \neq \hat{x}$ in the positive quadrant $\mathbb{R}^2_+$, we have $V(x) > 0$, and at points $x = x(t)$ along a trajectory of (68) we have

$$\frac{dV[x(t)]}{dt} = \frac{\partial V(x)}{\partial x_1}\dot{x}_1 + \frac{\partial V(x)}{\partial x_2}\dot{x}_2$$

$$= -\delta[x_1 - \hat{x}_1]u_1(x_1) - \beta[x_2 - \hat{x}_2]u_2(x_2) \ . \tag{69}$$

From (66), each term $[x_i - \hat{x}_i]u_i(x_i)$ in (69) is positive for $x_i \neq \hat{x}_i$, except for the case where $\mu[x_i - \hat{x}_i] \leq u_{i\min}$ and $u_{i\min} \leq 0$. In our application we have $u_{1\min} = u_{2\min} = 0$. Therefore $dV[x(t)]/dt < 0$ on $\mathbb{R}^2_+ - \{\hat{x}\}$, except in the region defined by $0 < x_i \leq \hat{x}_i$, $i = 1,2$. In this rectangular region we have

$V[\boldsymbol{x}(t)] =$ constant, corresponding to the system (48) for $\boldsymbol{u}(\boldsymbol{x}) = \boldsymbol{0}$. However, since the solutions to (48) are closed periodic orbits surrounding the point $\hat{\boldsymbol{x}}$ at the upper right–hand corner of the rectangle, every trajectory that enters this region ultimately leaves and enters a region where $dV[\boldsymbol{x}(t)]/dt < 0$. Thus we conclude that $V[\boldsymbol{x}(t)] \to 0$, hence $\boldsymbol{x}(t) \to \hat{\boldsymbol{x}}$, as $t \to \infty$.

In summary, $V(\boldsymbol{x})$ is a Lyapunov function on $\mathbb{R}^2_+ - \{\hat{\boldsymbol{x}}\}$, implying that $\hat{\boldsymbol{x}}$ is asymptotically stable and that its basin of attraction is all of $\mathbb{R}^2_+$. Thus, for every initial condition $\boldsymbol{x}(0)$ with positive populations, the corresponding solution $\boldsymbol{x}(t)$ to the continuous–time feedback control prey–predator system (68) converges asymptotically to the positive equilibrium point (50) as $t \to \infty$, under the action of the feedback controller given by (66).

## VIII. PREY–PREDATOR CONTROL SYSTEM SIMULATION

All of the results presented in this Section are for the discrete–time prey–predator feedback control system obtained by applying the pseudo-Euler integration method (53)–(54) to (67)–(68), using the saturating feedback control in (66), with $u_{1\min} = u_{2\min} = 0$ and $u_{1\max} = u_{2\max} = 1$. Figures 10–32 show state–space trajectories or attractors for various decreasing values of the controller gain $\mu$. In all cases the dynamic system parameters in (68) are $\alpha = \beta = \gamma = \delta = 1$, with the pseudo–Euler step sizes $\Delta t = 0.7$ and $\Delta \tau = 0.63$. The initial conditions are usually $x_1(0) = x_2(0) = 0.5$. Occasionally, to test alternate trajectories from the opposite side of an attractor, the initial conditions $x_1(0) = x_2(0) = 0.99$ are also used.

Figures 10 and 11 show pseudo–Euler state–space trajectories for controller strengths of $\mu = 1$ and $0.5$, respectively. In these two figures we have connected the points $\boldsymbol{x}(k)$ by straight lines to indicate the counter-clockwise motion, usually with nine points per revolution, that occurs in all of the results presented in this Section. The remaining trajectory and attractor figures show only the actual points $\boldsymbol{x}(k)$.

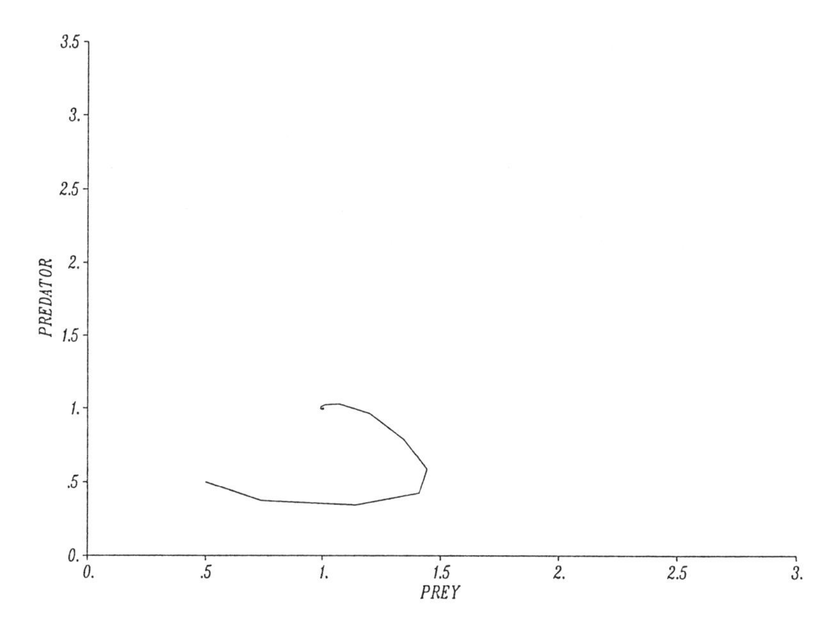


Figure 10. Pseudo–Euler trajectory → stable focus ($\mu = 1$).

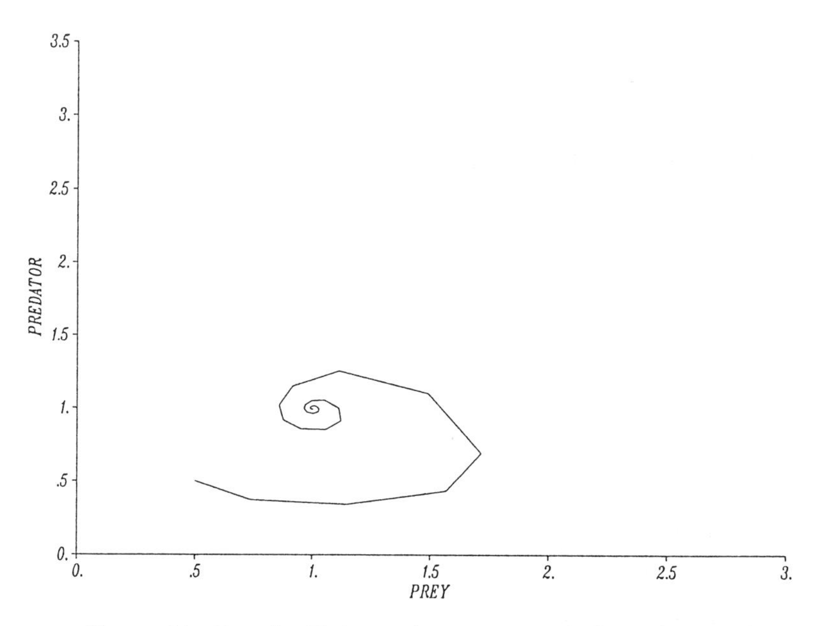


Figure 11. Pseudo–Euler trajectory → stable focus ($\mu = 0.5$).

Figures 12–15 show that as the strength $\mu$ of the controller is decreased the attractor changes (at $\mu \approx 0.1183$) from an asymptotically stable focus to what would be called an asymptotically stable limit cycle in a continuous-time system. For simplicity we will call this new attractor a "limit cycle", but for the discrete–time system the attractor, best illustrated by the oval in Figure 15, is a quasiperiodic orbit: with each counter–clockwise revolution points on the attractor are displaced slightly and do not appear to repeat.

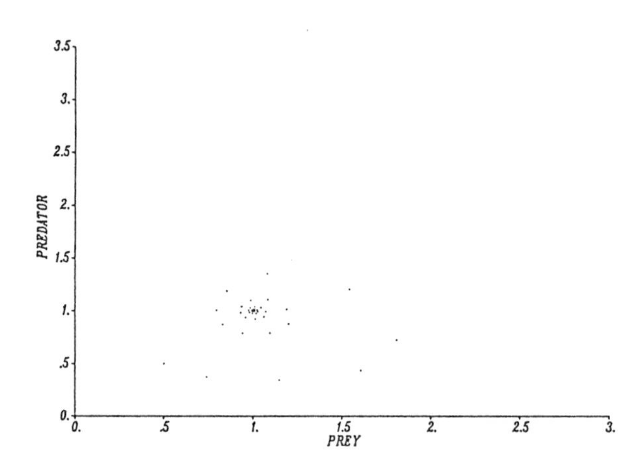


Figure 12. Pseudo–Euler trajectory → stable focus ($\mu = 0.4$).

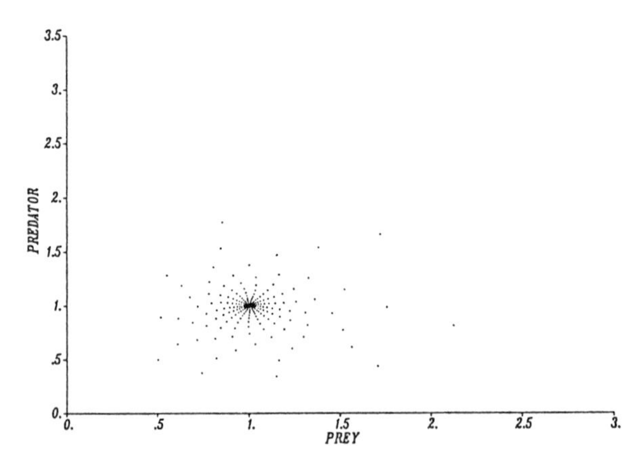


Figure 13. Pseudo–Euler trajectory → stable focus ($\mu = 0.15$).

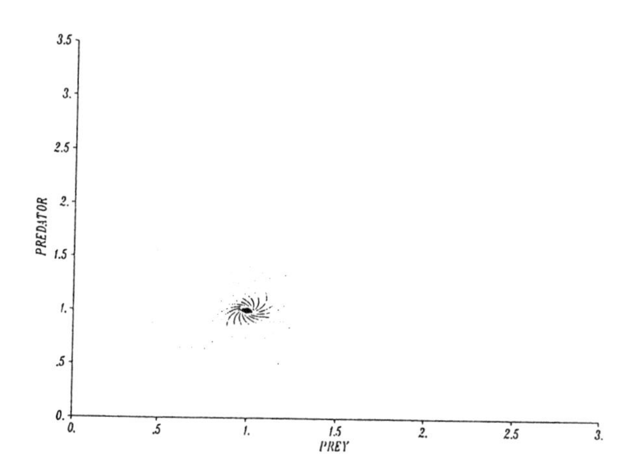


Figure 14. Pseudo–Euler trajectory → tiny "limit cycle" ($\mu = 0.118$).

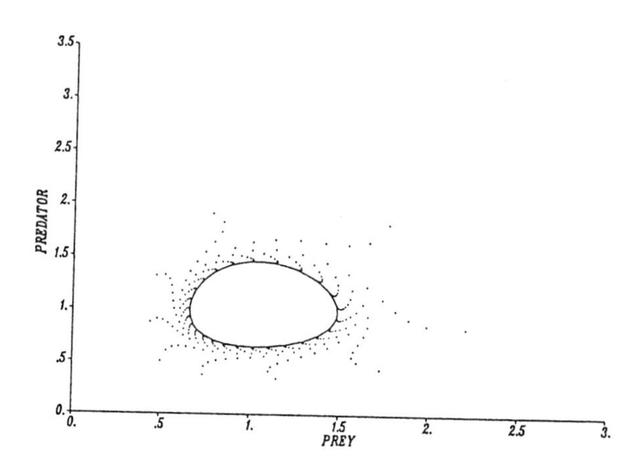


Figure 15. Pseudo–Euler trajectory → "limit cycle" ($\mu = 0.1$).

In a continuous–time system such a transition from a fixed–point attractor to a limit cycle would correspond to a Hopf bifurcation. As we shall see in the next section the transition to chaos for this system is not by way of Hopf bifurcations but rather via period doubling.

As illustrated in Figure 16, as the controller strength is decreased to a value of $\mu = 0.079774$ the trajectory from $x_1(0) = x_2(0) = 0.5$ converges to a 9–point cycle, while the trajectory from $x_1(0) = x_2(0) = 0.99$ converges to a quasiperiodic "limit cycle". The two attractors are shown in Figure 17.

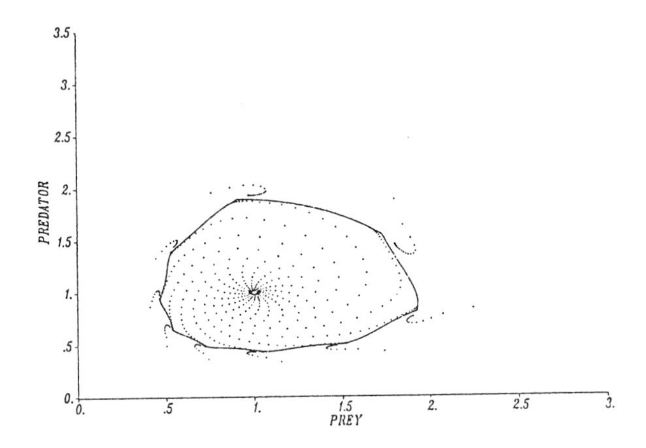


Figure 16. Trajectories → 9–cycle and limit cycle ($\mu = 0.079774$).

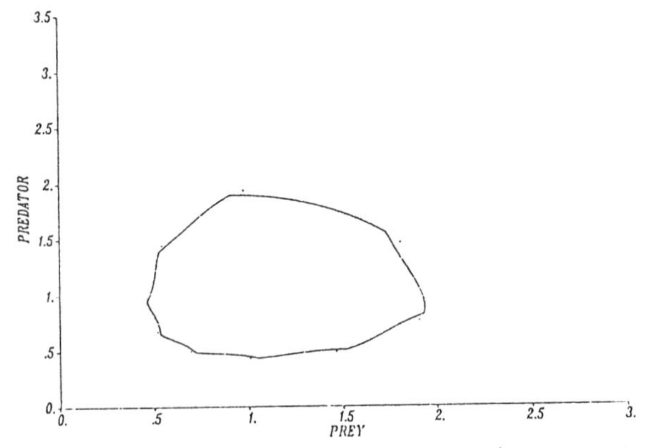


Figure 17. 9–cycle + limit cycle attractors ($\mu = 0.079774$).

The 9–cycle + limit cycle phenomenon in Figures 16–17 persists only for a very small range of $\mu$ values. At $\mu = 0.079772$ the quasiperiodic attractor vanishes, leaving only a 9–point cycle, as shown in Figure 18.

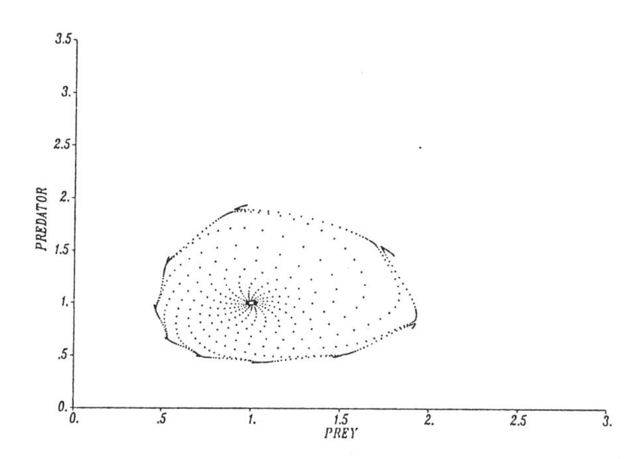


Figure 18. Trajectories → 9–cycle ($\mu = 0.079772$).

In Figure 19, for $\mu = 0.079$, trajectories from $x_1(0) = x_2(0) = 0.5$ and $x_1(0) = x_2(0) = 0.99$ converge to a 9–point cycle, shown in Figure 20.

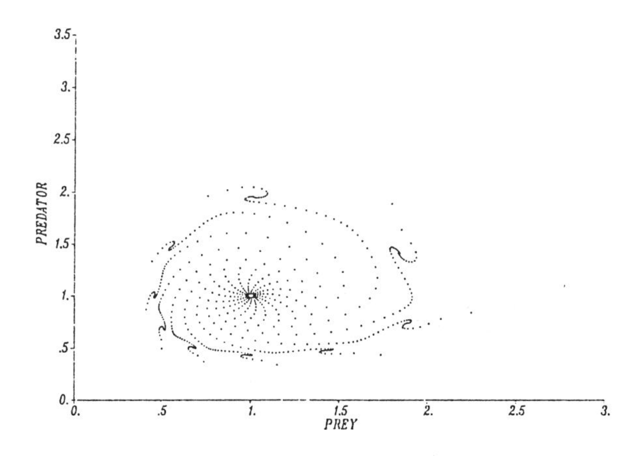


Figure 19. Trajectories → 9–cycle ($\mu = 0.079$).

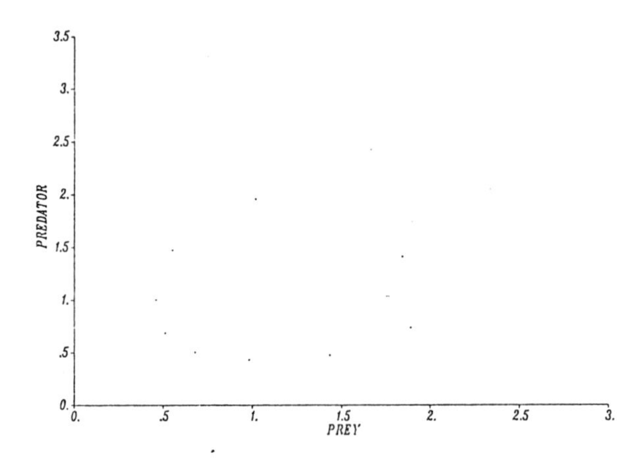


Figure 20. 9–cycle attractor ($\mu = 0.079$).

The 9–cycle in Figure 20 is not the only set of fixed points existing for $\mu = 0.079$. Figure 21 shows a *9–th iterate map* for this case, constructed by successively connecting each point $\boldsymbol{x}(9k)$, k = 3,4,... with the point $\boldsymbol{x}(9[k+1])$ on the next revolution of the trajectory, for various initial conditions. Figure 21 shows how the points $\boldsymbol{x}(9k)$ approach the 9–cycle attractor. Apparently there also exist nine unstable fixed points, or possibly an unstable 9–cycle.

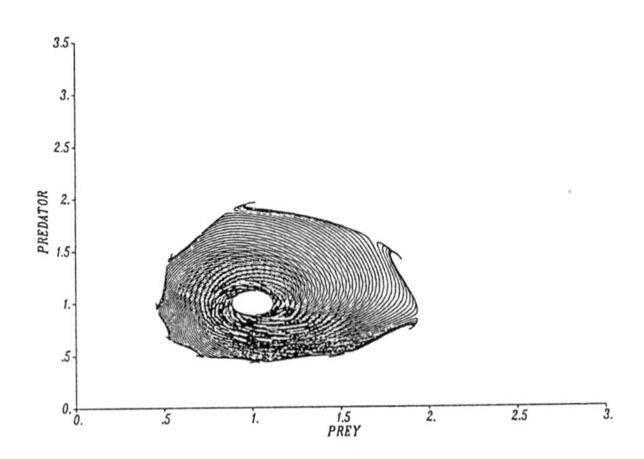


Figure 21. 9–th iterate map ($\mu = 0.079$).

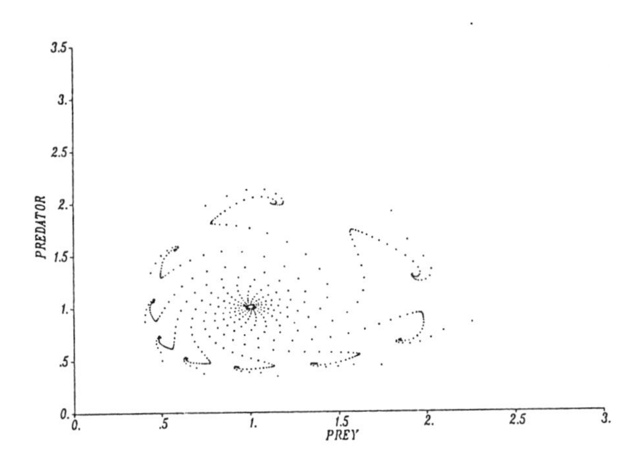


Figure 22. Trajectories → 9–cycle ($\mu = 0.075$).

Figure 22, for $\mu = 0.075$, also shows trajectories converging to a 9–cycle, with the points in the 9–cycle becoming more like foci than nodes.

In Figure 23, for $\mu = 0.05$, the 9–cycle has vanished and the attractor is once again a quasiperiodic "limit cycle".

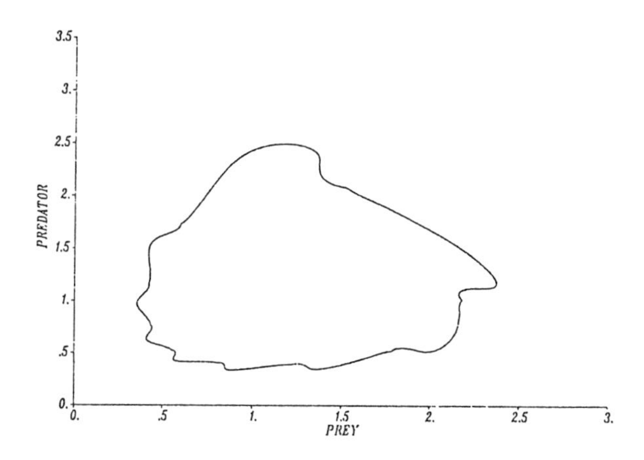


Figure 23. Quasiperiodic attractor ($\mu = 0.05$).

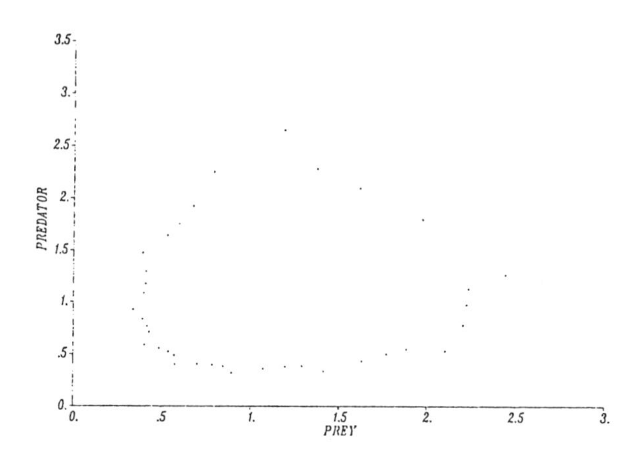


Figure 24. 37–cycle attractor ($\mu = 0.04$).

As illustrated in Figure 24, at $\mu = 0.04$ the quasiperiodic limit cycle has become (approximately) a 37–point cycle.

Figure 25 shows the trajectory from $x_1(0) = x_2(0) = 0.5$ for $\mu = 0.03$. While the transient portion of the trajectory may appear to be chaotic, the attractor, shown in Figure 26, is a nonchaotic 65–cycle (approximately).

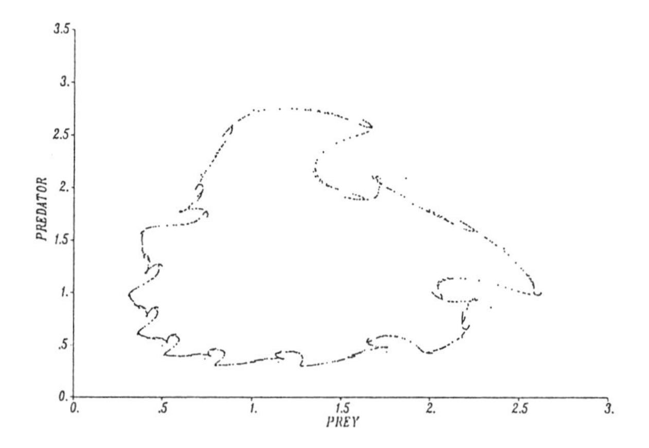


Figure 25. Trajectory → 65–cycle ($\mu = 0.03$).

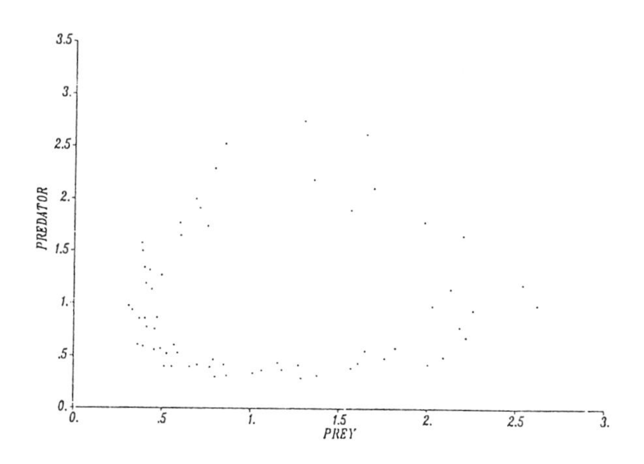


Figure 26. 65–cycle attractor ($\mu = 0.03$).

For $\mu = 0.02$ the attractor, shown in Figure 27, is a 28–point cycle.

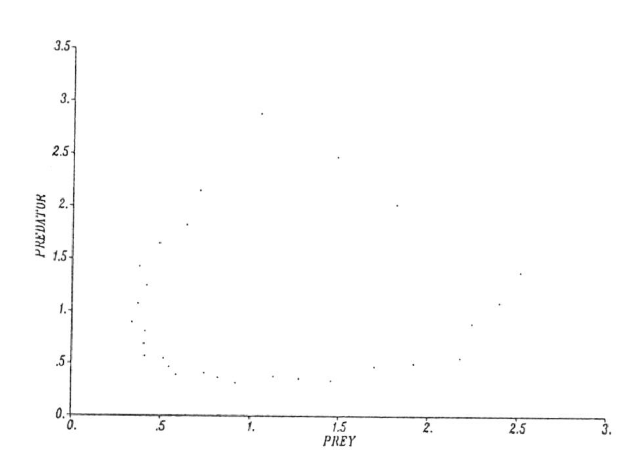


Figure 27. 28–cycle attractor ($\mu = 0.02$).

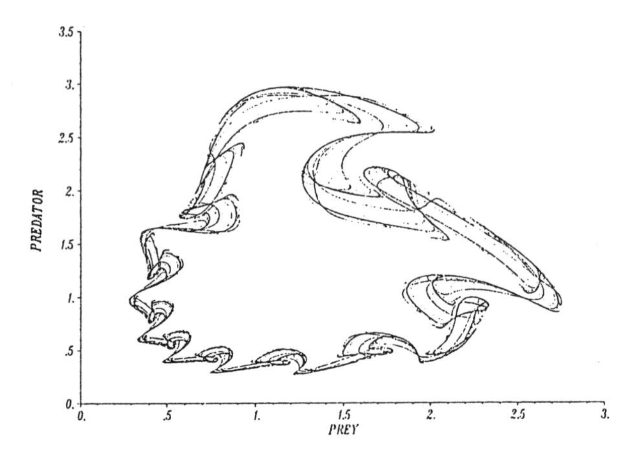


Figure 28. Chaotic attractor ($\mu = 0.016$).

For $\mu = 0.016$ the attractor, shown in Figure 28, has become chaotic. This figure was constructed by suppressing the first $10^4$ seconds of transient motion (with the two pseudo–Euler time steps being $\Delta t = 0.7$ sec and $\Delta\tau = 0.63$ sec) and then plotting $x(k)$ until $t_k = k\Delta t = 10^6$ seconds.

At $\mu = 0.01$ the trajectory shown in Figure 29 has a long ($10^4$ second) nearly chaotic transient, but then the trajectory converges to the 47–cycle attractor shown in Figure 30.

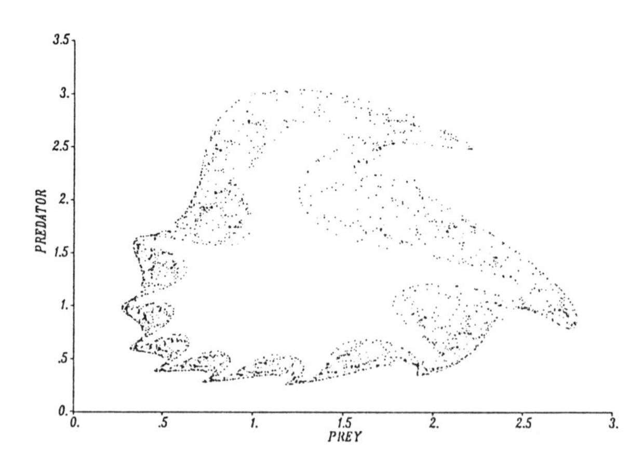


Figure 29. Trajectory $\rightarrow$ 47–cycle ($\mu = 0.01$).

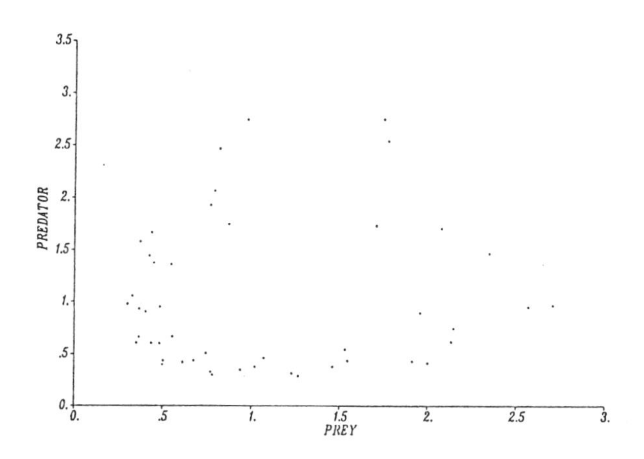


Figure 30. 47–cycle attractor ($\mu = 0.01$).

Even though the attractor in Figure 30 is a periodic cycle, its basin of attraction, shown as the dark region in Figure 31, apparently has a fractal basin boundary.

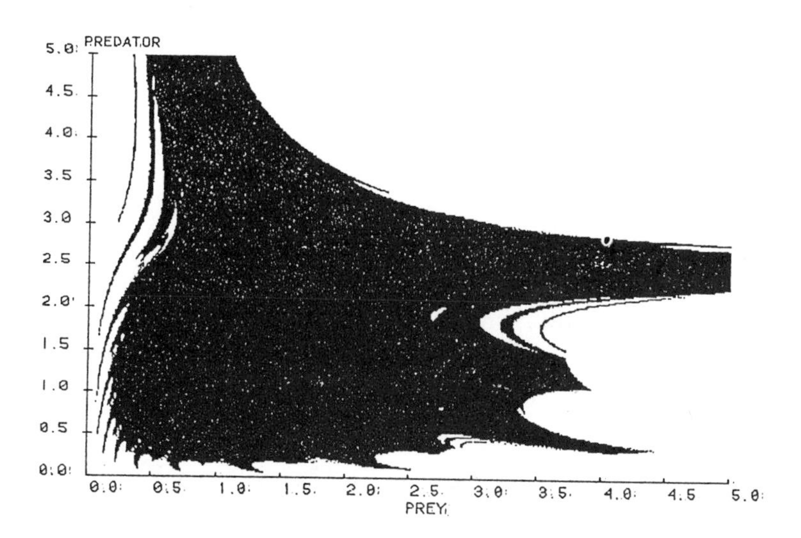


Figure 31. Fractal basin of attraction boundary ($\mu = 0.01$).

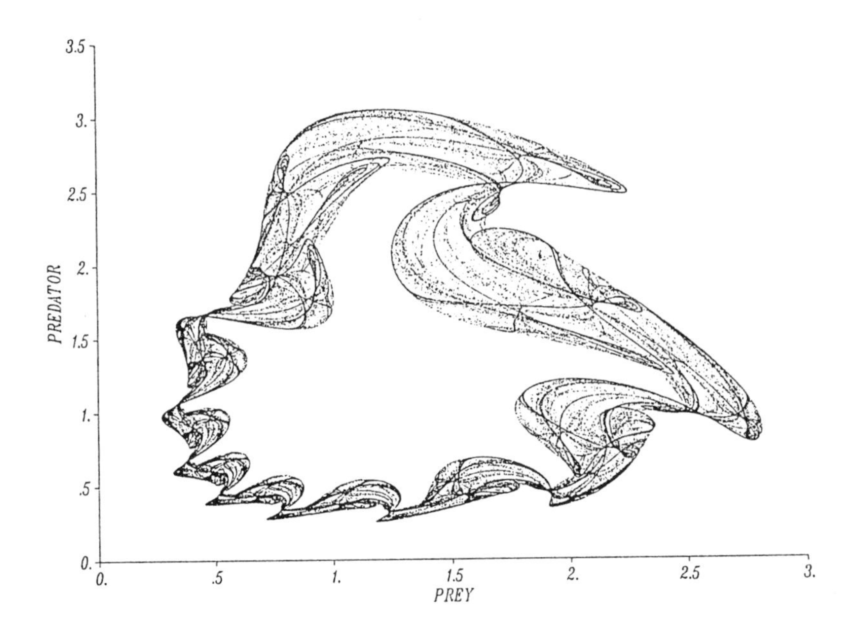


Figure 32. Chaotic Attractor ($\mu = 0.009$).

At a controller strength of $\mu = 0.009$ the attractor, shown in Figure 32, is again chaotic. This figure was constructed by plotting for $5{\times}10^5$ seconds, with the first $10^4$ seconds suppressed.

## IX. TRANSITION TO CHAOS

The simulation results in Section VIII do not indicated clearly the mechanism by which transition to chaos occurs in the pseudo–Euler simulation of the prey–predator feedback control system. The transition of the attractor from a fixed point to a quasiperiodic "limit cycle" suggests Hopf bifurcations and the Ruelle–Takens route to chaos [13], even though the "limit cycle" is not a true limit cycle. On the other hand, the occurrence of $m$–period cycles, with $m = 9$, 37, 65, suggests a period doubling route [13].

Part of the difficulty is that the simulation results do not cover all possible values for the controller gain $\mu$. Even if they did, state–space trajectory plots are not necessarily the best way to discover the occurrence of chaos. In this Section we will present two types of results that are very useful for determining and characterizing the transition to chaos.

## A. Bifurcation Diagrams

Figures 33–36 show a sequence of expanded views of the "steady–state" value of $x_1(k)$, $k$ large, for ranges of $\mu$ values in steps of $\Delta\mu = 10^{-4}$, $10^{-5}$, $10^{-6}$, and $5\times10^{-8}$, respectively. These figures were constructed by applying the pseudo–Euler integration algorithm to the feedback control system (67)-(68), using the controller in (66) with $u_{1\min} = u_{2\min} = 0$. The parameters in (68) are $\alpha = \beta = \gamma = \delta = 1$, the initial conditions are $x_1(0) = x_2(0) = 0.5$, and the pseudo–Euler time step parameters are $\Delta t = 0.7$ and $\Delta\tau = 0.63$.

For each value of $\mu$ the resulting values of $x_1(k)$ on the trajectory were plotted as points, until $k\Delta t = 10^4$ seconds ($4\times10^4$ seconds for Figure 36), with the initial transient points suppressed. In Figures 33, 34–35, and 36 the first 2000, 4000, 20000 seconds were suppressed, respectively.

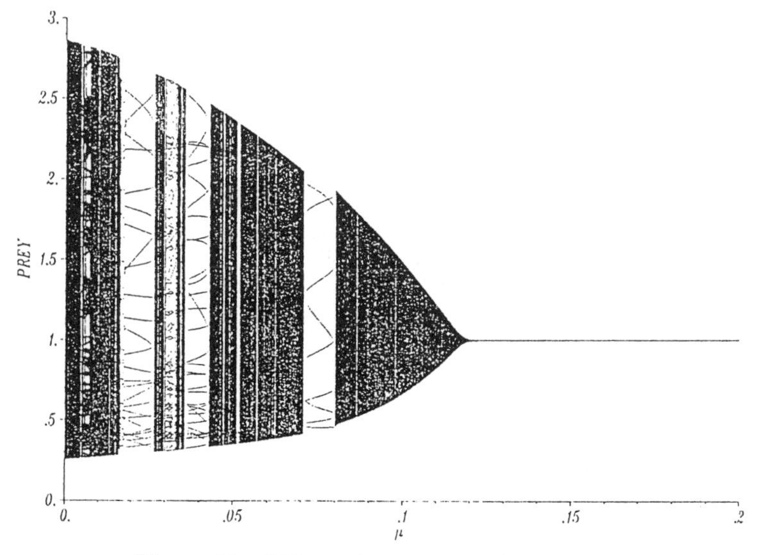


Figure 33. Bifurcation diagram ($\Delta\mu = 10^{-4}$).

Figure 33 clearly indicates the single fixed point attractor for $\mu$ values above about 0.12, which changes to an increasingly larger quasiperiodic "limit cycle" near this $\mu$ value, and then to a 9–cycle attractor in the approximate range $0.07 \leq \mu \leq 0.08$, with various other attractors corresponding to the vertical lines and points at other values of $\mu$. The "limit cycles", for example near the range $0.08 \leq \mu \leq 0.12$ in Figure 33, correspond to the dark vertical lines, indicating that the "steady–state" $x_1(k)$ values cover an interval rather than a single point or a set of discrete points.

The apparent transition from a fixed point attractor to a "limit cycle" at $\mu = 0.12$ in Figure 33, instead of at $\mu = 0.1183$, corresponding to Figure 14, is caused by the fact that the transient time of 2000 seconds is not long enough for the actual transient to die out: as plotting begins the trajectory is still slowly spiraling onto the limit cycle. For this reason subsequent bifurcation diagrams employ a 4000 second transient time, with an even longer transient time of 20000 seconds employed for Figure 36.

Figure 34 shows an expanded view of a region $0 \leq \mu \leq 0.02$ where chaos appears to occur in the state–space simulations of Section VIII. As we will see in the next Section, bands of chaos also occur for $0.025 < \mu < 0.037$.

In Figures 34–36, for $0.016 \leq \mu \leq 0.018$, we see expanded views of a sequence of period doubling bifurcations leading to a chaotic regime. The best view of this phenomenon occurs in Figure 36.

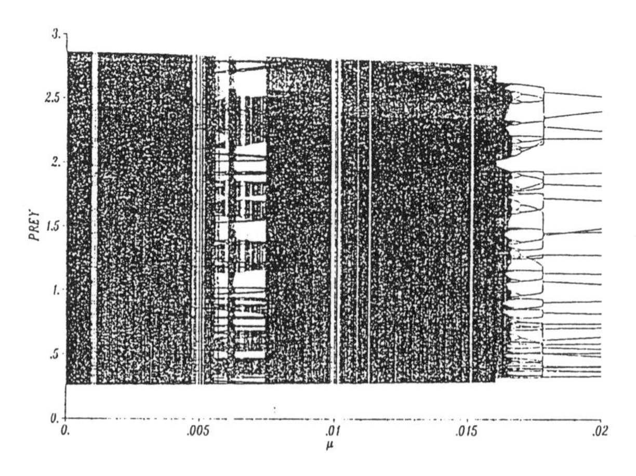


Figure 34. Bifurcation diagram ($\Delta\mu = 10^{-5}$).

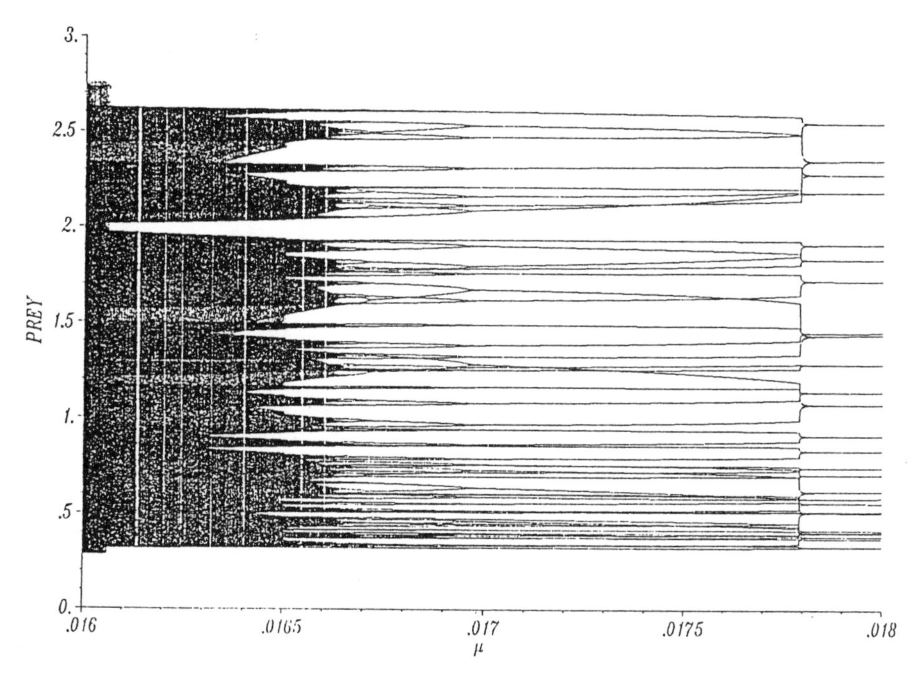


Figure 35. Bifurcation diagram ($\Delta\mu = 10^{-6}$).

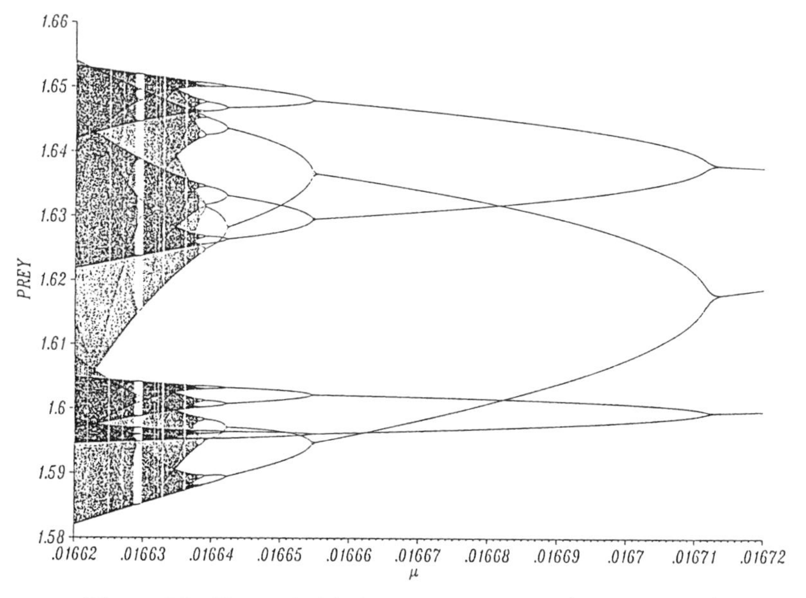


Figure 36. Expanded bifurcation diagram ($\Delta\mu = 5{\times}10^{-8}$).

From Figures 33–35 we see that as $\mu$ decreases below about 0.026 the attractor, which we shall see later is chaotic, apparently becomes a 27–cycle, although Figure 28 indicates a 28–cycle. Perhaps one of the lines in Figures 33–35 is actually two lines, or the discrepancy may be due to the fact that Figures 33–35 only account for the first $10^4$ seconds, whereas Figure 24 plots 1000 seconds, after suppressing the first $2\times10^4$ seconds of transient motion.

| $m$ | $\mu_m$ | $\rho_m$ |
|---|---|---|
| 0 | 0.01781 | - |
| 1 | 0.0167138 | 18.7385 |
| 2 | 0.0166553 | 4.5 |
| 3 | 0.0166423 | 3.8235 |
| 4 | 0.0166389 | 4.5333 |
| 5 | 0.01663815 | - |

Table I. Period Doubling Bifurcations.

Figure 36 shows a period doubling sequence for three of the 27 (or 28) branches in Figures 33–35 that lead to chaos. Table I shows the values $\mu_m$ at which the total number of cycles changes from $(2^m)N$ to $(2^{m+1})N$ as $\mu$ is decreased, where $N = 27$ (or 28). The values $\rho_m$ in Table I are the ratios of the distances between successive bifurcations:

$$\rho_m \triangleq \frac{\mu_m - \mu_{m-1}}{\mu_{m+1} - \mu_m}. \tag{70}$$

The value $\rho_4 = 4.533...$ in Table I differs by less than 3% from Feigenbaum's universal period doubling value [13] of

$$\rho = \lim_{m\to\infty} \{\rho_m\} = 4.6692016091...$$

Based on Feigenbaum's value for $\rho$, the bifurcations in Figure 36 scale as

$$\mu_m = \mu_\infty + c\rho^{-m} , \qquad m \gg 1 , \tag{71}$$

where $c = 0.000453629$ and $\mu_\infty = 0.016637946$. The value of $\mu_\infty$, below which one region of chaos occurs, agrees with the graphical results in Figure 36.

## B. Lyapunov Exponents and Fractal Dimension

In addition to bifurcation diagrams and state–space simulations, it is also important to examine the occurrence of chaos in terms of Lyapunov exponents and the (Kaplan–Yorke) fractal dimension of the attractors that result from various values of the controller gain parameter $\mu$.

The simulation results in Sections VIII and IX,A employed the pseudo-Euler integration algorithm (53)–(54) applied to the prey-predator system (55)–(56), with the feedback control given by (66). Thus the simulation results correspond to the three–dimensional discrete–time system

$$x(k+1) = G[x(k)]$$

$$\triangleq x(k) + \frac{\Delta t}{2}\Big[f\{x(k),u[x(k)]\} + f\{y(k),u[y(k)]\}\Big] , \tag{72}$$

where $x = [x_1,x_2,x_3]^{\mathrm{T}}$ with $x_3 = t$, $y(k) = y[x(k)]$ with

$$y(x) = x + f[x,u(x)]\Delta\tau , \tag{73}$$

$f(x,u) = [f_1(\cdot),f_2(\cdot),f_3(\cdot)]^{\mathrm{T}}$ with $f_1(\cdot)$ and $f_2(\cdot)$ defined by (56) for $\alpha = \beta = \delta = \gamma = 1$ and $f_3(\cdot) \equiv 1$, and $u(x)$ given by (66) with $u_{1\min} = u_{2\min} = 0$ and $u_{1\max} = u_{2\max} = 1$.

For this discrete–time system the three–dimensional linearized state perturbation equations

$$\eta(k+1) = J(k)\eta(k) \tag{74}$$

have a 3×3 Jacobian matrix given by

$$J(k) = \frac{dG[\boldsymbol{x}(k)]}{d\boldsymbol{x}} = \boldsymbol{I} + \frac{\Delta t}{2}\left[\frac{\partial \boldsymbol{f}(\boldsymbol{x},\boldsymbol{u})}{\partial \boldsymbol{x}} + \frac{\partial \boldsymbol{f}(\boldsymbol{x},\boldsymbol{u})}{\partial \boldsymbol{u}}\frac{\partial \boldsymbol{u}(\boldsymbol{x})}{\partial \boldsymbol{x}}\right] + \frac{\Delta t}{2}\left[\frac{\partial \boldsymbol{f}(\boldsymbol{y},\boldsymbol{v})}{\partial \boldsymbol{y}} + \frac{\partial \boldsymbol{f}(\boldsymbol{y},\boldsymbol{v})}{\partial \boldsymbol{v}}\frac{\partial \boldsymbol{u}(\boldsymbol{y})}{\partial \boldsymbol{y}}\right]\frac{\partial \boldsymbol{y}(\boldsymbol{x})}{\partial \boldsymbol{x}}, \quad (75)$$

where

$$\frac{\partial \boldsymbol{y}(\boldsymbol{x})}{\partial \boldsymbol{x}} = \boldsymbol{I} + \Delta\tau\left[\frac{\partial \boldsymbol{f}(\boldsymbol{x},\boldsymbol{u})}{\partial \boldsymbol{x}} + \frac{\partial \boldsymbol{f}(\boldsymbol{x},\boldsymbol{u})}{\partial \boldsymbol{u}}\frac{\partial \boldsymbol{u}(\boldsymbol{x})}{\partial \boldsymbol{x}}\right] \quad (76)$$

and the terms in (75)–(76) are evaluated at $\boldsymbol{x} = \boldsymbol{x}(k)$, $\boldsymbol{u} = \boldsymbol{u}[\boldsymbol{x}(k)]$, $\boldsymbol{y} = \boldsymbol{y}[\boldsymbol{x}(k)]$, and $\boldsymbol{v} = \boldsymbol{u}(\boldsymbol{y}) = \boldsymbol{u}\{\boldsymbol{y}[\boldsymbol{x}(k)]\}$.

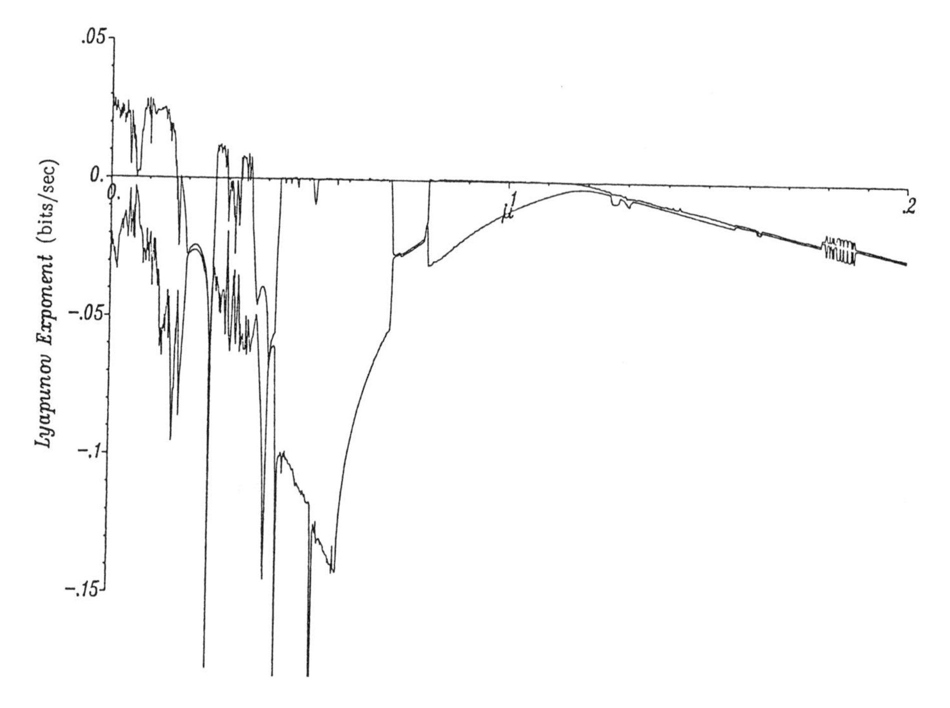


Figure 37. Lyapunov exponents for discrete–time prey–predator system.

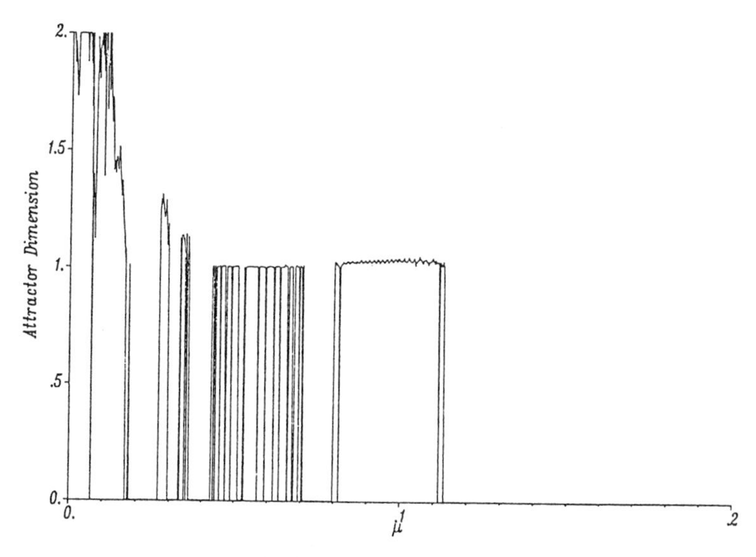


Figure 38. Fractal dimension of prey–predator attractors.

Figure 37 shows a plot of the Lyapunov exponents $\sigma_i$, i = 1,2,3, with one of the exponents, $\sigma_3$, corresponding to the time dimension being zero. The results in Figure 37 were computed, in steps of $\Delta\mu = 2\times10^{-4}$, by applying the perturbation equations (73) to a set of $n = 3$ initial condition vectors $[\eta_1(0),\eta_2(0),\eta_3(0)] = I_{3\times3}$. For each $\mu$ value pseudo–Euler integration was used to generate the reference trajectory $\boldsymbol{x}(k)$ starting from $x_1(0) = x_2(0) = 0.5$, $x_3(0) = 0$. The perturbation vectors $\boldsymbol{\eta}_i(k)$ were reorthonormalized after each $\Delta t = 0.7$ seconds, using the Gramm–Schmidt algorithm given by (30). The results were generated for 4000 seconds and the final values of the Lyapunov exponents were then plotted.

Figure 38 shows a plot of the Kaplan–Yorke fractal dimension computed from (44) using the Lyapunov exponents in Figure 37. Chaos occurs for $\mu$ values corresponding to a positive value for the largest Lyapunov exponent in Figure 37 and a fractional dimension in Figure 38.

Figures 37 and 38 give excellent agreement with the simulation and bifurcation diagram results presented previously. Any slight discrepancies are due to the fact that Figures 37 and 38 only employ a 4000 second transient time, whereas the actual transient time may be longer for some $\mu$ values, such as at $\mu = 0.01$.

It should also be noted that the generation of bifurcation diagrams, Lyapunov exponent plots, and plots of fractal dimension are all very intensive computationally and the study of chaos is a prime candidate for the use of supercomputers. For example, each of Figures 36–38 took approximately 36 hours to compute and plot, using an IBM PS/2 model 70–386 personal computer with an 80387 numeric coprocessor, both operating at 20 MHz.

### C. Eigenvalues, Open–Loop Controls, and Discontinuous Systems

In this Section we present some results on how *not* to compute Lyapunov exponents.

Recall from the discussion in Section IV,B that the Lyapunov exponents can be computed in terms of the eigenvalues of the state transition matrix as time approaches infinity. For our discrete–time system this matrix, given by (42), is the product of the Jacobian matrices $J(k)$.

Figures 39–41 show plots of the eigenvalues (with a slight abuse of notation) of the instantaneous Jabobian matrices $J(k)$ along the reference trajectory used for Figures 33–38 and for most of the simulation results. Clearly, the instantaneous eigenvalues bear little relation to the Lyapunov exponents. This situation is analogous to the case of a time–varying linear system $\dot{x} = A(t)x$, where the instantaneous eigenvalues of $A(t)$ have no general connection with the stability of the system.

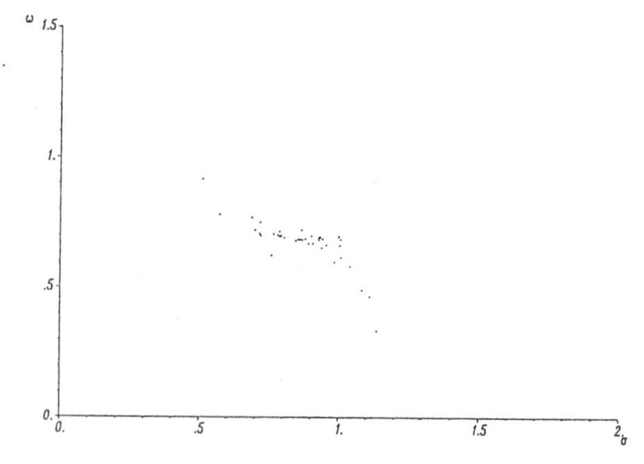


Figure 39. Instantaneous eigenvalues $\lambda = \sigma \pm i\omega$ of $J(k)$ $(\mu = 0.4)$.

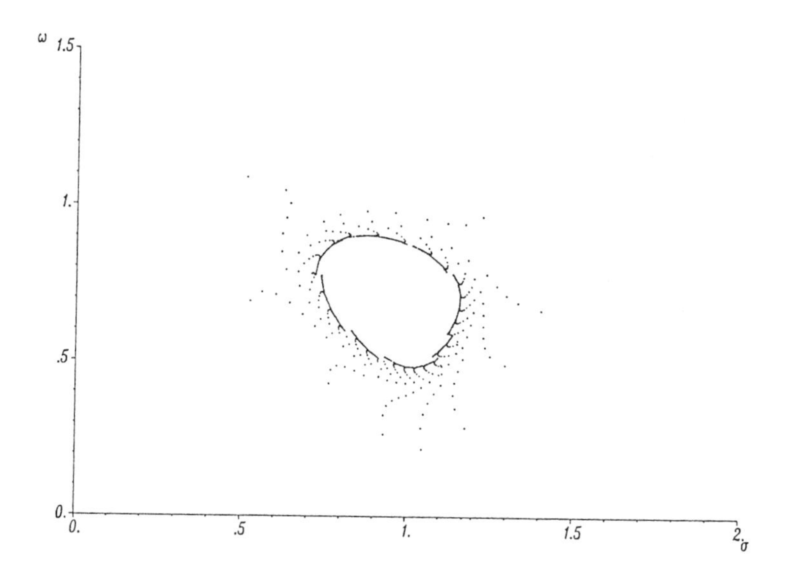


Figure 40. Instantaneous eigenvalues $\lambda = \sigma \pm i\omega$ of $J(k)$ $(\mu = 0.1)$.

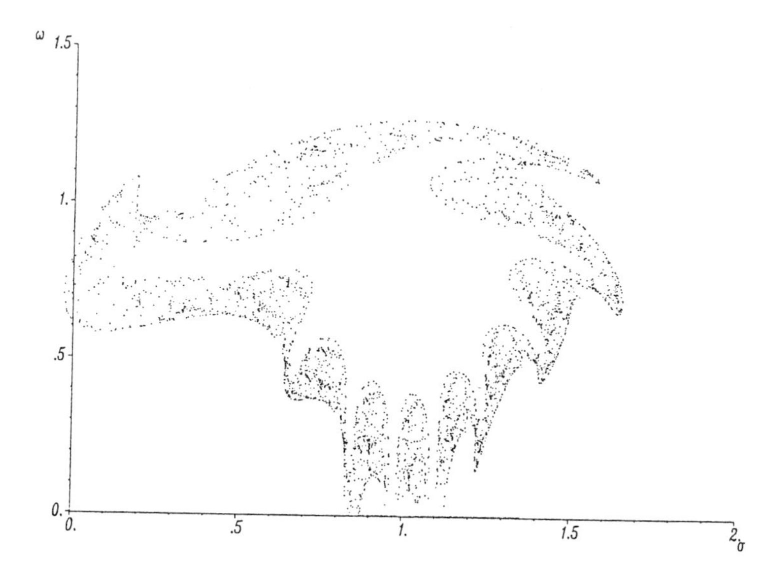


Figure 41. Instantaneous eigenvalues $\lambda = \sigma \pm i\omega$ of $J(k)$ $(\mu = 0.01)$.

Figures 42 and 43 show erroneous results for the Lyapunov exponents and fractal dimension, respectively, obtained by using an *open–loop control* $\boldsymbol{u}(k)$ instead of a closed–loop control $\boldsymbol{u}[\boldsymbol{x}(k)]$. These plots were generated by using the same reference trajectory $\boldsymbol{x}(k)$ and the same time sequence of control inputs $\boldsymbol{u}(k) = \boldsymbol{u}[\boldsymbol{x}(k)]$ as for Figures 37 and 38, but with $\partial\boldsymbol{u}(\boldsymbol{x})/\partial\boldsymbol{x} = \partial\boldsymbol{u}(\boldsymbol{y})/\partial\boldsymbol{y} \equiv \boldsymbol{0}$ in (75) and (76). Figures 42–43 show some qualitative agreement with Figures 37 and 38 for small values of $\mu$ corresponding to chaos. However, these open–loop results incorrectly indicate chaos for large values of $\mu$, where the actual attractor is an asymptotically stable fixed point.

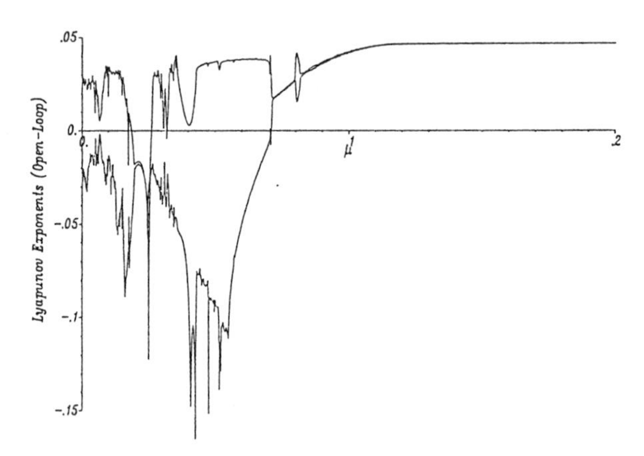


Figure 42. Open–loop Lyapunov exponents (bits/sec).

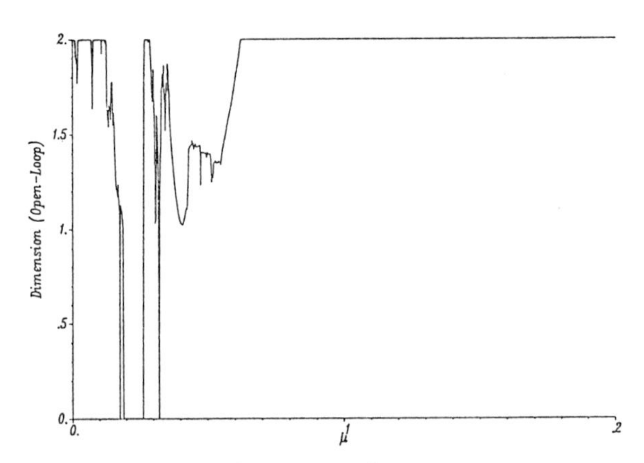


Figure 43. Open–loop fractal dimension.

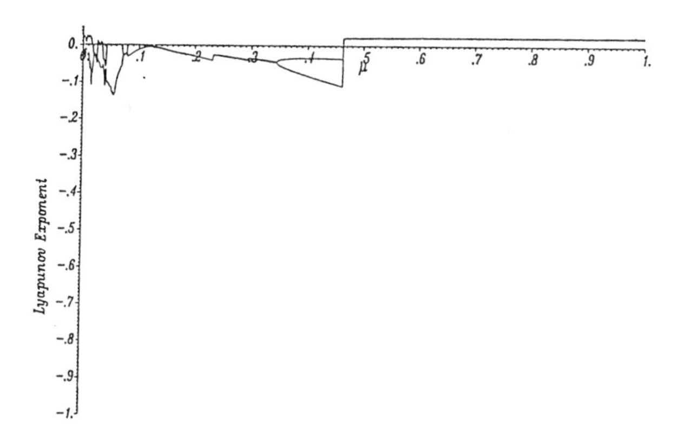


Figure 44. Lyapunov Exponents (bits/sec) for $u_{i\min} = 0$.

The fact that the feedback control $\boldsymbol{u}(\boldsymbol{x})$ in (66) is only piecewise differentiable can also create erroneous results. Figure 44 shows the closed-loop Lyapunov exponents generated as in Figure 37, but for a larger range of $\mu$ values and for a transient time of 2000 seconds. For values above $\mu \approx 0.2$ the results are not reliable, as indicated by the fact that Figure 44 has positive Lyapunov exponents for large values of $\mu$, corresponding to exponential separation of trajectories, instead of negative Lyapunov exponents corresponding to asymptotic stability.

For $u_{1\min} = u_{2\min} = 0$ and $\mu > 0.1183$ the $(x_1,x_2)$ attractor is an asymptotically stable fixed point at $x_1 = x_2 = 1$, where both $u_1(x_1) = 0$ and $u_2(x_2) = 0$ in (66) are continuous but not differentiable. This violates a basic premise underlying the method by which Lyapunov exponents were calculated. If the lower bounds on the control variables are replaced by $u_{1\min} = u_{2\min} = -\epsilon$, where $\epsilon > 0$ is small, then $u_1(x_1) = 0$ and $u_2(x_2) = 0$ become continuous and continuously differentiable at the attracting fixed point $x_1 = x_2 = 1$.

Figures 45 and 46 show the Lyapunov exponents and fractal dimension, respectively, for $\epsilon = 10^{-20}$ and a transient time of 2000 seconds. Even though this value of $\epsilon$ is smaller than the floating point accuracy of the computer, the results are improved significantly. The negative Lyapunov exponents for large $\mu$ in Figure 45 and the fractal dimension results in Figure 46 agree with the simulation results. The fluctuation in Lyapunov exponents in Figure 45 vanishes is $\epsilon$ is somewhat larger. Figure 47 shows the Lyapunov exponents for $\epsilon = 0.01$ and a transient time of 2000 seconds.

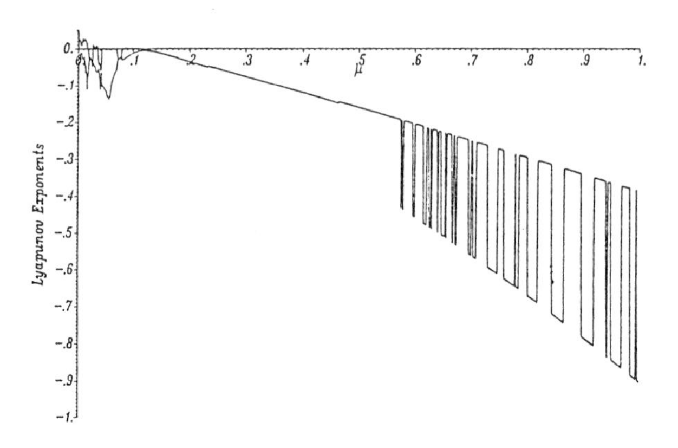


Figure 45. Lyapunov Exponents (bits/sec) for $u_{imin} = -10^{-20}$.

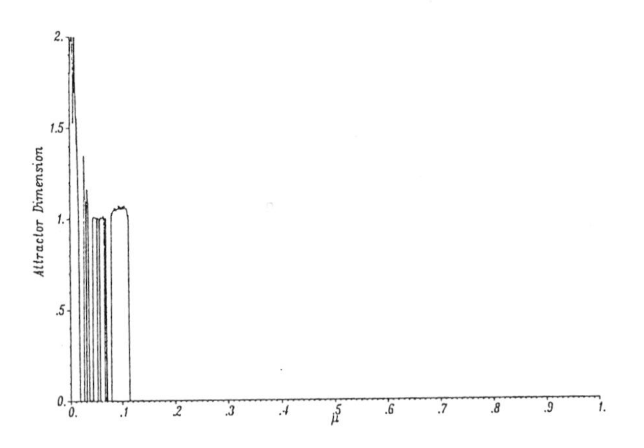


Figure 46. Fractal Dimension for $u_{imin} = -10^{-20}$.

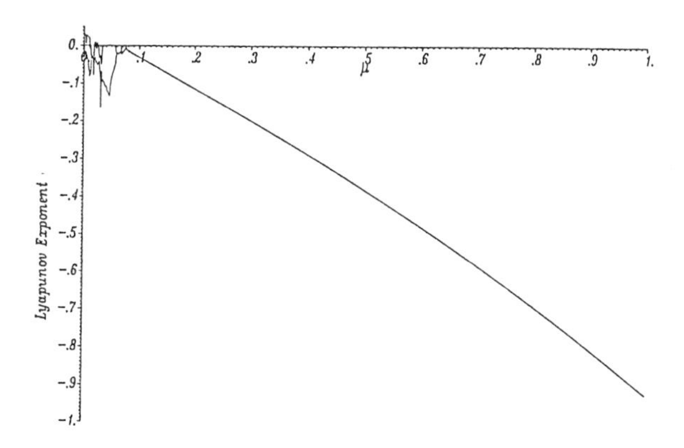


Figure 47. Lyapunov exponents (bits/sec) for $u_{imin} = -0.01$.

## X. SUMMARY

In this article we have presented an introduction to the analysis of chaos in both continuous–time and discrete–time feedback control systems. In addition we have presented two different approaches to obtaining a discrete-time model from a continuous–time system and have applied both discretization schemes to nonchaotic continuous–time systems.

An exponential discretization method was applied to an asymptotically stable Lotka-Volterra two–species competition model and it produced a chaotic trajectory, whereas the continuous–time system produced a corresponding trajectory that converged asymptotically to an equilibrium point. The second discretization method was a pseudo–Euler numerical integration procedure akin to second–order Runge–Kutta methods. When applied to a nonchaotic continuous–time Lotka–Volterra prey–predator system with periodic orbits, the pseudo–Euler integrator produced a chaotic attractor instead of periodic orbits.

When a saturating state feedback controller was added to the continuous-time prey–predator system, pseudo–Euler integration yielded trajectories that approach an equilibrium point asymptotically, but only if the strength (or gain) of the controller was sufficiently high. As the strength of the controller was decreased the equilibrium point became unstable under pseudo–Euler integration and the attractor became a quasiperiodic "limit cycle" curve. As the controller strength was decreased further, regions of chaotic motion began to occur, even though the continuous–time system still had an asymptotically stable equilibrium point as its attractor.

Bifurcation diagrams, Lyapunov exponents, and fractal dimension plots were used to show that the transition to chaotic motion in one of the regions of chaos for the pseudo–Euler prey–predator system is via a period doubling sequence that obeys Feigenbaum's universal rule for period doubling. In addition to chaotic motion, we have also shown that while the basin of attraction for the continuous–time prey–predator system is all of the positive quadrant, the basin of attraction under pseudo–Euler integration can have a fractal boundary, even for a nonchaotic periodic attractor.

In the computation of Lyapunov exponents, we have shown that open-loop control sequences can yield invalid results. In addition, we have shown that even discontinuities in control variable derivatives can yield invalid results for Lyapunov exponents and fractal dimension calculations.

There are several conclusions that can be drawn from the results presented in this article. One is that discretization schemes, such as the exponential discretization models often employed by population biologists or even certain numerical integration schemes, can produce drastically different results than those associated with continuous–time ordinary differential equations models. Indeed, the discrete–time system may be chaotic even though the continuous–time system is not.

A second conclusion is that, whether they occur naturally or as the result of applying a feedback controller, the properties of asymptotic stability and structural stability in a continuous–time system are *not* sufficient to eliminate chaotic motion in a corresponding discrete–time model of the system.

The results of this article also show that care must be taken when standard dynamical systems methods for the analysis of chaos are applied to control systems that may be discontinuous or nondifferentiable. The methods appear to be valid only if the system is continuous and continuously differentiable "almost everywhere" along a trajectory.

## REFERENCES


1. Mees, A.I. and Sparrow, C.T., "Chaos," ***IEE Proc. Part D***, Vol. 128, 1981, pp. 201–205.

2. Peitgen, H.–O. and Richter, P.H., ***The Beauty of Fractals***, Springer–Verlag, NY, 1986.

3. Lorenz, E.N., "Deterministic Nonperiodic Flows," ***J. Atmos. Sci.***, Vol. 20, 1963, p. 130.

4. Roux, J.C., Rossi, A., Bachelart, S. and Vidal, C., "Experimental Observation of Complex Behaviour During a Chemical Reaction," ***Physica 2D***, 1981, p. 395.

5. Moon, F.C. and Holmes, P.J., "A Magnetoelastic Strange Attractor," ***J. Sound and Vibration***, Vol. 65, No. 2, 1979, pp. 275–296.

6. Pezeshki, C. and Dowell, E.H., "On Chaos and Fractal Behavior in a Generalized Duffing's System," ***Physica D***, Vol. 32, 1988, pp. 194–209.

7. May, R.M., "Biological Populations with Nonoverlapping Generations: Stable Points, Stable Cycles, and Chaos," ***Science***, Vol 186, Nov. 1974, pp. 645–647.

8. May, R.M., "Simple Mathematical Models with Very Complicated Dynamics," ***Nature***, Vol. 261, 1976, pp. 459–467.

9. Inoue, M. and Kamifukumoto, H., "Scenarios Leading to Chaos in a Forced Lotka–Volterra Model," ***Progress in Theoretical Physics***, Vol. 71, No. 5, May 1984, pp. 930–937.

10. Poincaré, H., ***Les Méthodes Nouvelles de la Méchanique Celeste***, Gauthier–Villars, Paris, 1892; in English: ***N.A.S.A. Translation TT F–450/452***, U.S. Fed. Clearinghouse, Springfield, VA, 1967.

11. Devaney, R.L., ***An Introduction to Chaotic Dynamical Systems***, Revised 2nd Printing, Benjamin/Cummings Pub. Co., Menlo Park, CA, 1988.

12. Guckenheimer, J. and Holmes, P., ***Nonlinear Oscillations, Dynamical Systems, and Bifurcations of Vector Fields***, Revised 2nd Printing, Applied Mathematical Sciences series, Vol. 42, Springer–Verlag, NY, 1983.

13. Schuster, H.G., ***Deterministic Chaos***, 2nd Revised Ed., VCH Pub., NY, 1988.

14. Hsu, C.S., ***Cell–to–Cell Mapping***, Applied Mathematical Sciences series, Vol. 64, Springer–Verlag, NY, 1987.

15. Wiggens, S., ***Global Bifurcations and Chaos***, Applied Mathematical Sciences series, Vol. 73, Springer–Verlag, NY, 1988.

16. Ushio, T. and Hirai, K., "Chaos in Nonlinear Sampled–Data Control Systems," ***Int. J. Control***, Vol. 38, No. 5, 1983, pp. 1023–1033.

17. Cook, P.A., "Chaotic Behaviour in Feedback Systems," ***Proc. 25th Conf. on Decision and Control***, Athens, Greece, December 1986, pp. 1151–1154.

18. Cook, P.A., "Simple Feedback Systems with Chaotic Behaviour," ***Systems and Control Letters***, Vol. 6, 1985, pp.223–227.

19. Rubio, F.R., Aracil, J., and Camacho, E.F., "Chaotic Motion in an Adaptive Control System," ***Int. J. Control***, Vol. 42, No. 2, 1985, pp. 353–360.

20. Grantham, W.J. and Athalye, A.M., "A Chaotic System: Discretization and Control," in ***Dynamics of Complex Interconnected Biological Systems***, A.I. Mees and T.L. Vincent, eds., Birkhäuser Boston, New York, 1989 (in press).

21. Libchaber, A. and Maurer, J., "A Rayleigh–Bénard Experiment: Helium in a Small Box," in ***Nonlinear Phenomena at Phase Transitions and Instabilities***, T. Riste, ed., Plenum Press, New York, 1982, pp. 259–286.

22. Coddington, E.A. and Levinson, N., ***Theory of Ordinary Differential Equations***, McGraw–Hill, N.Y., 1955.

23. Filippov, A.F., "Differential Equations with Discontinuous Right–Hand Side," ***Amer. Math. Soc. Translations***, Series 2, Vol. 42, 1964, pp. 199–231.

24. Fisher, M.E. and Grantham, W.J., "Estimating the Effect of Continual Disturbances on Discrete–Time Population Models," ***J. of Math. Biology***, Vol. 22, 1985, pp. 199–207.

25. Grantham, W.J. and Fisher, M.E., "Generating Reachable Set Boundaries for Discrete–Time Systems," in ***Modeling and Management of Resources Under Uncertainty***, Vincent, T.L., *et. al.*, eds, Springer–Verlag, New York, 1987, pp. 152–166.

26. Fröberg, C.–E., ***Introduction to Numerical Analysis***, 2nd Ed., Addison-Wesley, Reading, MA, 1969.

27. Ruelle, D. and Takens, F., "On the Nature of Turbulence," ***Comm. Math. Phys.***, Vol. 20, pp. 167–192, Vol. 23, pp. 343–344, 1971.

28. Feigenbaum, M.J., "Universal Behavior in Nonlinear Systems," ***Physica 7D***, 1983, pp. 16–39.

29. Minorsky, N., ***Nonlinear Oscillations***, Van Nostrand, Princeton, NJ, 1962.

30. Grebogi, C., Ott, E., and Yorke, J.A., "Attractors on an N–Torus: Quasiperiodicity Versus Chaos," ***Physica 15D***, 1985, pp. 354–373.

31. Grebogi, C., Ott, E., Pelikan, and Yorke, J., "Strange Attractors That Are Not Chaotic," ***Physica 13D***, 1984, pp. 261–268.

32. Grebogi, C., Ott, E., and Yorke, J.A., "Chaos, Strange Attractors, and Fractal Basin Boundaries in Nonlinear Dynamics," ***Science***, Vol. 238, 1987, pp. 632–638.

33. Mandelbrot, B.B., ***The Fractal Geometry of Nature***, Freeman Pub., San Francisco, 1982.

34. Farmer, J.D., Ott, E., and Yorke, J.A., "The Dimension of Chaotic Attractors," ***Physica 7D***, 1983, pp. 153–180.

35. McDonald, S.W., Grebogi, C., Ott, E., and Yorke, J.A., "Fractal Basin Boundaries," ***Physica 17D***, 1985, pp. 125–153.

36. Wolf, A., Swift, J.B., Swinney, H.L., and Vastano, J.A., "Determining Lyapunov Exponents from a Time Series," ***Physica 16D***, 1985, pp. 285–317.

37. Shimada, I. and Nagashima, T., "A Numerical Approach to Ergodic Problem of Dissipative Systems," ***Progress of Theoretical Physics***, Vol. 61, No. 6, June 1979, pp. 1605–1616.

38. Haken, H., "At Least One Lyapunov Exponent Vanishes if the Trajectory of an Attractor Does Not Contain a Fixed Point," ***Physics Letters 94A***, 1983, p. 71.

39. Kaplan, J.L. and Yorke, J.A., "Chaotic Behavior of Multidimensional Difference Equations," in ***Functional Differential Equations and Approximation of Fixed Points***, H.O.Peitgen and H.O. Walther, eds., Lecture Notes in Mathematics, Vol. 730, Springer–Verlag, New York, 1979, pp. 228–237.

CONTROL AND DYNAMIC SYSTEMS, VOL. 34

# ON A CLASS OF NONSTANDARD DYNAMICAL SYSTEMS: SINGULARITY ISSUES

**Ramesh S. Guttalu and Pedro J. Zufiria**

Department of Mechanical Engineering
University of Southern California
Los Angeles, CA 90089-1453

## I. INTRODUCTION

This paper is concerned with the general behavior and the role of the singularities possessed by the dynamical system given by

$$\dot{x}(t) = F(x(t)) = -J^{-1}(x)f(x), \quad J(x) = \nabla_x f, \quad x \in \mathbf{R}^\mathbf{N}, \quad t \in \mathbf{R} \tag{1}$$

$$f : \mathbf{R}^\mathbf{N} \to \mathbf{R}^\mathbf{N}, \quad f \in C^1$$

where $\dot{x}$ is the time derivative of the vector $x$. The vector function $f(\cdot)$, which generates the vector field $F(\cdot)$, is in general nonlinear and $J(\cdot)$ is the Jacobian matrix of $f$. The purpose of constructing the differential equation (1) for a given $f$ is to assist in determining all the solutions to the algebraic system

$$f(x) = 0 \tag{2}$$

Location of all zeros has application in determining equilibria and periodic solutions of nonlinear systems, bifurcation of solutions, global optimization, inverse kinematics, etc. The zeros of $f$ and the equilibria of (1) (or zeros of $F$) are closely related. Trajectories of the dynamical system (1) can provide useful information regarding the location of the zeros of $f$. Besides the system (1), a great variety of dynamical systems can be formulated for the purpose of finding zeros of $f$. We further study the type of systems described by (1) which has been extensively considered in the literature. The aim of this paper is to analyze the behavior of the system (1), especially to identify the type of singularities the system possesses and how they affect its local and global dynamics.

Several investigations have been carried out in an attempt to characterize the dynamical system (1), see [1, 2, 3, 4, 8, 9, 10, 15, 20, 22]. The system (1) appears often in techniques based on the homotopic and continuation methods where the main issue deals with defining paths leading to zeros of $f$, see [7, 17]. Other investigations have primarily focused on the analysis of the discrete system which results from numerically integrating (1). The major objective of these investigations has been to study the convergence of algorithms, obtained from such discrete systems, for locating zeros.

In the algorithms obtained by numerically integrating (1), a class of discrete systems is generated. Only a local analysis is developed to study the behavior of the trajectories of the discrete systems. Most of the properties derived from these studies are related to the specific scheme used for the integration of the dynamical system. In general, the references mentioned above do not exploit the concept of domain of attraction to globally characterize the dynamics of (1). The roots of $f$ are related to the equilibria of the system (1). Every equilibrium point of (1) is asymptotically stable and is a root of $f$. For the purpose of characterizing convergence of the trajectories of the system (1), Zufiria and Guttalu [22] have used the general concept of domain of attraction associated with an asymptotically stable equilibrium point of (1). This concept is natural in dynamical systems theory and attraction domains can be computed numerically, for example, using the method of cell mapping, see Hsu [16]. Results about the stability of equilibria of (1) are made possible due to the existence of a strict Lyapunov function which is applicable at once to all these equilibria and which can also be used to provide a rough estimate of the size of the attraction domains.

A systematic analysis of domains of attraction is presented in [22]. In addition, the importance of the role of the singularity of the Jacobian $J$ is also emphasized. Certain manifolds, referred to as singular and barrier manifolds, play crucial role in delineating the global behavior of the dynamical system (1). Their existence is strongly related to the global structure of the Lyapunov function. However, when a root of $f$ falls on the singular manifold the dynamics of the system (1) can be very difficult to analyze. This aspect is not considered in [22].

The present paper provides additional results to characterize the effect of singular manifolds on the global behavior of the system (1). The role of the local minima (which do not correspond to roots of $f$) of the Lyapunov function is elaborated. The relationship between roots of $f$ and zeros of $F$ (equilibria of (1)) is further characterized for the case when the root of $f$ lies on a singular manifold. Some additional issues regarding the effect of singularities on the local and global dynamics of the system (1) are also provided. The concepts of singular manifolds, barrier manifolds and isolated regions introduced in [22] have been found quite useful in further pursuing the analysis of the local and global behavior of the dynamical system (1).

In order to illustrate the effect of singularities beyond the analysis provided by Zufiria and Guttalu [22], a pathological example of the vector function $f$ is provided. The associated dynamical system exhibits different types of singularities as a parameter is varied. A detailed and careful analysis of special cases, when the theory developed in [22] is not applicable, is carried out. To illustrate further the effect of singular manifolds on the dynamics of (1), they are computed for two examples by making use of the algorithm provided in [23]. In addition, another formulation of a dynamical system is presented which is applicable to the numerical integration of mechanical systems with algebraic constraints.

The organization of this paper is as follows. Known global results for the dynamical system (1) are summarized in Section II. New issues related to the singularities associated with (1) are brought out and a pathological example is analyzed in detail in Section III. The algorithm for locating singular manifold is applied for some specific problems in Section IV. Another formulation of a dynamical system which has more general applicability than the system (1) is discussed in Section V. Concluding remarks appear in Section VI.

# II. ANALYTICAL RESULTS

Generic properties of the autonomous dynamical system (1) constructed by using the vector function $f$ are studied in Zufiria and Guttalu [22]. This section provides a summary of both local and global results. Within this framework stability is to be interpreted in the sense of Lyapunov. The properties of the Lyapunov function $V = f^T f$ for (1) are also given. The reader is referred to [22] for a proof of the lemmas and theorems quoted in this section.

## A. RELATIONSHIP BETWEEN THE ZEROS OF $f$ AND $F$

The Jacobian matrix $J$ is called *regular* at $x \in \mathbf{R}^\mathrm{N}$ if $det\ J(x) \neq 0$, otherwise it is referred to as *singular* or *irregular* at $x$. Let $x^* \in \mathbf{R}^\mathrm{N}$ denote a root of $f$, i.e. $f(x^*) = 0$. Then, similar definitions apply to regular and singular roots of $f$. A root $x^*$ is regular or singular if $J(x^*)$ is regular or singular, respectively. The relation between $F$ and $f$ is given in the following lemmas:

**Lemma 1**

*(a) $F(x_d)$ is defined if and only if $f(x_d)$ belongs to the image of $J(x_d)$.*

*(b) Let $F(x_u)$, $x_u \in \mathbf{R}^\mathrm{N}$, be undefined. Then $J(x_u)$ is singular.*

Note that in some cases $F(x_d)$ could be defined but not uniquely. In this case, sometimes it is possible to choose a value of $F(x_d)$ that preserves the continuity of $F$ at that point. If this happens, then we will consider that the function $F(\cdot)$ is defined at such a point.

**Lemma 2** *Suppose that $J(x^*)$, $x^* \in \mathbf{R}^\mathbf{N}$, is nonsingular. Then $f(x^*) = 0$ if and only if $F(x^*) = 0$. In other words, $x^*$ is an equilibrium point of (1).*

**Lemma 3** *Let $F(x^*) = 0$, $x^* \in \mathbf{R}^\mathbf{N}$. Then $f(x^*) = 0$.*

**Lemma 4** *$F(x_0)$, $x_0 \in \mathbf{R}^\mathbf{N}$, is defined if and only if $F$ is continuous at $x = x_0$.*

The following definitions are used later for the analysis of the dynamical system (1).

**Definition 1** *$F(x)$ is* directionally defined *at $x = x_0$ if there exists a sequence of points $x_k$ approaching $x_0$ such that $\lim_{x_k \to x_0} F(x_k)$ is defined.*

**Definition 2** *$F(x)$ is* directionally undefined *at $x = x_0$ if it is not directionally defined.*

We note that when $F(x)$ is directionally undefined at $x = x_0$

$$\lim_{x \to x_0} \frac{1}{\|F(x)\|} = 0 \tag{3}$$

where $\|\cdot\|$ is any norm in $\mathbf{R}^\mathbf{N}$.

## B. LYAPUNOV STABILITY ANALYSIS

The indirect method of Lyapunov for studying the relationship between the zeros of $f$ and the equilibrium points $x^*$ of the dynamical system (1) leads to the following theorem.

**Theorem 1** *Suppose that $J$ is regular and $\nabla_x J^{-1}$ is defined at a zero $x^*$ of $f$. Then $x^*$ is an asymptotically stable equilibrium point of (1).*

Accordingly, each equilibrium point has eigenvalues $\lambda = -1$ with multiplicity $N$. Theorem 1 turns out to be a weaker form of the result provided by the direct method. It can be proved that the function $V = f^T f$ is a strict Lyapunov function of (1) (for the definition of strict Lyapunov function see Hirsch and Smale [14]) and that all the zeros of $f$ are asymptotically stable equilibria of the system (1). These results can be stated as follows.

**Theorem 2** *Given $f$, consider the scalar function*

$$V(x) = f^T(x)f(x) = f(x) \cdot f(x), \; V : \mathbf{R}^\mathbf{N} \to \mathbf{R} \tag{4}$$

*where $f^T(x)$ is the transpose of $f(x)$. Suppose that $F(x^*) = 0$, $x^* \in \mathbf{R}^\mathbf{N}$, that is, $x^*$ is an equilibrium point of (1). Then $V$ is a strict Lyapunov function of (1) for $x^*$.*

**Theorem 3** *Let $x^*$ be a zero of $f$. If $F(x^*) = 0$, then $x^*$ is an asymptotically stable equilibrium point of (1).*

## 1. Local considerations of analytic $V$

It is instructive to study the case when $f$ is analytic in a neighborhood of an equilibrium point $x^*$. In this case $V(x)$ can be approximated by

$$V(x) = (x - x^*)^T J^T(x^*)J(x^*)(x - x^*) + \mathcal{O}(\|x - x^*\|^3) \tag{5}$$

One observes that the positive definiteness of $V(x)$ is indicated by the second order term in (5) whenever $J(x)$ is nonsingular at $x^*$. Higher order terms are needed in (5) to arrive at the positive definiteness of $V$ when $J$ is singular at $x^*$.

On the other hand, referring to Theorem 1 and [21], we know that there exists a Lyapunov function of the form

$$V_{x^*}(x) = (x - x^*)^T P(x - x^*) \tag{6}$$

where the Lyapunov matrix equation

$$PA + A^T P = -Q \tag{7}$$

is to be satisfied by any pair of positive definite matrices $P$ and $Q$. Under the regularity conditions mentioned in Theorem 1, for any $x^*$, $A = -I$ and we can choose $Q = 2P$ to satisfy (7). This means that any positive definite matrix $P$ can be used in (6) to make $V_{x^*}$ a Lyapunov function in a small neighborhood of $x^*$. Referring to (5), the Lyapunov function $V$ may be approximated by taking only second order terms. Notice that for each $x^*$ the matrix $P$ in (6) can be obtained from equation (5) by setting $P = J^T(x^*)J(x^*)$. In this case $P$ would depend on the equilibrium point $x^*$ of the dynamical system (1). Hence, it confirms the fact that the original Lyapunov function $V$ given by (4) is general and applicable at once to every equilibrium point $x^*$.

## 2. Global considerations of $V$

The following result explains the relationship between the regularity of $J$ and the nature of the extrema of the Lyapunov function $V$.

**Theorem 4** *Assume that $J$ is regular in a region $\Phi \subset \mathbf{R}^\mathbf{N}$. Then, the Lyapunov function $V$ given by (4) can only have minima (not a maximum or a saddle point). Besides, a point $x_m \in \Phi$ can be a minimum of $V(x)$ if and only if $x_m$ is a zero of $f$.*

Suppose that $D_{x^*}$ denotes the domain of attraction or the region of attraction of an asymptotically stable equilibrium point $x^*$ of the system (1). The boundary of the level curves $V(x) = f^T f = l$ for appropriate constant $l > 0$ and for a given zero $x^*$ of $f$ defines a region $\Psi$. It can be shown that there always exists a constant $l = l_\epsilon$, $l_\epsilon > 0$, for which $\Psi$ is a subset of $D_{x^*}$. The following theorem provides an improved estimate of the domain of attraction $D_{x^*}$ (that is, an estimate $l > l_\epsilon$) by making use of the Lyapunov function $V$.

**Theorem 5** *Let $\Psi = \{x \mid V(x) < l\}$ be a bounded region with the boundary $\partial\Psi$ on which $V(x) = l$, $l \in \mathbf{R}^+$. Suppose that $\Psi$ contains only one root $x^*$ of $f$. If $F$ is defined in $\Psi$, then $\Psi$ is included in the domain of attraction $D_{x^*}$ associated with $x^*$ of (1).*

### 3. Practical significance of $V$

What we have established so far is that $V$ serves as a suitable Lyapunov function for the dynamical system associated with $F = -J^{-1}f$. One can also make use of this function to serve as a novel theoretical tool to address computational issues involved in finding all the zeros of $f$.

An estimate of $l = l_\epsilon$ in Theorem 5, obtained by considering the eigenvalues of the matrix $A$ defined in (7), may prove useful for determining either partially or completely the extent of the domain of attraction associated with a root of $f$. Such an estimate of $l$ may be improved by considering level curves of the Lyapunov function $V$. This estimate of $l$ can be employed to determine the complete domain of attraction associated with $x^*$ by using a numerical method developed by [13] which is based on the backward mapping of trajectories of (1). The estimate provided by the Lyapunov function $V$ constructed above can be related to the radius of convergence of the region of initial guess associated with Newton-like methods of finding zeros.

Another computational method to determine the domains of attraction is the cell mapping technique which is based on discretization of the state space, see [16]. This method can potentially provide all the roots of $f$ to a given accuracy. Later, we discuss results obtained for a pathological example using the cell mapping method.

## C. GLOBAL BEHAVIOR OF THE DYNAMICAL SYSTEM

The global dynamics of the system (1) depend naturally on the function $f$, specifically on its gradient $J$. When $f$ has only isolated roots, as the following theorem shows, the system trajectories behave in a predictable way.

**Theorem 6** *Suppose that $f$ has only isolated zeros. Then, every trajectory of the dynamical system (1) can only have one of the following behaviors:*

*I.* $\| x(t) \| \to \infty$ *as* $t \to T \in \mathbf{R}^+$
*(If $T$ is finite we have a finite* escape time, *see [18])*

*II.* $x(t) \to x_S$ *as* $t \to T$ *where* $F(x_S)$ *is not defined and* $x(t)$ *is not defined for* $t > T$

*III.* $x(t) \to x^*$ *as* $t \to \infty$ *where* $F(x^*) = 0$ *(that is* $x^*$ *is a root of* $f$*)*

One important conclusion of Theorem 6 is that the dynamical system (1) cannot possess periodic solutions when roots of $f$ are isolated. Even in higher dimensions, the system cannot possess complicated behavior such as chaotic motions.

The Lyapunov function constructed previously, in addition to providing local stability character and estimates of domains of attraction, can also give information about the behavior of system trajectories near degenerate points as stated in the following theorem.

**Theorem 7** *Let $d(x_k, x_S)$ denote the distance between a point $x_k$ and the point $x_S$ where $F(x_S)$ is directionally undefined. Consider any trajectory of (1) which contains the sequence of points $x_k$, $k = 1, 2, \ldots$, for which $d(x_k) \to 0$ as $x_k \to x_S$. Then the angle between the trajectory of (1) and $\nabla_x V$ at $x_k$ tends to $\pi/2$ as $d(x_k) \to 0$.*

## D. DOMAINS OF ATTRACTION AND THE ROLE OF SINGULARITIES OF J

From previous analysis, we have already observed that the regularity assumptions are not sufficient to provide a complete description of the global behavior of (1). The critical case results when the Jacobian matrix $J$ is singular. It is then important to ascertain how the sets of points where $J$ is singular are related to attraction domains. The following concepts of singular manifolds, barrier manifolds and isolated regions are needed for enhancing our understanding of the global dynamical behavior of (1).

**Definition 3** *A* singular manifold $S_i$, $i = 1, 2, \ldots$, *is defined as a* $N-1$ *dimensional manifold*

$$S_i = \{x \mid rank\ J(x) \leq N - 1\} \tag{8}$$

In general, singular manifolds separate the state space region into two different regions (denoted by $\Phi_1$ and $\Phi_2$, each a subset of $\mathbf{R}^\mathbf{N}$). If we consider two points $x_1 \in \Phi_1$ and $x_2 \in \Phi_2$ then one can define a special class of singular manifolds as follows:

**Definition 4** *A* barrier manifold $B_i$, $i = 1, 2, \ldots$ *is defined as a singular manifold for which any arbitrary pair* $x_1$ *and* $x_2$ *of directional limits of system trajectories satisfy*

$$\lim_{x_1, x_2 \to x_B} (\vec{n} \cdot F(x_1))\, (\vec{n} \cdot F(x_2)) < 0 \tag{9}$$

*or at least one of the conditions*

$$\lim_{x_1 \to x_B} F(x_1) = 0, \quad \lim_{x_2 \to x_B} F(x_2) = 0 \tag{10}$$

*where* $\vec{n}$ *is a vector normal to the* $B_i$ *manifold at* $x = x_B$, $x_B \in B_i$.

For example, the system (1) with $f = x^2 - 1$, $x \in \mathbf{R}$ has the barrier manifold given by $x = 0$.

**Definition 5** *An* isolated region $\Omega_i \subset \mathbf{R}^{\mathbf{N}}$, $i = 1, 2, \ldots$, *is any region in which*

1. $\Omega_i$ *is connected;*

2. $\Omega_i \cap B_j = \emptyset$, $j = 1, 2, \ldots$;

3. $\partial\Omega_i \subset \cup_j B_j$, *where* $\partial\Omega_i$ *is the boundary of* $\Omega_i$.

A barrier manifold is a special class of singular manifold and a dynamical system of the type (1) may possess several such manifolds. The isolated regions are the regions separated by the barrier manifolds. These definitions lead to the following corollaries indicating the behavior of system trajectories.

**Corollary 1** *Let* $\Omega_j$ *be an isolated region. Then the dynamical system (1) does not have any equilibrium point in* $\Omega_j$ *other than the zeros* $x_i^*$ *of* $f$ *for* $i = 1, 2, \ldots$.

**Corollary 2** *Assume that* $J$ *is regular in an isolated region* $\Omega_j$. *Then, a point* $x_m \in \Omega_j$ *is an extremum or a saddle point of the Lyapunov function* $V(x)$ *given by (4) if and only if* $x_m$ *is a zero of* $f$.

**Corollary 3** *If the Lyapunov function* $V(x)$ *given by (4) has either a saddle point or a local minimum (which is not a root of* $f$*) at* $x = x_S$, *then* $J(x_S)$ *is singular.*

**Corollary 4** *Assume that* $J$ *is regular and* $f$ *is analytic in an isolated region* $\Omega_j$. *Then, the Lyapunov function* $V(x)$ *given by (4) can only have minima (not a maximum or a saddle point). Besides, a point* $x_m \in \Omega_j$ *can be a minimum of* $V(x)$ *if and only if* $x_m$ *is a zero of* $f$.

**Corollary 5** *Let* $d_i(x_k, B_i)$ *denote the distance of a point* $x_k$ *to the manifold* $B_i$ *defined by (4). Consider any trajectory of (1) which contains the sequence of points* $x_k$, $k = 1, 2, \ldots$ *such that* $d_i(x_k, B_i) \to 0$ *as* $x_k \to x_S$. *Then, the angle between the trajectory of (1) and* $\nabla_x V$ *at* $x_k$ *tends to* $\pi/2$ *as* $d(x_k) \to 0$.

Corollary 1 corresponds to Lemma 3, Corollaries 2 and 3 to Theorem 5, Corollary 4 to Theorem 4, and Corollary 5 to Theorem 7. The following theorems further characterize isolated regions and the domains of attraction associated with $x^*$.

**Theorem 8** *Trajectories of (1) cannot cross the barrier manifolds.*

**Theorem 9** *Every trajectory of (1) that starts in any isolated region $\Omega_j$ evolves only within $\Omega_j$.*

**Theorem 10** *Let $x^*$ be a zero of $f$ such that $F(x^*) = 0$. If $D_{x^*}$ is the domain of attraction associated with $x^*$ of the dynamical system (1), then $D_{x^*} \subseteq \Omega_j$ for some specific $j$.*

These results allow one to separate the study of the global dynamics of the system (1) into individual study of different isolated regions. An important question arises as to the number of roots of $f$ that may exist in $\Omega_i$ and how domains of attraction are related to them.

**Theorem 11** *Suppose that $J$ is regular in an isolated region $\Omega_j$.*

*(a) Consider the case $N = 1$, $x \in \mathbf{R}$. Then, $\Omega_j$ contains* at most *one asymptotically stable equilibrium point $x_k^*$ of the system (1) (with the corresponding domain of attraction given by $D_{x_k^*}$).*

*(b) Consider the case $N \geq 2$, $x \in \mathbf{R}^{\mathbf{N}}$. Then, it is possible that $\Omega_j$ may contain more than one asymptotically stable equilibrium points $x_k^*$ of (1), $k = 1, 2, \ldots$ (with their domains of attraction $D_{x_k^*}$).*

Theorem 8 is very important, because when trajectories of (1) are obtained by numerical integration they may tend to cross the barrier manifolds leading to misinterpretation of the global behavior. Theorems 9–11 indicate how isolated regions bound and restrict the extent of domains of attraction associated with the zeros of $f$. An estimate of the sizes of domains of attraction is provided by the Lyapunov function $V(x)$. In order to obtain domains of attraction completely (at least from a numerical standpoint), one resorts to global computational analysis such as the cell mapping method. In this case, one can expect inaccuracies in the computed domains of attraction when Theorem 8 is violated by trajectories which may cross barrier manifolds during numerical computation. For an additional use of Theorem 11, see [24].

# III. OTHER SINGULARITY ISSUES

## A. Local minima of the Lyapunov function

When globally characterizing the Lyapunov function $V$, Corollary 3 stated that the local minima of $V$ which are not roots of $f$ must be at points where the Jacobian $J(x)$ is singular. Based on this result we present the following theorem.

**Theorem 12** *Suppose that $x_L$ is a strict local minimum of the Lyapunov function $V$ given by (4) and $f(x_L) \neq 0$. Then there exists a neighborhood $B$ of $x_L$ such that every trajectory starting in $B$ will terminate at a point where $J$ is singular and $F$ is not defined.*

*Proof:* As $f(x_L) \neq 0$, then $V(x_L) = c_L > 0$. Since $x_L$ is a strict local minimum, one can always define a neighborhood $B$ of $x_L$ as a connected bounded set of points given by $B = \{x | \; c_L \leq V(x) < c_L + \epsilon, \; \epsilon > 0\}$ for sufficiently small $\epsilon$. Following Lemma 3 and the definition of $V$, $F(x) = 0 \Rightarrow f(x) = 0 \Rightarrow V(x) = 0$ implying that $B$ does not contain any equilibria of the dynamical system (1). From the properties of the Lyapunov function, $\dot{V}(x) = -2V(x) \leq -2c_L < 0, \; x \in B$, meaning that every trajectory starting in $B$ remains in $B$ forever. Hence, only case II of Theorem 6 can occur. This concludes the proof. ■

In the next few sections we will address the issue of the nature of the vector field $F(\cdot)$ at points belonging to the singular manifolds.

## B. Definition of the vector field $F$

The theory developed by Zufiria and Guttalu [22] states that when the Jacobian $J(x)$ is regular at a given point then the vector field $F(x)$ is defined at that point. On the other hand, this is not always the case when $J$ is singular. The singularity of the Jacobian $J(x)$ can lead to the following possibilities:

1. Lemma 1 shows that there may be points on the singular manifolds where $F(x) = -J(x)^{-1}f(x)$ is defined although the Jacobian matrix $J(x)$ is singular,
2. It is possible that the vector field $F$ is directionally defined at a point, meaning that its value at that point depends on the direction of approach of the point. The vector field $F$, being discontinuous, is not properly defined in such a case. This leads to nonstandard and pathological dynamics for the system (1).
3. There may be points where the vector field $F$ is directionally undefined.

The following theorems characterize the various cases stated above.

**Theorem 13** *Suppose that a point $x_S$ belongs to a singular manifold $S_i$. If the vector field $F(x)$ is defined at $x_S$ then there exists a neighborhood $B_r$ of $x_S$ along the singular manifold for which $adj(J(x))f(x) = 0, \; \forall x \in B_r$. Also $adj(J(x_S)f(x_S)) = 0$.*

*Proof:* By definition, $F = -\frac{adj(J)f}{det\ J}$, where $adj(J)$ is the adjoint of the matrix $J$. If $F(x)$ is defined at $x_S$, following Lemma 4, it is continuous at $x = x_S$. Hence, if we take $\lim_{x \to x_S} F(x)$ along the singular manifold (where $det\ J(x) = 0$), this limit must exist. Accordingly, $adj(J(x))f(x) = 0$ must be satisfied in a neighborhood of $x_S$ along the singular manifold $S_i$. Finally, since $f \in C^1$, $adj(J(x))f(x)$ is continuous and the result follows. ■

**Theorem 14** *Suppose that a point $x_S$ belongs to a singular manifold $S_i$. The vector field $F$ is defined along some direction (directionally defined) at $x_S$ if and only if $adj(J(x_S))f(x_S) = 0$.*

*Proof:* From the definition of the vector field $F$ and following arguments similar to that of Theorem 13, $\lim_{x \to x_S} adj(J(x))f(x) = 0$ along the direction mentioned in the theorem. The proof concludes noting the continuity of $adj(J(x_S)f(x_S))$. ■

Related to the previous theorems, it is worth mentioning that there may be points which satisfy $adj(J(x))f(x) = 0$ but which do not satisfy $f(x) = 0$. These points are termed *extraneous singularities* by Branin [2]. Using Theorems 13 and 14, the following corollaries can be derived.

**Corollary 6** *Let $x_S$ be a point on a singular manifold $S_i$. Suppose that $F(x_S)$ is defined or at least directionally defined. Then one of the following two cases must exist:*

1. *$x_S$ is a root of $f$.*
2. *$x_S$ is an extraneous singularity.*

Note that we can have cases where a root of $f$ leads to $F$ being defined as well as cases for which $F$ is directionally defined. The same applies to an extraneous singularity.

**Corollary 7** *Suppose that $F(x)$ is defined at a point $x = x_S$ on a singular manifold $S_i$. Then there exist a submanifold of $S_i$ formed by either the roots of $f$ or by extraneous singularities.*

**Corollary 8** *Let $x_S$ be a point on a singular manifold $S_i$. If $x_S$ is neither a root of $f$ nor an extraneous singularity, then $F(x_S)$ is directionally undefined.*

If we consider a given singular manifold $S_i$, the nature of the vector field $F(x)$ along it can be represented by a combination of the cases mentioned above. For instance, the following cases can occur:

1. $F(x)$ is defined on the entire singular manifold, except maybe at a countably finite number of points which may or may not correspond to the roots of $f$.
2. $F(x)$ is not defined on the singular manifold. In addition, $F(x)$ is directionally undefined except on a set of points where it is directionally defined. This set of points may contain some of the roots of $f$ or extraneous singularities. Even here the behavior of the dynamical system (1) may be further classified according to the topological nature of trajectories asymptotically approaching these points.

These cases will be illustrated in the example provided at the end of this section.

## C. Zeros and equilibria

Following Lemma 2, when the Jacobian $J(x)$ is regular, there is a one-to-one correspondence between the roots of $f$ and the equilibrium points of the dynamical system (1). However, a characterization of this relationship is deeply involved at points where the Jacobian $J(x)$ is singular. This relation can be summarized as follows:

1. Following Lemma 3, every equilibrium point of the dynamical system (1), that is $F(x^*) = 0$, corresponds to a root of the vector function $f$.

2. A root of $f$ need not always correspond to an equilibrium point of (1), that is $f(x^*) = 0 \not\Rightarrow F(x^*) = 0$.

In the previous section defining the vector field $F$, we mentioned that when a root of $f$ lies on a singular manifold $F$ may be defined or directionally defined at the root. The following theorem provides additional information regarding this issue.

**Theorem 15** *Suppose that $x^*$ is a point on a singular manifold $S_i$ such that $f(x^*) = 0$ and $F(x^*)$ is defined. If $\nabla_x(det\ J) \not\perp ker(J)$, then $F(x^*) = 0$ implying that $x^*$ is an equilibrium point of (1).*

*Proof:* By definition, $F$ may be expressed in the form $F = -\frac{adj(J)f}{det\ J}$. A Taylor series expansion of $f$ and $det\ J$ around $x^*$ leads to

$$F(x) = -\frac{adj(J(x))\, J(x^*)\,(x - x^*) + \mathcal{O}\|x - x^*\|^2}{\nabla_x(det\ J(x^*))) \cdot (x - x^*) + \mathcal{O}\|x - x^*\|^2} \tag{11}$$

Since $J(x)$ is singular at $x = x^*$, it has a null space $ker(J)$ such that $dim(ker(J)) \geq 1$. So we can take the directional limit $x \to x^*$ such that $(x - x^*) \in ker(J)$. This implies that $\lim_{x \to x^*} \frac{\|J(x^*)\,(x-x^*)\|}{\|x-x^*\|} = 0$. The expression for $F$ in (11) then becomes

$$\lim_{x \to x^*} F(x) = -\frac{adj(J(x^*))\,\left(\frac{J(x^*)\,(x-x^*)}{\|x-x^*\|}\right) + \mathcal{O}\|x - x^*\|}{\nabla_x(det\ J(x^*)) \cdot \frac{(x-x^*)}{\|x-x^*\|} + \mathcal{O}\|x - x^*\|} = 0,$$

since the denominator is nonzero. Finally, as $F(x)$ is defined at $x = x^*$, the value of any directional limit is the value $F(x^*)$. ∎

If $f$ has roots on the singular manifolds, then these roots seem to play a prominent role in characterizing dynamic behavior near the manifolds. In some cases, $F$ can be defined at these roots which would imply a regular behavior of the dynamical system (1). In other cases, $F$ may be just directionally defined at the roots. Then, the classical notions of equilibrium points, stability and attractivity are not applicable.

Unlike the case of smooth dynamical systems which have been intensively studied and often arise from describing the behavior of physical systems, the dynamics described by (1) do not seem to be related to any known physical system. Since $f$ is any given vector function, there exist some forms of $f$ which lead to pathological cases for (1). Following the analysis provided in [22], the obvious step is to investigate in great detail some of the pathological cases. To enhance our understanding of this situation, one such pathological example is provided in the next section and its global behavior is examined in detail.

## D. A PATHOLOGICAL EXAMPLE

Consider the following vector field $f$ in $\mathbf{R}^2$:

$$\begin{aligned} f_1 &= x_1^2(x_1 - 1) \\ f_2 &= x_1^4 + x_2^2 - a, \quad a \in \mathbf{R} \end{aligned} \tag{12}$$

This example possesses a wide range of dynamic behavior depending on on the value of the parameter $a$. The extent of applicability of the theory developed in [22] will be considered. By direct computation

$$J = \begin{bmatrix} x_1(3x_1 - 2) & 0 \\ 4x_1^3 & 2x_2 \end{bmatrix}, \quad det\ J = 2(3x_1 - 2)x_1x_2$$

$$J^{-1} = \begin{bmatrix} \frac{1}{x_1(3x_1-2)} & 0 \\ \frac{-2x_1^2}{(3x_1-2)x_2} & \frac{1}{2x_2} \end{bmatrix}, \quad F = \begin{bmatrix} \frac{-x_1(x_1-1)}{3x_1-2} \\ \frac{4(x_1-1)x_1^4-(3x_1-2)(x_1^4+x_2^2-a)}{2(3x_1-2)x_2} \end{bmatrix}$$

$$\dot{x}(t) = F(x(t)) \tag{13}$$

Note that the function $V$ defined by (4) becomes

$$V(x) = [x_1^2(x_1 - 1)]^2 + [x_1^4 + x_2^2 - a]^2 \tag{14}$$

and is defined everywhere and it satisfies the relation $\dot{V} = -2V$. However, one can only consider $V(\cdot)$ as a proper Lyapunov function of (1) only at points which correspond to the equilibrium points of (1) in the classical sense.

Clearly, $J$ is singular on the three manifolds defined by $x_1 = 0$, $x_2 = 0$, and $x_1 = \frac{2}{3}$. On the first manifold $x_1 = 0$, $F(\cdot)$ is defined everywhere except at $x_2 = 0$ which is the intersection point of the two manifolds $x_1 = 0$ and $x_2 = 0$. Hence, following Corollary 6, the set of points $x_1 = 0$, $x_2 \neq 0$ must only contain either roots of $f$ or extraneous singularities. On the second manifold $x_2 = 0$, $F(\cdot)$ is directionally undefined except at $x_1^4 = a$, $a \geq 0$, where the value of $F$ depends on how this point is approached. Again, the point $x_1^4 = a$ must be either a root of $f$

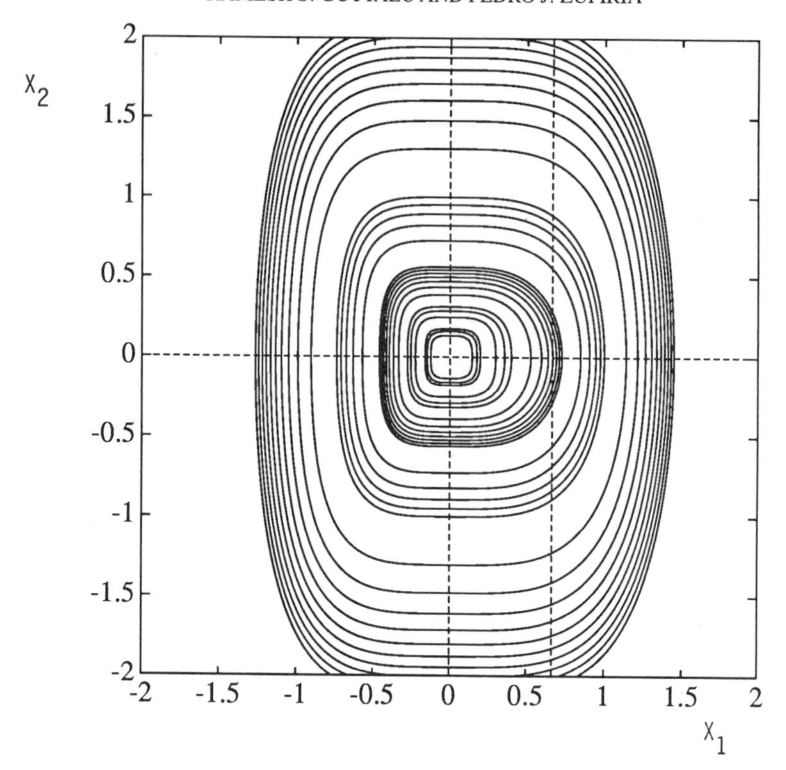


Figure 1: Level curves of the Lyapunov function $V(x)$ given by (14) for the case $a = 0$.

or an extraneous singularity. On the third manifold $x_1 = \frac{2}{3}$, $F(\cdot)$ is directionally undefined.

The analysis of the system (12) can be classified in the following five cases.

*Case I: $a < 0$*

The algebraic system (12) has no roots. The manifolds $x_2 = 0$ and $x_1 = \frac{2}{3}$ are barrier manifolds and hence they restrict the evolution of trajectories of the dynamical system (13) to be contained in the isolated regions $\Omega_i$ defined by them. The function $V$ has a minimum at $x = (0,0)$. Note that it satisfies Corollary 3. In addition, following Theorem 12, every trajectory starting close enough to this point will terminate at a point on a singular manifold where the trajectory is not defined any more.

*Case II: $a = 0$*

The algebraic system (12) has only one root given by $x^* = (0,0)$ (intersection point of the two manifolds $x_1 = 0$ and $x_2 = 0$) where $F$ is not defined. Hence, $x^* = (0,0)$ is not an equilibrium point of the dynamical system (13) in the classical sense. Following Theorem 14, it is easy to verify that $adj(J(x_S))f(x_S) = 0$. By using the transformation $x_1 = r\cos\theta$ and $x_2 = r\sin\theta$, one can be deduce that

$$\lim_{r\to 0} F_1(r,\theta) = 0, \quad \forall\, \theta$$

$$\lim_{r\to 0} F_2(r,\theta) = \begin{cases} 0, & \sin\theta \neq 0 \\ \infty, & \sin\theta = 0 \end{cases}$$

Using polar coordinates, the dynamical system (13) in a small neighborhood $U - \{(0,0)\}$ of the root $x^* = (0,0)$ may be expressed as

$$\dot{r} = -\frac{r}{2} + \mathcal{O}(r^2)$$

$$\dot{\theta} = \frac{r}{4}\sin\theta\cos^2\theta + \mathcal{O}(r^2)$$

These equations indicate that any trajectory starting at a point in $U$, excluding the manifold $x_2 = 0$, will eventually terminate at the point $x^* = (0,0)$.

Figure 1 shows $V-$level curves together with the barrier manifolds shown by the dashed lines. The cell mapping results are provided in Figures 2-3. We divide $x_1-$axis covering $-2 \leq x_1 \leq 2$ into $N_1 = 300$ intervals with interval size $h_1 = 1/75$, and $x_2$–axis covering $-2 \leq x_2 \leq 2$ into $N_2 = 300$ intervals with interval size $h_2 = 1/75$. The state space region $(-2,2) \times (-2,2)$ in this case contains a total of $90,000$ cells. An associated cell mapping of (13) is obtained by integrating (13) over a time period of $T = 1$ unit with integration time step size $\Delta T = 1/15$. Figure 2 gives the location of periodic cells discovered by the cell mapping method. The locations of periodic cells serve also the purpose of providing excellent initial guesses for refining the roots by means iterative methods. The symbol "+" refers to periodic cells corresponding to the equilibrium point $x^* = (0,0)$. The symbol "." corresponds to cells lying on the line $x_1 = 1$ but these locations do not contain any equilibria of (13). If trajectories are started on the line $x_1 = 1$, they reach the state $x = (1,0)$ in time $T = \ln(1 + x_2^2(0))$. Once this point is reached, the trajectories are not defined any more. Figure 3 provides the domains of attraction of periodic cells where the white region refers to cells which map to the sink cell, that is, they map outside the state space region considered. Obviously, the periodic cells corresponding to the point $x^* = (0,0)$ attract a large number of cells indicating that this equilibrium point attracts every point except the line $x_2 = 0$. Note that this can be interpreted as a "cone-like" attraction set with the cone angle approaching $\pi$. Its region of attraction is demarcated by the barrier manifold $x_1 = 2/3$. Some cells close to the barrier manifold $x_2 = 0$ are seen to map to the sink cell. It has been verified that trajectories starting inside these cells actually map to sink cell but eventually return to be mapped to the equilibrium point. The periodic cells corresponding to the line

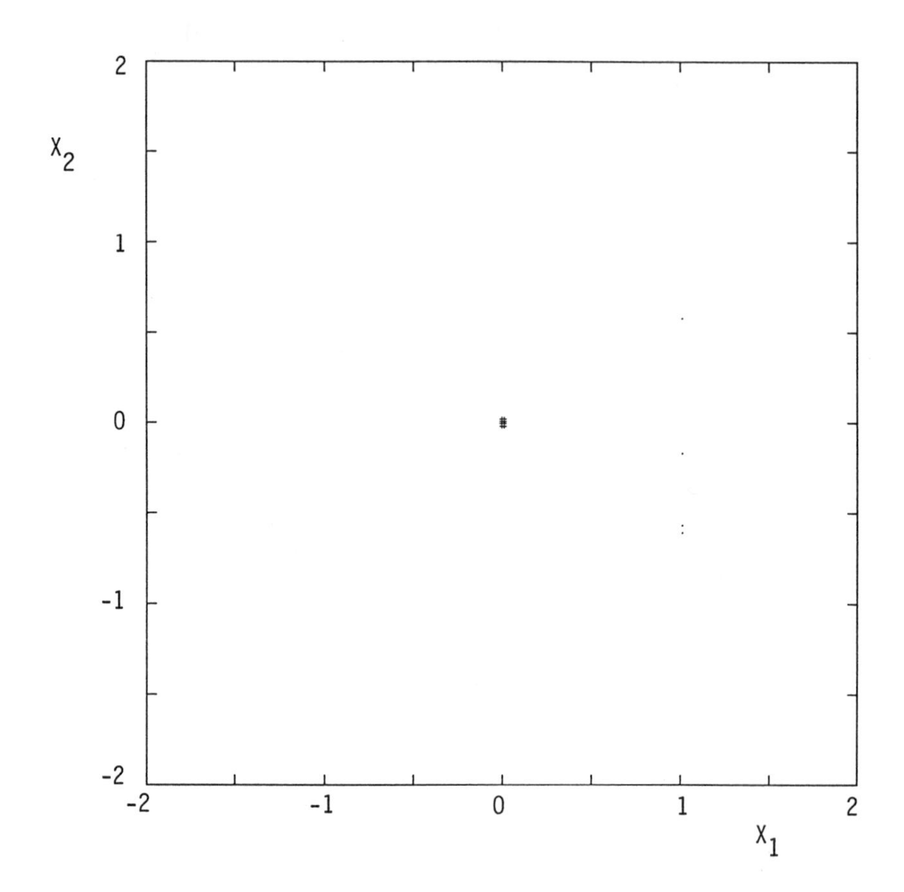


Figure 2: Location of periodic cells for the pathological example with $a = 0$.

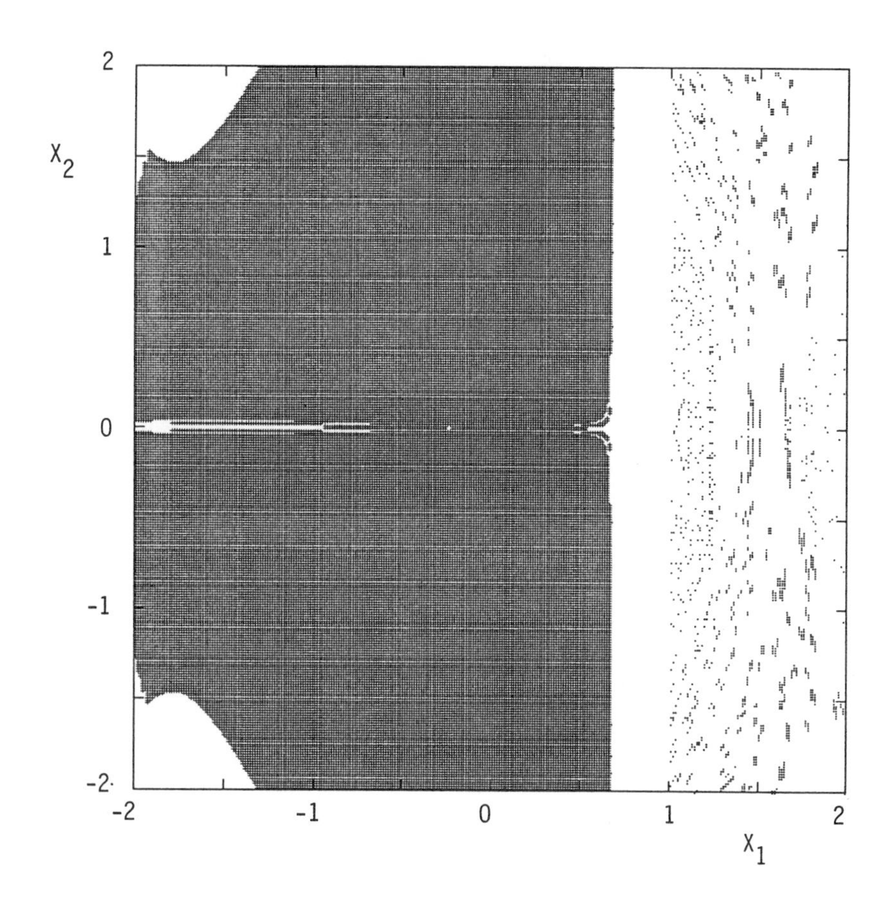


Figure 3: Domains of attraction of periodic cells for the pathological example with $a = 0$.

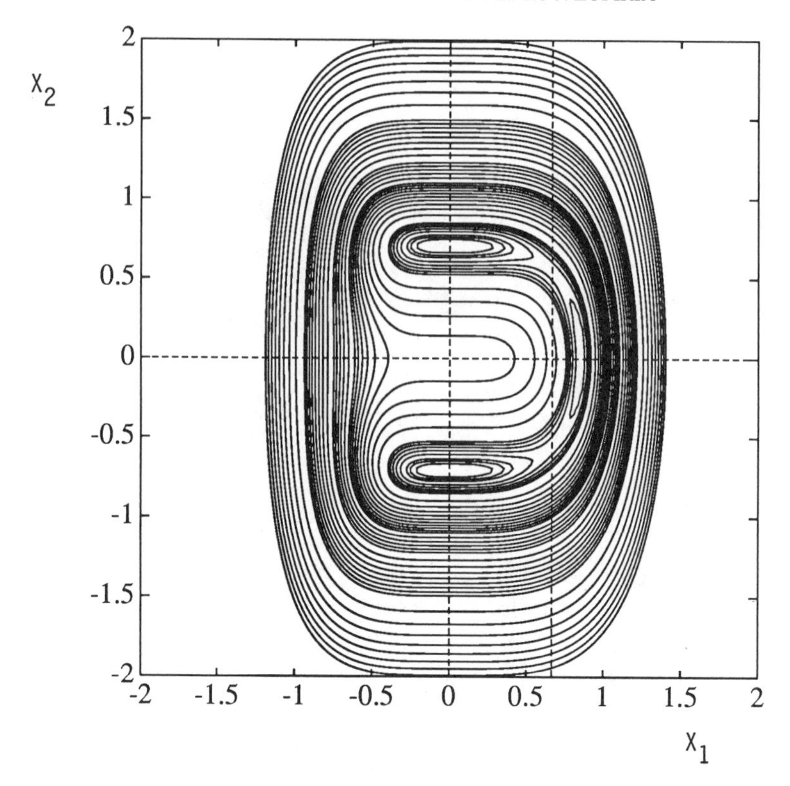


Figure 4: Level curves of the Lyapunov function $V(x)$ given by (14) for the case $a = 0.5$.

$x_1 = 1$ also attract some cells. This may be explained by the fact that if we start at a point where $x_1 > 2/3$ then $x_1(t) \to 1$.

*Case III:* $0 < a < 1$

The single root of Case II splits into two zeros of $f$ given by $x_a^* = (0, -\sqrt{a})$ and $x_b^* = (0, \sqrt{a})$. Notice that these are singular roots of $f$ since they lie on the singular manifold $x_1 = 0$. Hence, Theorem 1 is not applicable in this case. However, $F$ is defined at these roots. Hence, Theorem 13 and Corollary 7 are satisfied. In addition, a very important result follows from Theorem 15 which implies that $F(x_a^*) = 0$, and $F(x_b^*) = 0$. Therefore, by Theorem 3, it is known that the roots of $f$ are asymptotically stable equilibrium points of the dynamical system (13). Note that the eigenvalues of $J$ at both the roots are $\lambda_1 = -\frac{1}{2}$ and $\lambda_2 = -1$ (as opposed to the case of regular roots for which $\lambda = -1$ is the only eigenvalue with multiplicity 2). The eigenvalues are not equal to $-1$ because $J$ does not satisfy the conditions

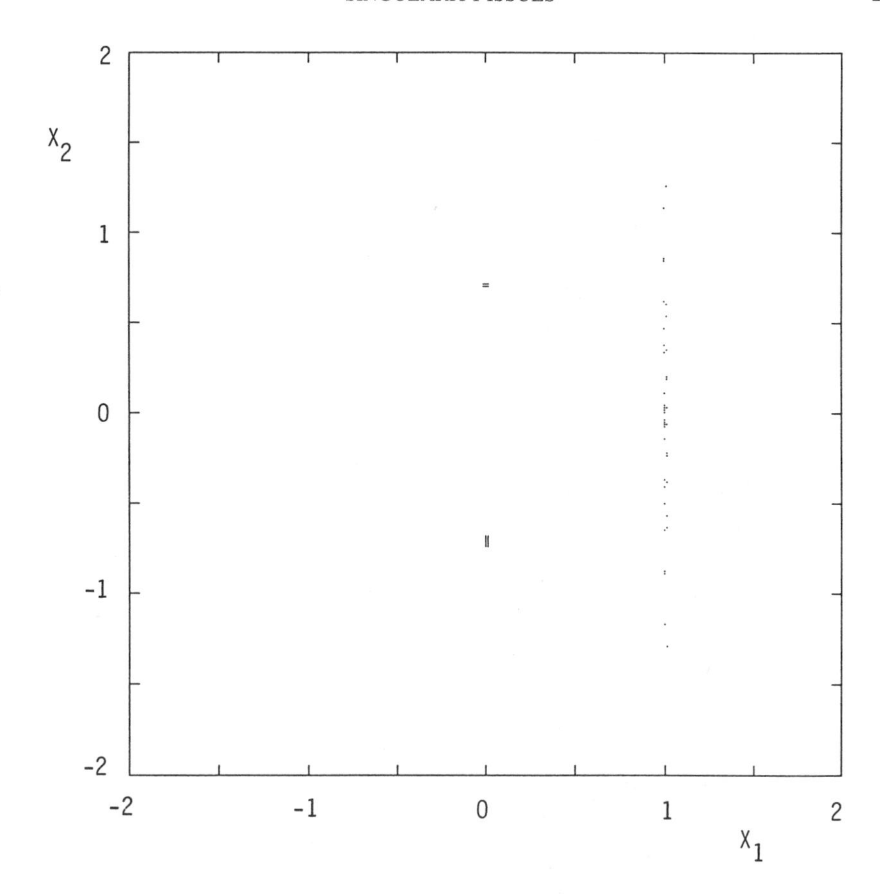


Figure 5: Location of periodic cells for the pathological example with $a = 0.5$.

of Theorem 1 exactly.

In this case, $V$ defined by (14) can be considered as a Lyapunov function of both the zeros. Figure 4 shows level curves of the Lyapunov function $V(x)$ for (12) with $a = \frac{1}{2}$. The regions of attraction near both the roots are depicted in this figure. The saddle points of the Lyapunov function are located on the barrier manifold $x_1 = \frac{2}{3}$ as predicted by Corollary 3. Also, there exists a strict local minimum of $V(x)$ at $x_S = (0.8494427, 0)$ which is not a root of $f$ and it lies on the singular manifold $x_2 = 0$. It follows from Theorem 12 that every trajectory starting close enough to the minimum will terminate at a point on a singular manifold where it will not be defined any more.

This particular case illustrates broader applicability of Theorem 3 proved using the direct method as compared with Theorem 1 proved by indirect method.

The cell mapping results are given in Figures 5-6. The state region $(-2, 2) \times$

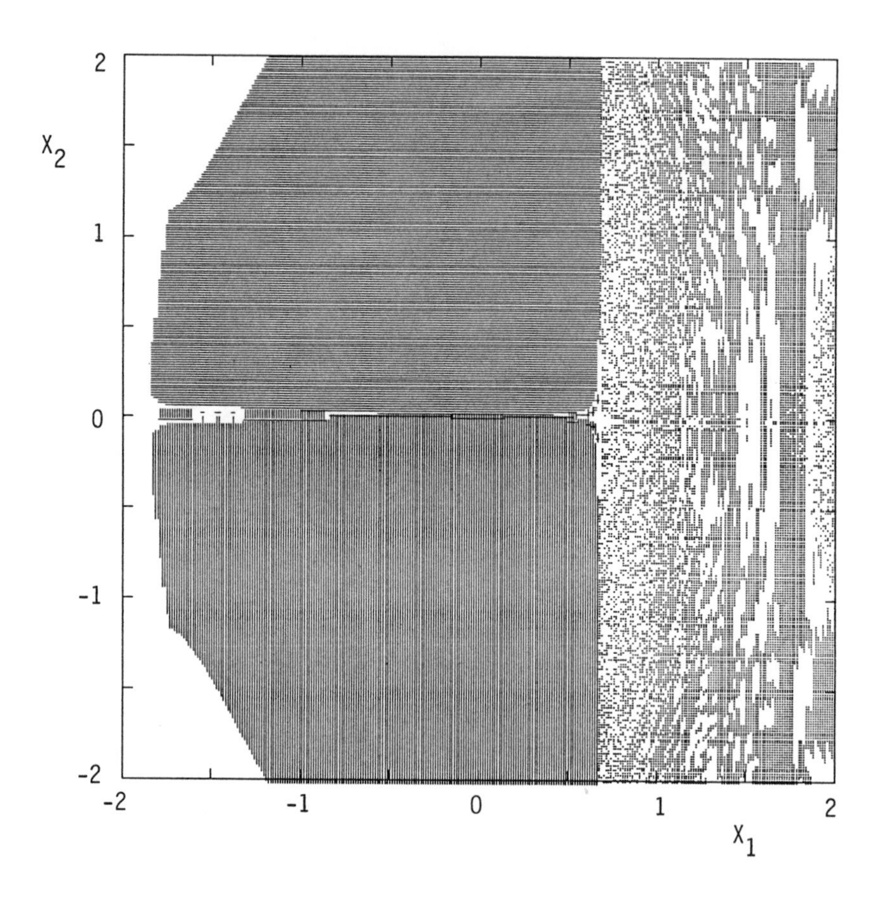


Figure 6: Domains of attraction of periodic cells for the pathological example with $a = 0.5$.

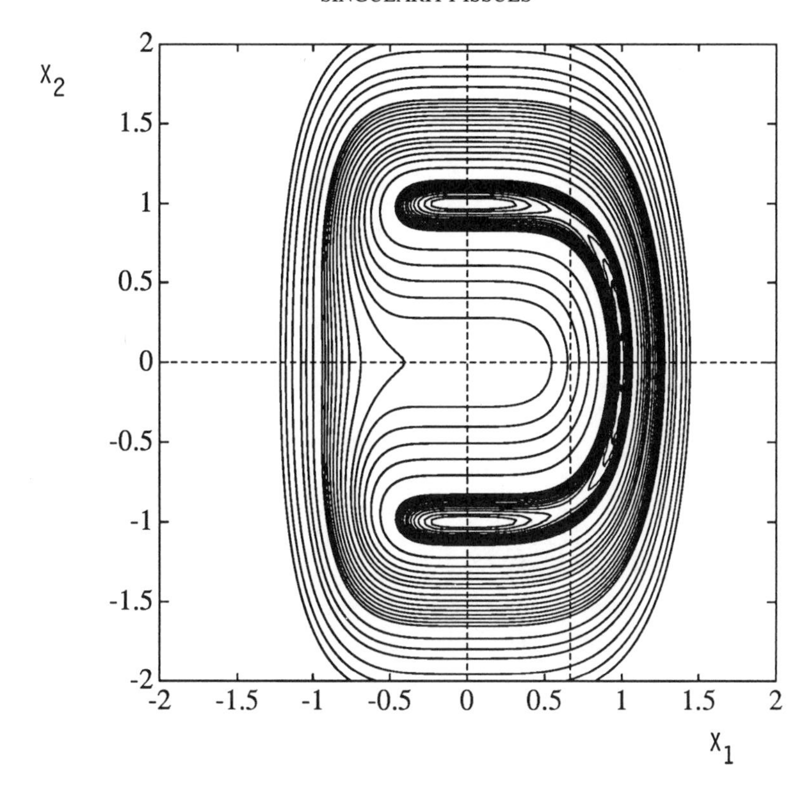


Figure 7: Level curves of the Lyapunov function $V(x)$ given by (14) for the case $a = 1$.

$(-2, 2)$ is divided into $N_1 = 300$ and $N_2 = 300$ intervals with 90,000 cells in it. For generating an associated cell mapping of (13), we use $T = 1$ and $\Delta t = 1/15$. The equilibrium point $x_a^*$ in Figure 5 is represented by periodic cells denoted by the symbol "|" and $x_b^*$ by the symbol "–". The periodic cells represented by the symbol "." do not correspond to an equilibrium point and their presence is explained as indicated in Case II. The domains of attraction of the periodic cells are shown in Figure 6. The two attraction regions are separated by the barrier manifolds $x_1 = 2/3$ and $x_2 = 0$. The cells in the white areas are mapped to the sink cell.

*Case IV: $a = 1$*

The roots located at $x_a^* = (0, -1)$ and $x_b^* = (0, 1)$ behave as stated in Case III. The function $f$ has a new root given by $x_c^* = (1, 0)$ (which lies on the manifold $x_2 = 0$) where $F$ is not defined. By using the transformation $x_1 = 1 + r\cos\theta$ and

$x_2 = r\sin\theta$, it can be observed that

$$\lim_{r\to 0} F_1(r,\theta) = 0, \quad \forall\,\theta$$

$$\lim_{r\to 0} F_2(r,\theta) = \begin{cases} 0, & \sin\theta \neq 0 \\ \infty, & \sin\theta = 0 \end{cases}$$

Using polar coordinates, the dynamical system (13) in a small neighborhood $U - \{(1,0)\}$ of $x_c^* = (1,0)$ can be approximated by

$$\dot{r} = -\frac{r}{2}(1 + 3\cos^2\theta) + \mathcal{O}(r^2)$$

$$\dot{\theta} = \frac{1}{2\tan\theta}\left(1 - 3\cos^2\theta\right) + \mathcal{O}(r)$$

For $r$ small enough, the evolution in the $\theta$ variable can be studied separately because the coupling term is of $\mathcal{O}(r)$. The differential equation in $\theta$ has six equilibrium points given by $\theta_{1,2}^* = \pm\frac{\pi}{2}$ which are stable and $\theta_{3,4,5,6}^* = \pm\cos^{-1}(\pm\frac{1}{\sqrt{3}})$ which are unstable. Note also that the domains of attraction for the stable equilibria are respectively defined by the strips $\cos^{-1}(\frac{1}{\sqrt{3}}) < \theta < \cos^{-1}(-\frac{1}{\sqrt{3}})$ and $-\cos^{-1}(-\frac{1}{\sqrt{3}}) < \theta < -\cos^{-1}(\frac{1}{\sqrt{3}})$. In addition, trajectories starting in regions other than these domains of attraction or the equilibria go to either $\theta = 0$ or $\theta = \pi$ in finite time, where they are not defined any more.

These results reveal that the state space $(r,\theta)$ near the point $x_c^* = (1,0)$ has a cone-like region of attraction for $x_c^*$. Trajectories inside the cone converge to $x_c^*$ and trajectories outside the cone terminate on the singular manifold $x_2 = 0$. For an illustration, consider the case of any initial state of the form $x(t_0) = (1,c)$, where $c$ is an arbitrary constant. These trajectories map to the point $x_c^*$. On the other hand, every trajectory starting at $x(t_0) = (1+\delta,\delta)$, for small $\delta$, hits the singular manifold $x_2 = 0$. Notice that, since $\dot{r}$ is of order higher than $\dot{\theta}$ for small r, the trajectories tend to hit the singular manifold orthogonally in this case.

Figure 7 illustrates $V-$level curves. Notice that the local minimum at $x_2 = 0$ of Figure 3 has now become a root of $f$. Again, the saddle points of the Lyapunov function are on the barrier manifold located at $x_1 = \frac{2}{3}$.

The cell mapping results for the state space region $(-2,2)\times(-2,2)$ with $N_1 = 300$, $N_2 = 300$, $T = 1$ and $\Delta t = 1/15$ are provided in Figures 8-9. In Figure 8, the periodic cells with the symbol "–" correspond to the point $x_a^*$, the symbol "|" to the point $x_b^*$ and the symbol "." to the point $x_c^*$. Notice that the periodic cells representing the point $x_c^*$ are disconnected. The domains of attraction of periodic cells are shown in Figure 9. As in case IV, some cells cross the manifold $x_2 = 0$.

*Case V: $a > 1$*

The roots of $f$ located at $x_a^* = (0,-\sqrt{a})$ and $x_b^* = (0,\sqrt{a})$ behave as given in Case III. The single root of Case IV splits into two zeros of $f$ and are given by

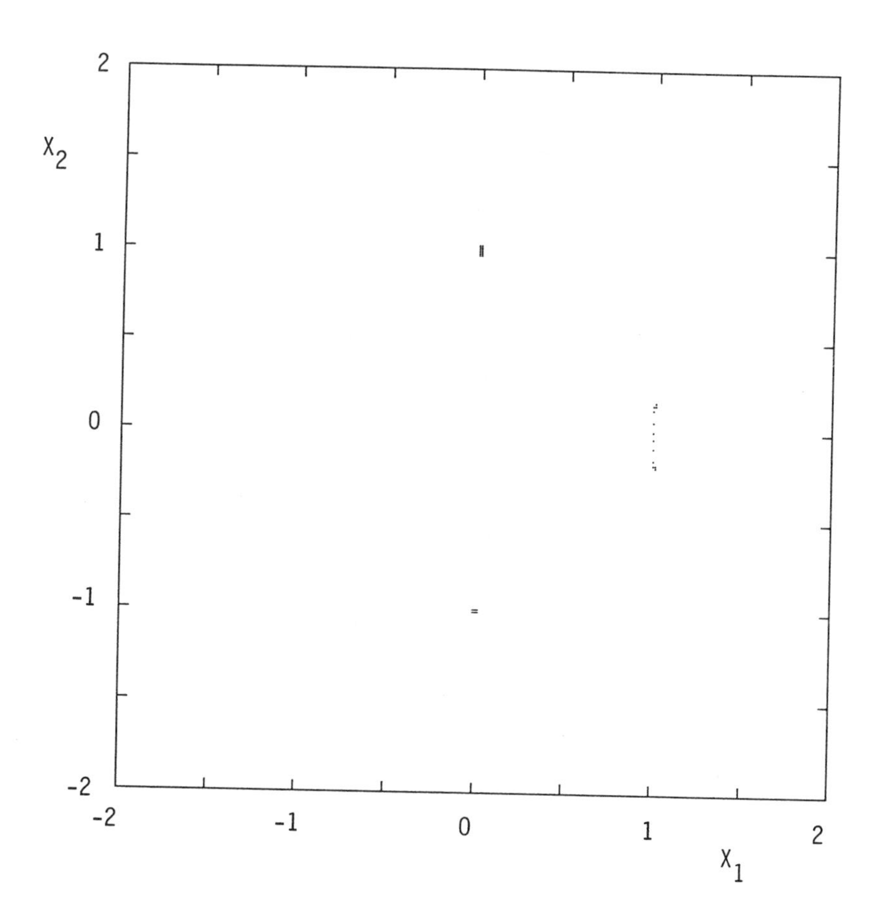


Figure 8: Location of periodic cells for the pathological example with $a = 1$.

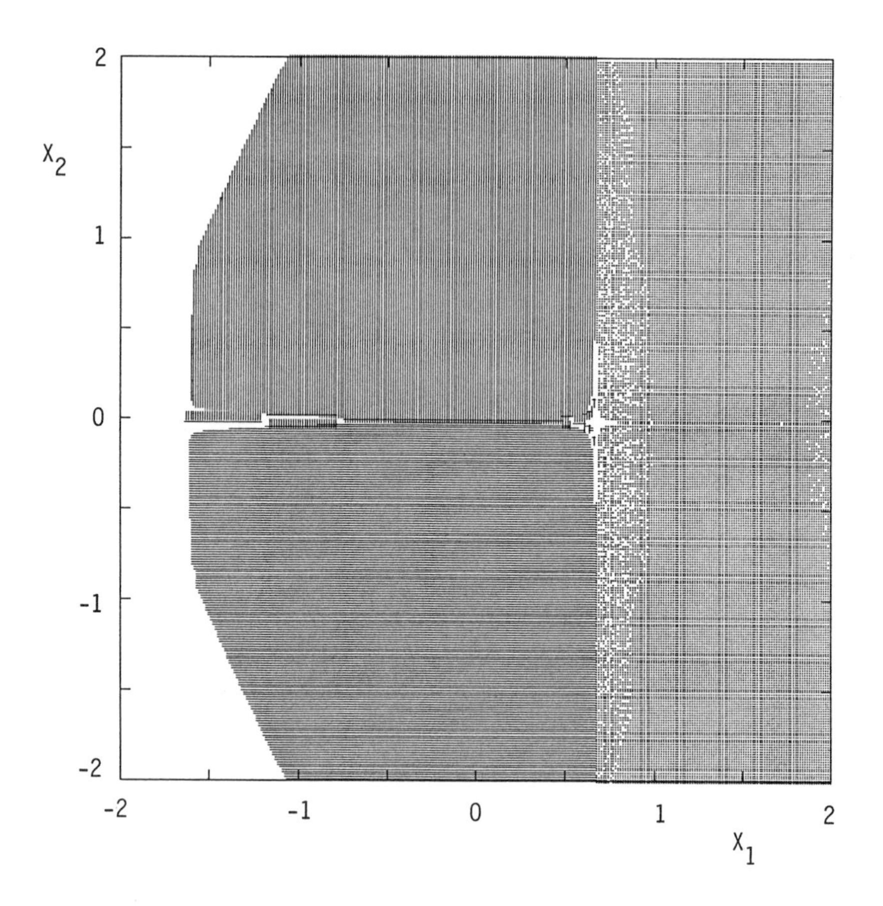


Figure 9: Domains of attraction of periodic cells for the pathological example with $a = 1$.

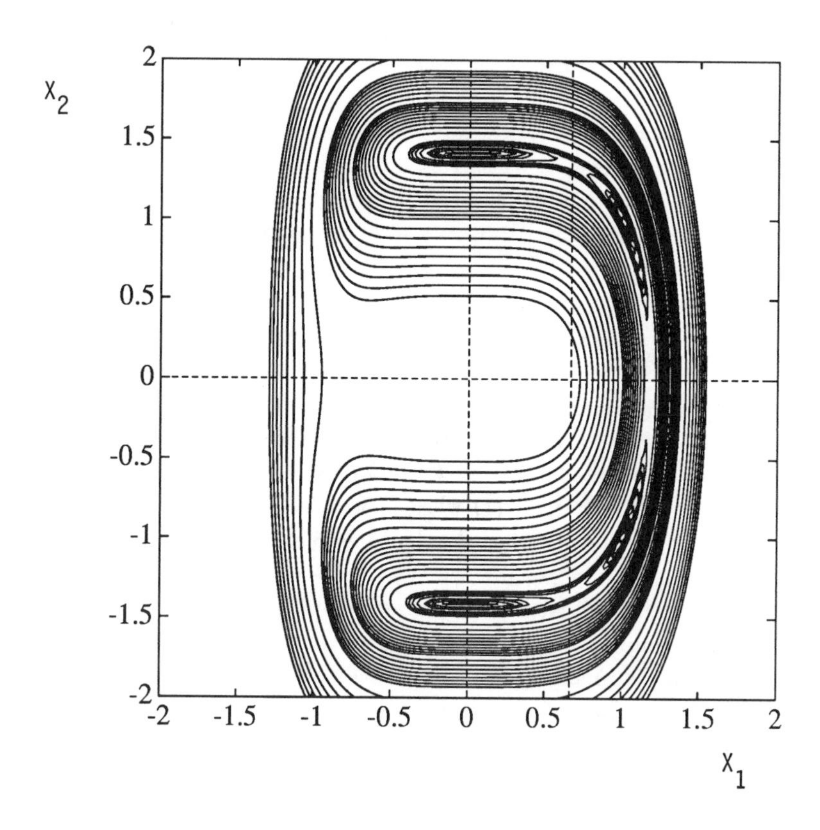


Figure 10: Level curves of the Lyapunov function $V(x)$ given by (14) for the case $a = 2$.

$x_c^* = (1, -\sqrt{a-1})$ and $x_d^* = (1, \sqrt{a-1})$. These zeros are equilibrium points of the dynamical system (13) defined by $F$ and satisfy all the regularity conditions of the theory presented in [22]. It is already known that they are asymptotically stable equilibrium points of (13) and that the eigenvalues of $J$ at both the roots are $\lambda = -1$ with multiplicity two. Of course, $V$ defined by (14) is a Lyapunov function at the new roots and its level curves are shown in Figure 10 for $a = 2$.

In this section several pathological cases have been studied for different parameter values. Note that in the cases where $F = 0$ is defined only directionally, cone-like shapes of region of attraction are obtained. This type of result has been reported in the literature for an associated discrete form which is defined by the Newton method, see [6, 11, 12].

# IV. EFFECT OF SINGULAR MANIFOLDS

Recall that the singular manifolds of the dynamical system (1) are defined as $(N-1)$-dimensional sets of points $x$ where $det\ J(x) = 0$. A trajectory of the system may not be defined on such manifolds since $F(\cdot)$ may not be defined. From a numerical point of view, a trajectory of (1) when approaching a singular manifold may evolve unpredictably especially when $det\ J(x)$ becomes extremely small. Singular manifolds also seem to influence the extent of domains of attraction. Thus, it is important to determine the location of singular manifolds. One should note that the problem of finding singular manifolds can be as difficult as finding the roots of $f(x)$ since again the problem is reduced to solving for the zeros of a nonlinear scalar algebraic equation $det\ J(x) = 0$. In this section, we present some analytical results and provide a computational approach based on them for locating singular manifolds of (1).

One way to construct an algorithm for locating the singular manifolds is to discretize the state space $\mathbf{R^N}$ by dividing each state variable $x_i$ into intervals denoted by $z_i$. The collection of intervals will be called *cells*, see Hsu [16]. The cell vector $z$ is said to belong to the cell space $S$ and satisfies the condition

$$\left(z_i - \frac{1}{2}\right) h_i \le x_i < \left(z_i + \frac{1}{2}\right) h_i, \quad i = 1, 2, \ldots, N \tag{15}$$

where $h_i$ is the cell size along the $x_i$–axis. We define a new cell state space $\bar{S}$ in which the cells are closed (that is, each cell contains its boundary). It should be noted that the way in which we have defined the cells in $\bar{S}$, these cells are compact and convex in $\mathbf{R^N}$. The points where $det\ J(x) = 0$ represent, in general, a hypersurface of dimension $(N-1)$ in $\mathbf{R^N}$ whose sign changes as this surface is crossed. Intuitively, if a cell is traversed by the singular manifold then $det\ J(x)$ must change sign within the cell. By making use of the cellular state space, Zufiria and Guttalu [23] have derived the following results.

**Theorem 16** *Let $z_0$ be a cell occupying the region $x^L \leq x \leq x^U$ in the state space where det $J(x)$ is continuous. Suppose that $x_a$ and $x_b$ are two vertices of $z_0$ for which det $J(x_a) \leq 0 \leq$ det $J(x_b)$. Then there exists a point $x_0$ belonging to the cell $z_0$ such that det $J(x_0) = 0$.*

Note that in our case the continuity of *det* $J(x)$ is guaranteed since $f \in C^1$.

**Theorem 17** *Let $x_0 \in \mathbf{R^N}$ be a point on the singular manifold and det $J(x)$ be analytic at $x = x_0$. Let $z_0$ be the cell in $\bar{S}$ containing $x_0$ as an interior point. Suppose that the vector $\nabla_x(\det J(x_0)) \neq 0$. Then for cell size small enough there exist two vertices $x_a$ and $x_b$ of $z_0$ such that det $J(x_a) > 0 >$ det $J(x_b)$.*

**Theorem 18** *Let $\Phi \subset \mathbf{R^N}$ be any subset of the state space. Consider that there exists an interior point $x_0 \in \Phi$ which belongs to a singular manifold. Assume that the cell size in $\mathbf{R^N}$ can be taken to be as small as we wish. Then one of the following cases must occur:*

*(a) There exists a point $x \in \Phi$ which is on the singular manifold such that $x$ is an interior point of a cell.*

*(b) There exists a point $x \in \Phi$ which is on the singular manifold such that $x$ is a vertex of a cell.*

Theorems 16–18 provide a framework to develop a computational algorithm for locating singular manifolds of the dynamical system (1) with a given accuracy. Suppose that we consider the following algorithm: For each cell $z \in \bar{S}$, check if the condition

$$\det J(x_a) \det J(x_b) \leq 0,$$

is satisfied; where $x_a$ and $x_b$ represent the coordinates of any two of the vertices of $z$. The question when this cell contains a point belonging to a singular manifold is addressed by the theorem given below.

**Theorem 19** *If the cell size in $\mathbf{R^N}$ is small enough, then the above algorithm locates at least one cell in any state space region which is a subset of $\mathbf{R^N}$ which intersects a singular manifold.*

In practice, the above algorithm is implemented to evaluate the *det* $J(x)$ at each of the $2^N$ vertices of every cell $z$ in $\bar{S}$. If the sign of *det* $J(x)$ differs at any two of the vertices of $z$, then this cell contains a point on a singular manifold. There is no need to compute the *det* $J(x)$ for all the vertices in this case. The computation is terminated for each cell as soon as the difference in sign of *det* $J(x)$ is detected for any two of the vertices. The procedure is repeated for all the cells in the cell state space.

In the next section, we apply the singular manifold algorithm to two examples to illustrate their effect on the global behavior of the dynamical system (1).

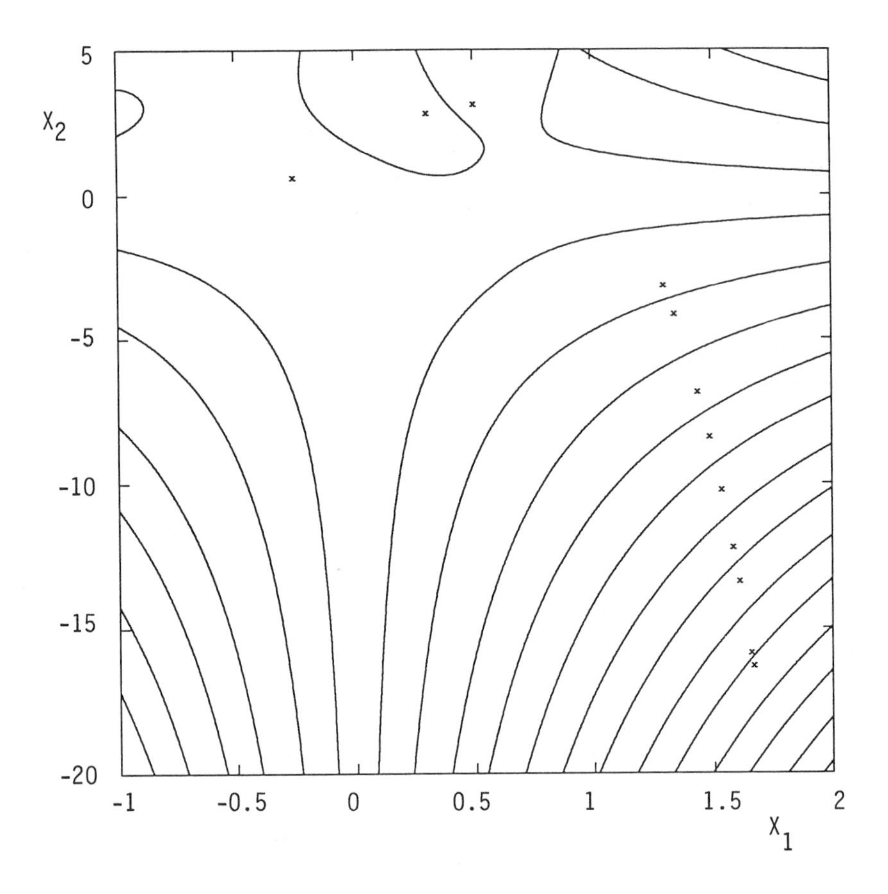


Figure 11: Computed singular manifolds for Brown's example. Location of all the zeros of $f$ are also shown.

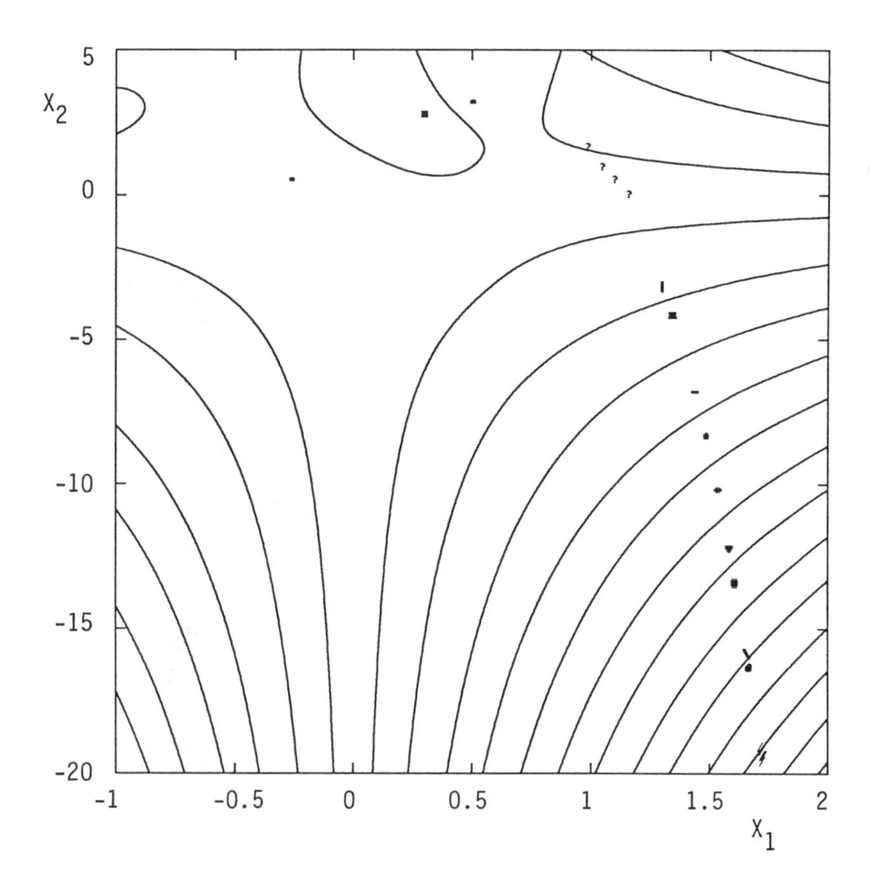


Figure 12: Location of periodic cells for Brown's example.

## A. BROWN'S EXAMPLE

Consider the algebraic system in $\mathbf{R}^2$ taken from [5] and [19]:

$$\begin{aligned} f_1 &= \frac{1}{2}\sin(x_1x_2) - \frac{x_2}{4\pi} - \frac{x_1}{2} \\ f_2 &= \left(1 - \frac{1}{4\pi}\right)(e^{2x_1} - e) + \frac{ex_2}{\pi} - 2ex_1 \end{aligned} \tag{16}$$

where $e$ denotes the base of the natural logarithms. This problem was analyzed by Zufiria and Guttalu [22] to determine all of its roots as well the domains of attraction associated with them.

We now apply the singular manifold algorithm developed above to the region $-1.0 \leq x_1 \leq 2.0$ and $-20.0 \leq x_2 \leq 5.0$ by taking $N_1 = 2000$ and $N_2 = 2000$ intervals in the region (totaling four million cells) with cell sizes $h_1 = 0.0015$ and

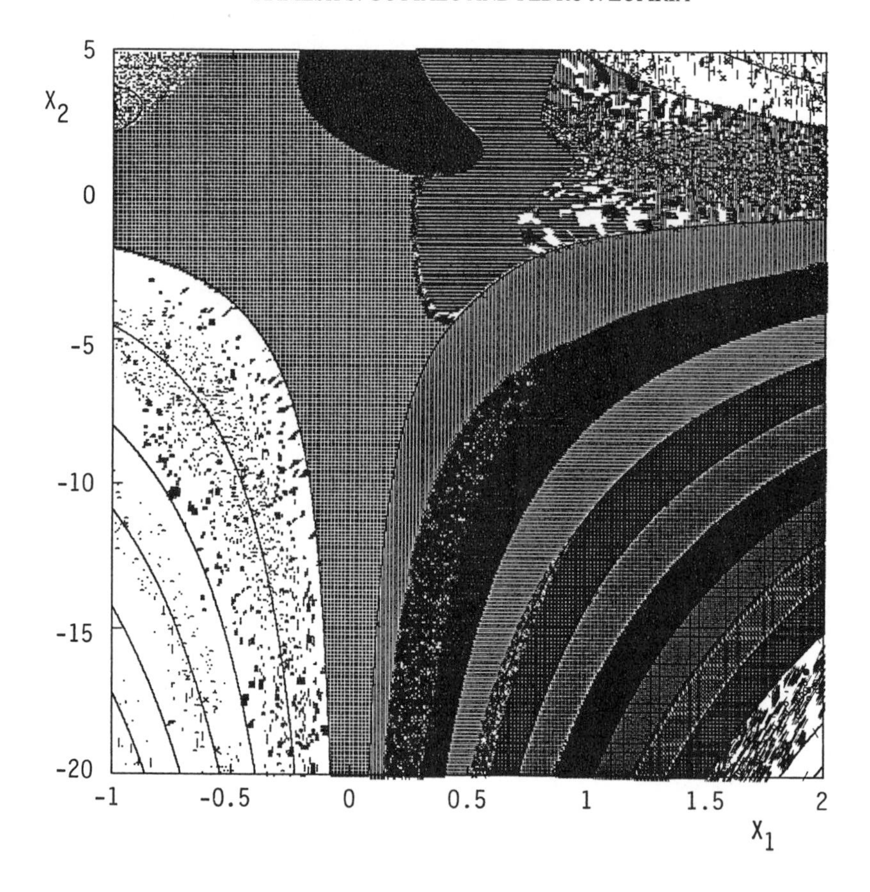


Figure 13: Domains of attraction of the roots of $f$ for Brown's example.

$h_2 = 0.0125$. Figure 11 shows the cells discovered by the algorithm in which $det\ J(x)$ changes sign. Also shown in this figure are all the zeros of (16) whose locations are indicated by "x". For the chosen cell size, the singular manifolds are seen to be continuous curves in the plane. This figure compares excellently with that provided by [22] where the singular manifolds were computed by an exhaustive application of Newton's method. The results of cell mapping analysis applied to the same region with $N_1 = 400$, $N_2 = 400$, $T = 0.6$, $\Delta t = T/30$ is shown in Figure 12-13. In Figure 12, except for the periodic cells represented by the symbol "?" (which corresponds to a local minimum of the Lyapunov function given by (4)), all other periodic cells have a direct correspondence with the roots of $f$. The importance of the role played by the barrier manifolds, strongly suggested by theory, is clearly demonstrated in this example. The extent of the domains of attraction is demarcated by the isolated regions in a theoretical sense which is verified numerically by the cell mapping

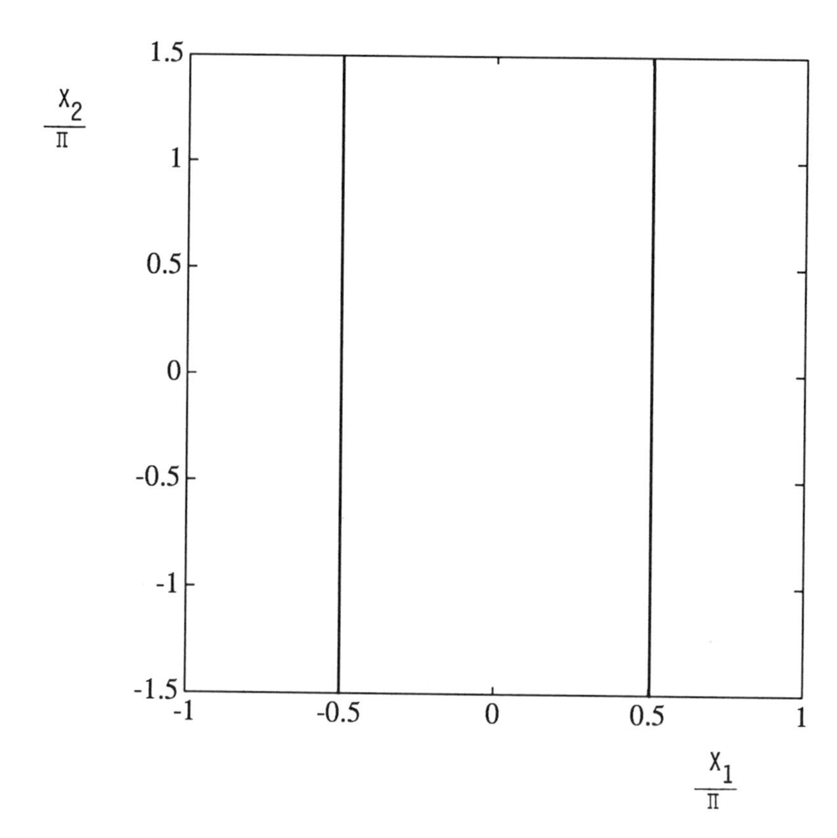


Figure 14: Singular manifolds associated with $f(x) = G(x) - x$ for the mechanical system described by (17).

method. In addition, it should be noted that some points on the barrier manifolds act like an attractor as seen by a large number of cells which straddle across them. It is likely that these points are extraneous singularities.

## B. A MECHANICAL SYSTEM

Consider a rigid bar hinged at one end and subjected to periodic impulse load at the other end, see Hsu [16]. The bar is restrained at the hinged end by a linear rotational spring and a damper. The bar undergoes large rotational motion in the plane and its dynamics are described exactly by means of a two-dimensional period-to-period mapping. We first consider the case when there exists no elastic restraint. In this case the map describing the motion of the bar is given by

$$x(n+1) = G(x(n)), \quad G : \mathbf{R}^2 \to \mathbf{R}^2, \quad x \in \mathbf{R}^2 \tag{17}$$

$$x_1(n+1) \;=\; x_1(n) + a\left(x_2(n) - \alpha \sin x_1(n)\right), \quad -\pi \leq x_1 \leq \pi$$

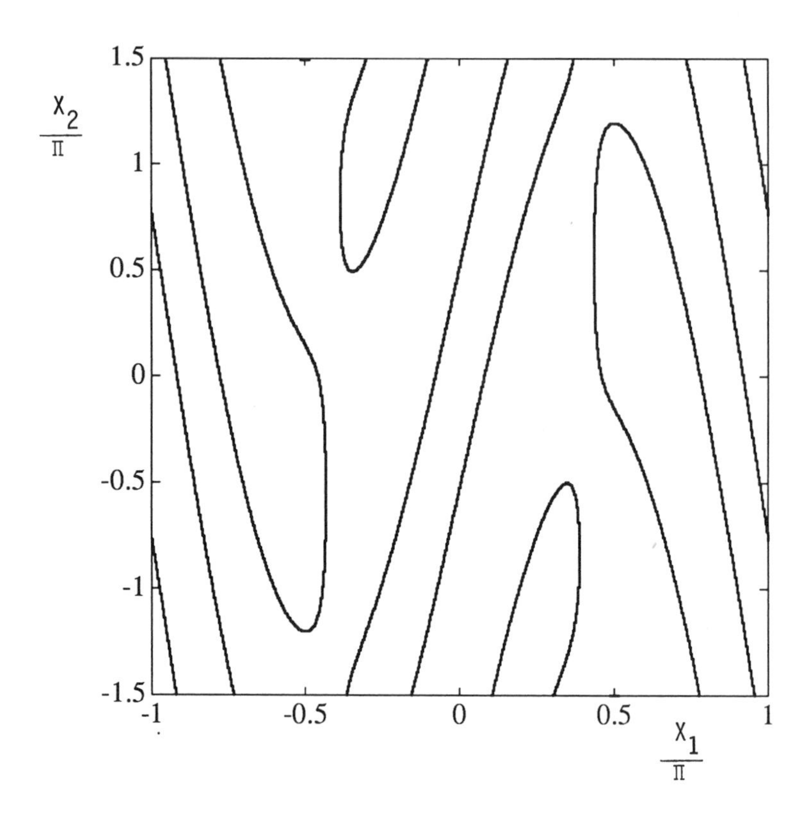


Figure 15: Singular manifolds associated with $f(x) = G^2(x) - x$ for the mechanical system described by (17).

$$x_2(n+1) = b(x_2(n) - \alpha \sin x_1(n))$$
$$a = \frac{1-e^{-2\mu}}{2\mu}, \quad b = e^{-2\mu}$$

where $\mu$ is the nondimensional rotational damping coefficient and $\alpha$ is the strength of the impulse. The dynamics of such maps are often controlled by fixed points of order $K$ defined to be the roots of the function $f(x) = G^K(x) - x$. Since the dynamical system (1) can be used to locate fixed points, it is important to ascertain the nature of singular manifolds for this problem. We choose the system parameters to be $\mu = 0.1\pi$ and $\alpha = 8$. The algorithm for computing the singular manifolds is applied to the state space region $(-\pi, \pi) \times (-1.5\pi, 1.5\pi)$ which is divided into $N_1 = N_2 = 500$ intervals. There are 250,000 cells in this region.

Figure 14 shows the singular manifolds for the system (1) associated with the vector function $f(x) = G(x) - x$ which is used for determining fixed points of the map $G$. In this case *det* $J(x) = \alpha a \cos x_1$ and only two singular manifolds exist which are located at $x_1 = \pm\frac{\pi}{2}$.

Figure 15 shows the singular manifolds for the system (1) associated with the vector function $f(x) = G^2(x) - x$ which determines the fixed points of both order one and two. Obviously, more solutions exists in this case than the previous one. There are 6 singular manifolds. Notice that the symmetry which exists in the map $G$ (that is, $G$ is symmetric with respect to the transformation $x_1 \to -x_1$ and $x_2 \to -x_2$) is reflected also in the symmetry of singular manifolds.

Figure 16 shows the singular manifolds for the system (1) associated with the vector function $f(x) = G^3(x) - x$ which determines the fixed points of both order one and three. The large number of singular manifolds may be attributed to the fact that a large number of zeros of the vector function $f$ exist.

Figure 17 displays the singular manifolds for the system (1) associated with $f(x) = G^6(x) - x$ which determines fixed points of order one, two, three and six. Almost all the state space is covered by the singular manifolds indicating that the computation of fixed points of even moderately high order can be extremely difficult. It is known that the above mechanical system possesses chaotic behavior for the parameter values chosen, Hsu [16]. Thus it is to be expected that as $K$ increases, the pattern of singular manifolds associated with $f(x) = G^K(x) - x$ in the state space becomes highly complex. Accordingly, determination of fixed points of high orders is computationally challenging when the dynamical system formulation (1) is employed.

We now consider the case when, in addition to the damper, an elastic restraint is also present in the system. The map $G$ takes the form

$$x_1(n+1) = e^{-\mu}\left\{\left(\cos\omega + \frac{\mu\sin\omega}{\omega}\right)x_1(n) + \frac{\sin\omega}{\omega}A(x)\right\} \tag{18}$$
$$x_2(n+1) = e^{-\mu}\left\{-\left(\omega + \frac{\mu^2}{\omega}\right)\sin\omega\, x_1(n) + \left(\cos\omega - \frac{\mu\sin\omega}{\omega}\right)A(x)\right\}$$

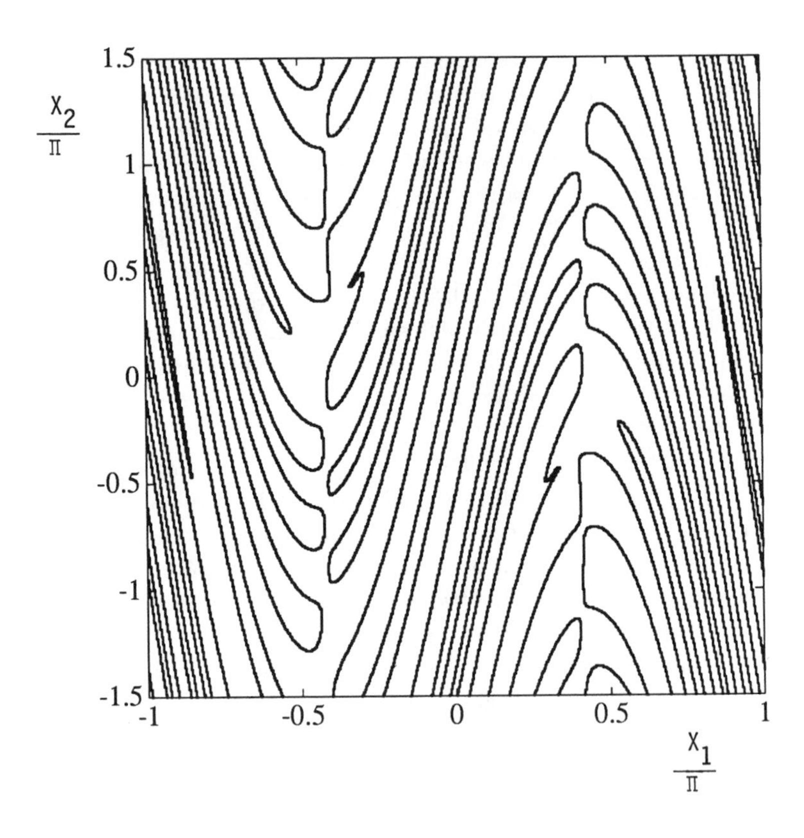


Figure 16: Singular manifolds associated with $f(x) = G^3(x) - x$ for the mechanical system described by (17).

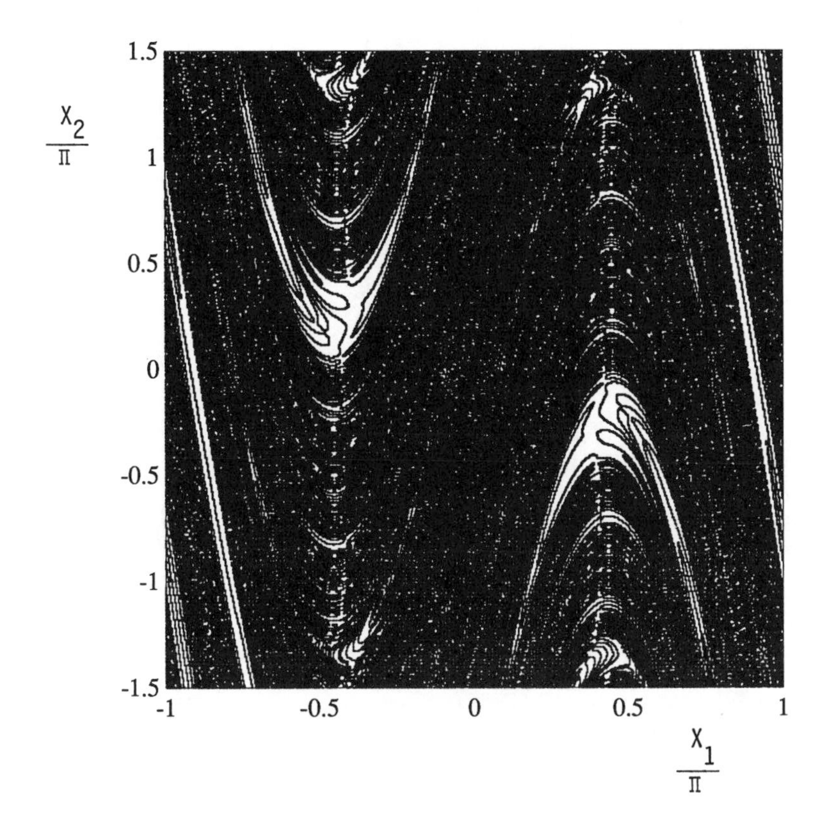


Figure 17: Singular manifolds associated with $f(x) = G^6(x) - x$ for the mechanical system described by (17).

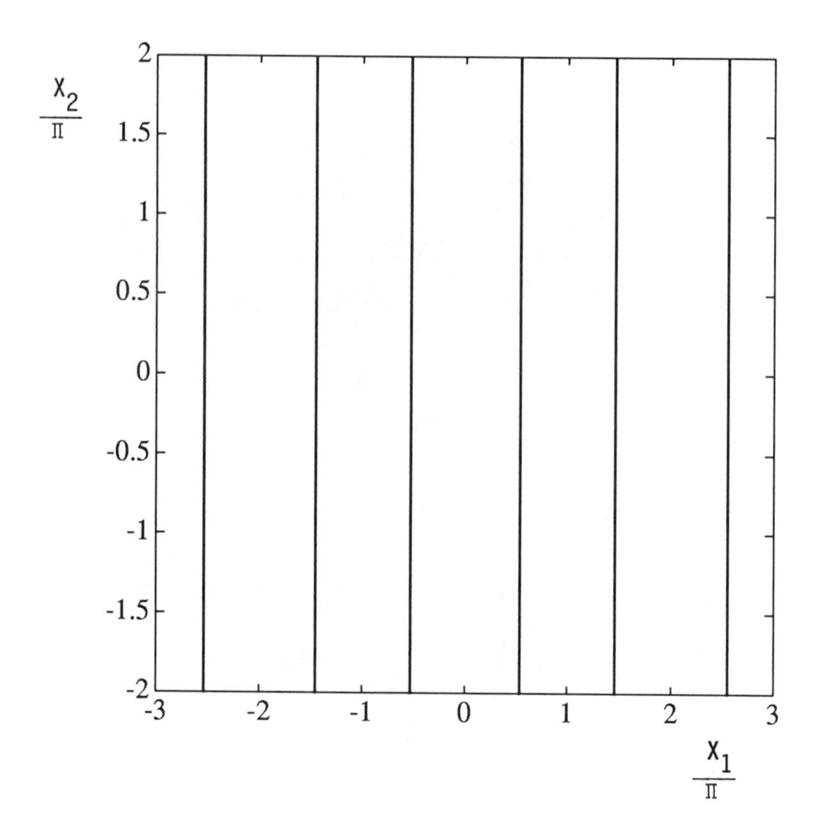


Figure 18: Singular manifolds associated with $f(x) = G(x) - x$ for the mechanical system with an elastic restraint described by (18).

$$A(x) = x_2(n) - \alpha \sin x_1(n)$$

where $\omega^2 = \omega_0^2 - \mu^2$ and $\omega_0$ represents the nondimensional natural frequency of the system.

We treat the case for which the parameters are $\alpha = 3.0$, $\mu = 0.1$ and $\omega = 0.5$. The state space region $(-3\pi, 3\pi) \times (-2\pi, 2\pi)$ is divided into $N_1 = N_2 = 500$ intervals to contain 250,000 cells. The singular manifolds associated with $f(x) = G^K(x) - x$ for $K = 1, 2, 3, 4, 6, 12$ are respectively shown in Figures 18-23. It can be observed that the number of singular manifolds increases as $K$ is increased. However, the singular manifolds do not seem to cover all of the state space region under consideration as compared to the case when the elastic restraint was absent.

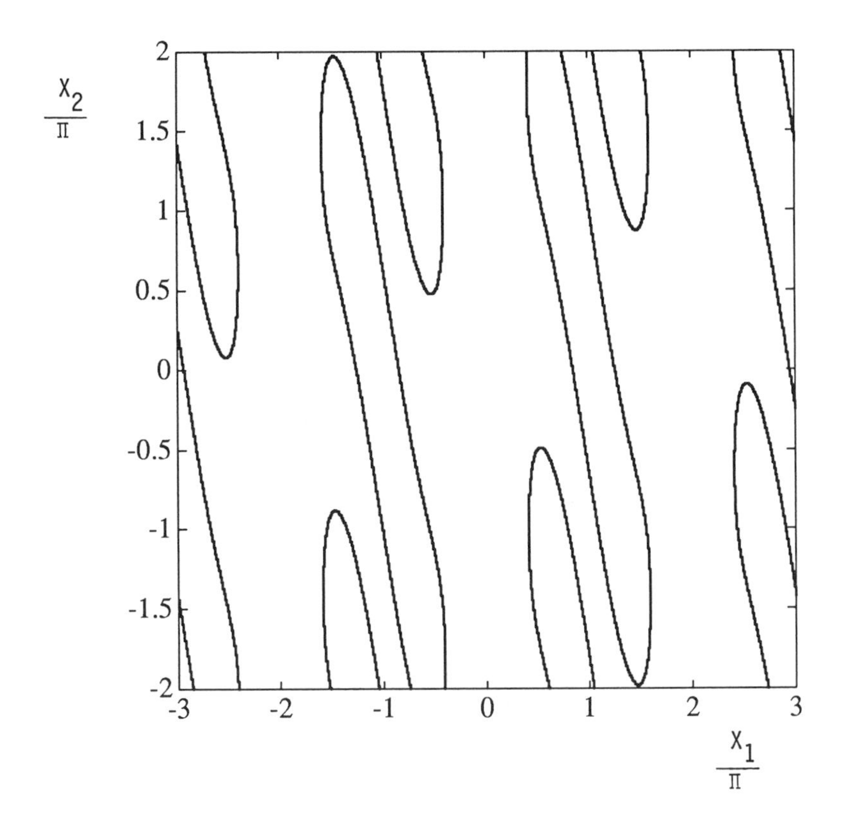


Figure 19: Singular manifolds associated with $f(x) = G^2(x) - x$ for the mechanical system with an elastic restraint described by (18).

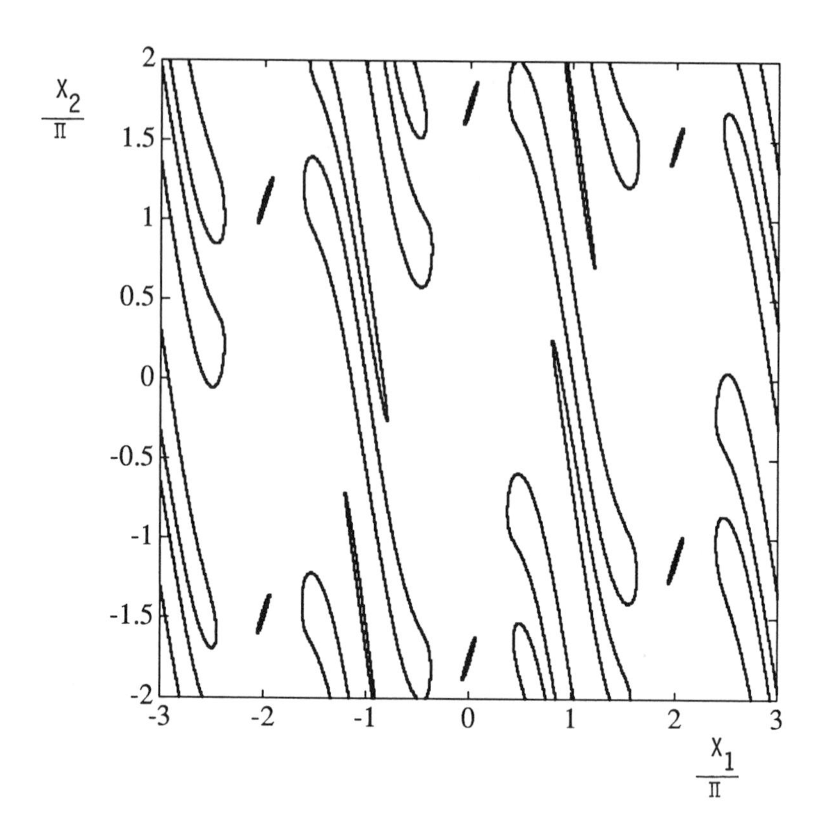


Figure 20: Singular manifolds associated with $f(x) = G^3(x) - x$ for the mechanical system with an elastic restraint described by (18).

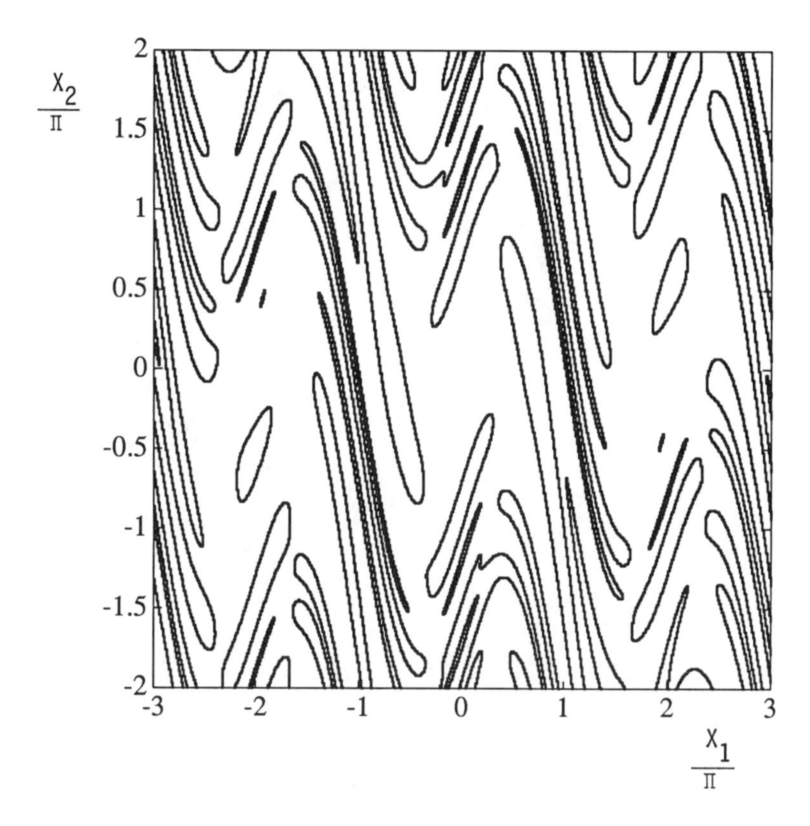


Figure 21: Singular manifolds associated with $f(x) = G^4(x) - x$ for the mechanical system with an elastic restraint described by (18).

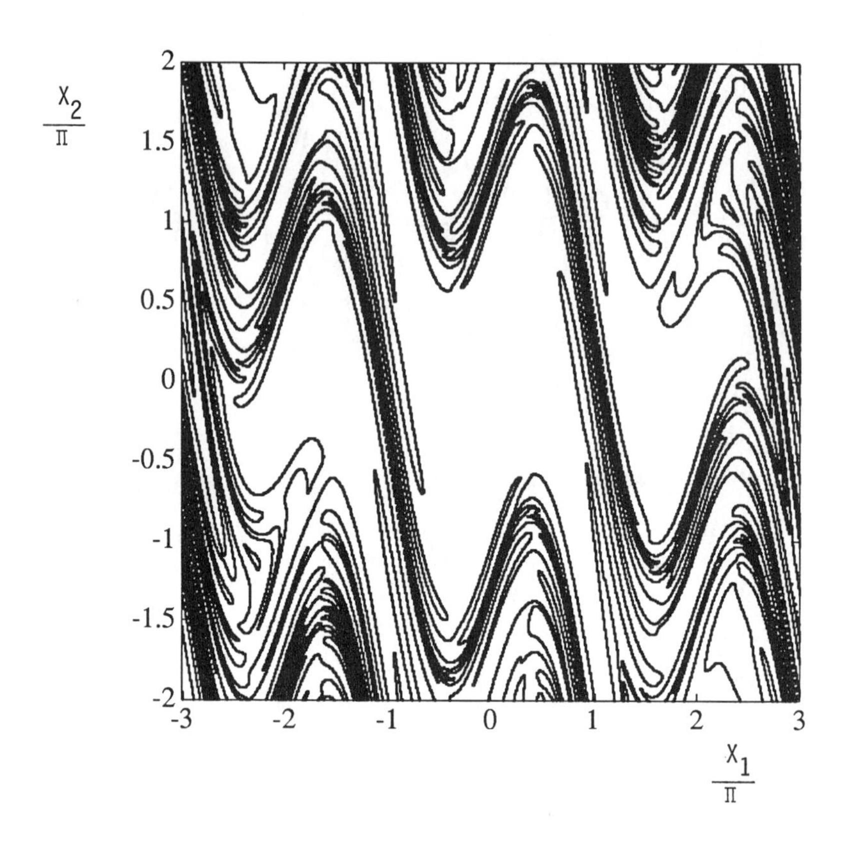


Figure 22: Singular manifolds associated with $f(x) = G^6(x) - x$ for the mechanical system with an elastic restraint described by (18).

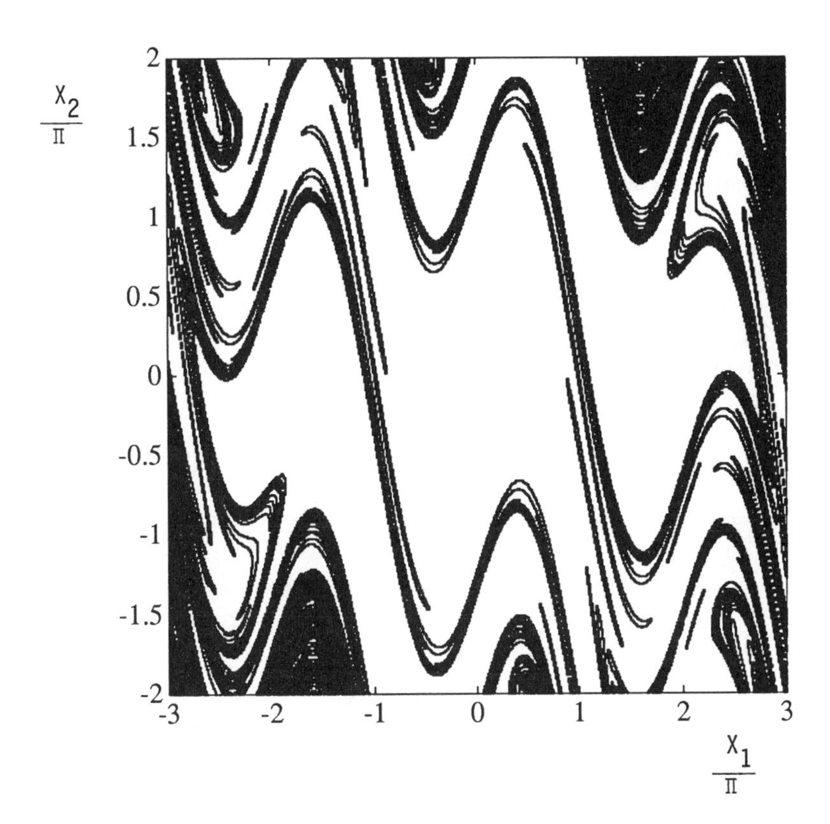


Figure 23: Singular manifolds associated with $f(x) = G^{12}(x) - x$ for the mechanical system with an elastic restraint described by (18).

# V. REDUCED ORDER SYSTEMS

The dynamical system (1) has been constructed to provide valuable information on the distribution of the solution of a determinate set of nonlinear algebraic equations (2), that is there are as many equations as unknowns. However, there are many practical instances where one is required to solve an underdetermined set of algebraic equations. In this case, the solution is sought on a manifold. We consider one such application in this section and apply some of the theory developed by Zufiria and Guttalu [22].

Suppose that a system of nonlinear algebraic equations of the form

$$f(x) = 0, \quad f : \mathbf{R}^{\mathrm{N}} \to \mathbf{R}^{\mathrm{M}} \tag{19}$$

where $M < N$ needs to be solved. In general, the solutions of (19) will be manifolds whose state space dimension will be of order $N - M$. Notice that the Jacobian of $f(x)$ is no longer a square matrix and the dynamical system (1) cannot be used either for locating or for approaching these solution manifolds. Instead of (1), we propose to study the following dynamical system:

$$\dot{x}(t) = H(x(t)) = -J^T(x)f(x), \quad J(x) = \nabla_x f(x), \tag{20}$$

where the Jacobian $J(\cdot)$ is an $M \times N$ matrix. No restriction need to be placed on the value of $M$.

Some stability results concerning the solution manifolds of (20) may be stated following an approach similar to the one provided in [22].

**Theorem 20** *Given $f$, consider the scalar function*

$$V(x) = f^T(x)f(x) = f(x) \cdot f(x), \; V : \mathbf{R^N} \rightarrow \mathbf{R} \tag{21}$$

*where $f^T(x)$ is the transpose of $f(x)$. Let $\Pi$ be a manifold of dimension $N - M$. Suppose that $H(x^*) = 0$, $\forall\, x^* \in \Pi$, that is, $\Pi$ is an equilibrium manifold of (20). Then $V$ is a strict Lyapunov function of (20) for* $\Pi$.

*Proof:* By definition, $H$ is continuous everywhere. Hence the existence of a solution of (20) is guaranteed everywhere. Also, $x^*$ is a root of $f$. Then,

$$V(x^*) = f^T(x^*)f(x^*) = 0 \tag{22}$$

$$V(x) = \sum_{k=1}^{N} f_k^2 = \|f(x)\|^2 > 0 \quad if\ x \notin \Pi \tag{23}$$

Let $x_i^* \in \mathbf{R^N}$ be one of the equilibrium points of (20). From equations (22) and (23), there always exists a neighborhood $U_i \subset \mathbf{R^N}$ of $x_i^*$ for which $V(x_i^*) = 0$ and $V(x) > 0$ for all $x \in U_i - \Pi$.

The gradient of $V$ defined by (21) is

$$\nabla_x V = J^T \, \nabla_f V = 2J^T f \tag{24}$$

Since $f$ is continuously differentiable, $V$ is also differentiable everywhere (and, consequently in every $U_i$).

Along the trajectories of the system (20), the time derivative of $V$ is given by

$$\begin{aligned} \dot{V}(x) &= (\nabla_x V)^T \cdot \dot{x} \\ &= (2f^T J) \cdot (-J^T f) \\ &= -2f^T J J^T f \\ &= -2H^T H \end{aligned} \tag{25}$$

where $H^T H > 0 \iff x \notin \Pi$.

In this case $\dot{V}$ is unique, even if the uniqueness of the solution of (20) is not guaranteed. Referring to (23), $\dot{V} < 0$ in $U_i - \Pi$. This proves that $V$ is a strict Lyapunov function for $\Pi$ where $\Pi$ is any equilibrium manifold of (20). ■

**Example:** *A Constrained Mechanical System*
The above formulation finds an application in integrating differential equations with algebraic constraints (for instance, one can cite Lagrange multipliers in Lagrangian formulation of mechanical systems). Consider the dynamical system

$$\begin{aligned} \dot{x} &= f(x,t,\Lambda), \quad f: \mathbf{R^N} \times \mathbf{R} \times \mathbf{R^L} \to \mathbf{R^N} \\ g(x) &= 0, \quad g: \mathbf{R^N} \to \mathbf{R^M} \\ g(x_0) &= 0, \quad x_0 \in \mathbf{R^N} \end{aligned} \tag{26}$$

where $\Lambda(x,t)$ are the Lagrange forces which guarantee that the motion of the system remains on the manifold defined by $g(x) = 0$. Usually, $L = M$ in the Lagrangian formulation, however, we can consider $L$ and $M$ to take on independent values for without loss of generality.

The system (26) may be converted into a singular perturbation problem where the fast dynamics corresponds to the algebraic constraints and the slow dynamics proceeds along the constraint manifolds. The fast dynamics system could be constructed via the dynamical system (20) independent of the number $M$ of constraints. The asymptotic stability of the manifolds where the constraints are satisfied would be guaranteed. The fast dynamics of this subsystem would guarantee a rapid convergence to the values of $\Lambda$ that satisfy the algebraic constraints. The state of the system (26) would evolve very close to the manifolds where the constraints are satisfied. Hence, the solution of (26) would be searched for on the manifolds as desired.

If trajectories of (26) remain on the manifold, then the following equation must be satisfied:

$$\begin{aligned} \dot{g} &= \nabla_x g \, \dot{x} \\ &= \nabla_x g \, f(x,t,\Lambda) \\ &= 0 \end{aligned} \tag{27}$$

This equation defines $\Lambda$ implicitly as a function of $x$. For a fixed value of $x = x_c$, we get a system of $M$ equations with $L$ unknowns which can be solved for $\Lambda$ via (20):

$$\dot{\Lambda} = -K^T \nabla_x g f(x,t,\Lambda)$$

where $K = \nabla_\Lambda(\nabla_x g f(x,t,\Lambda))$.

In summary, in order to integrate the original constrained problem, the following type of singular perturbation problem may be posed:

$$\begin{aligned}\dot{x} &= f(x,t,\Lambda) \\ g(x_0) &= 0, \quad x_0 \in \mathbf{R}^{\mathbf{N}} \\ \epsilon\dot{\Lambda} &= -K^T \nabla_x g f(x,t,\Lambda)\end{aligned}$$

Here $\epsilon \ll 1$. Again, in the usual case where $L = M$, one can construct system (1) of the form $\dot{\Lambda} = -K^{-1}\nabla_x g f(x,t,\Lambda)$ and follow the same singular perturbation procedure presented above. Note that the general procedure is based on guaranteeing $\dot{g} = 0$ so that any numerical error which forces a trajectory to leave the manifold is not detected. Therefore, it is necessary to have some feedback based on evaluating $g(x)$. This feedback can be performed via an intermediate adjusting procedure, after every prescribed number of integration steps, by integrating

$$\dot{x} = -(\nabla_x g(x))^T g(x)$$

where the state vector $x(t)$ goes to the manifold provided it starts close enough.

# VI. CLOSING REMARKS

A class of nonstandard dynamical systems constructed for the purpose of locating zeros of a vector function is analyzed. New results concerning the role of singularities possessed by the systems in characterizing its general behavior are provided. Local results concerning the relation between zeros of $f$ and equilibria of the system are further developed. When the zeros of $f$ happen to lie on the singular manifolds of the dynamical system, it is possible that the classical notions of equilibrium, stability and attractivity are no longer applicable in some cases. The role played by the singular manifolds in controlling the local and global behavior is highlighted by providing a pathological example. An algorithm for locating the singular manifolds is applied to two examples to illustrate their effect on the global behavior of the system. Another dynamical system formulation has been provided to extend the applicability of the system (1). Its application to numerical integration of a class of mechanical systems has been presented. These mechanical systems are constrained via a Lagrange formulation.

**ACKNOWLEDGMENTS**

The research reported here is partially supported by a grant from the National Science Foundation. P. J. Zufiria wants also to thank the support of a Formación de Postgrado fellowship of the Programa Nacional de F.P.I. provided by the D.G.I.C.T. of the Ministerio de Educación y Ciencia of Spain.

# References


[1] P. T. Boggs, "The solution of nonlinear systems of equations by A-stable integration techniques," *SIAM J. Numer. Anal.* **8**, 767-785 (1971).

[2] F. H. Branin, "Widely convergent method for finding multiple solutions of simultaneous nonlinear equations," *IBM J. Res. Develop.* **16**, 504-522 (1972).

[3] R. Brent, "Some efficient algorithms for solving systems of nonlinear equations," *SIAM J. Numer. Anal.* **10**, 327-344 (1973).

[4] R. Brent, "On the Davidenko-Branin method for solving simultaneous nonlinear equations," *IBM J. Res. Develop.* **16**, 434-436 (1972).

[5] K. N. Brown, "A Quadratically Convergent Method for Solving Nonlinear Equations," Ph.D. Thesis, Purdue University, 1966.

[6] D. W. Decker, H. B. Keller and C. T. Kelley, "Convergence rates for Newton's method at singular points," *SIAM J. Numer. Anal.* **20**, 296-314 (1983).

[7] C. B. García and W. I. Zangwill, "Pathways to Solutions, Fixed Points, and Equilibria," Prentice-Hall, Englewood Cliffs, New Jersey, 1981.

[8] I. Gladwell, "Globally convergent methods for zeros and fixed points" *in* "Numerical Solutions of Nonlinear Problems" (T. H. Baker and C. Phillips, eds.), Oxford University Press, New York, 1981.

[9] J. Gomulka, "Remarks on Branin's method for solving nonlinear equations," *in* "Towards Global Optimisation" (I. C. W. Dixon and G. P. Szego, eds.), North-Holland/American Elsevier, 1975.

[10] J. Gomulka, "Two implementations of Branin's method: numerical experience," *in* "Towards Global Optimisation 2" (I. C. W. Dixon and G. P. Szego, eds.), North-Holland, 1978.

[11] A. Griewank and M. R. Osborne, "Analysis of Newton's Method at Irregular Singularities," *SIAM J. Numer. Anal.* **20**, 747-773, (1983).

[12] A. Griewank and M. R. Osborne, "Newton's Method for singular problems when the dimension of the null space is $> 1$," *SIAM J. Numer. Anal.* **18**, 145-149 (1981).

[13] R. S. Guttalu and H. Flashner, "A Numerical Method for Computing Domains of Attraction for Dynamical Systems," *Int. J. Num. Meth. Eng.* **26**, 875-890, 1988.

[14] M. W. Hirsch and S. Smale, "Differential Equations, Dynamical Systems, and Linear Algebra," Academic Press, 1974.

[15] M. W. Hirsch and S. Smale, "On algorithms for solving $f(x) = 0$," *Comm. Pure Appl. Math.* **32**, 281-312 (1979).

[16] C. S. Hsu, "Cell-to-Cell Mapping: a method of global analysis of nonlinear systems," Springer-Verlag, New York, 1987.

[17] H. B. Keller, "Global homotopies and Newton methods," *in* "Symposium on Recent Advances in Numerical Analysis" (C. de Boor and Golub, eds.), 1978.

[18] J. LaSalle and S. Lefschetz, "Stability by Lyapunov's Direct Method with applications," Academic Press, New York, 1961.

[19] A. Morgan, "Solving polynomial systems using continuation for engineering and scientific problems," Prentice-Hall, Englewood Cliffs, New Jersey, 1987.

[20] J. M. Ortega and W.C. Rheinboldt, "Iterative solution of nonlinear equations in several variables," Academic Press, New York, 1970.

[21] M. Vidyasagar, "Nonlinear Systems Analysis," Prentice-Hall, Englewood Cliffs, New Jersey, 1978.

[22] P. J. Zufiria and R. S. Guttalu, "On an application of dynamical system theory to determine all the zeros of a vector function," *J. of Math. Anal. and Appl.* **151** (1990) (tentative).

[23] P. J. Zufiria and R. S. Guttalu, "A computational method for finding all the roots of a vector function," *Appl. Math. Comp.* **34** (1989) (tentative).

[24] P. J. Zufiria and R. S. Guttalu, "The role of singularities in Branin's method" (submitted for publication).

# INDEX

E

## F



## G



## H



## I



## J



## K



## L

**M**



**N**

T



U